Spatial Database Systems

T0143018

The GeoJournal Library

VOLUME 87

Managing Editor: Max Barlow, Toronto, Canada

Founding Series Editor:
Wolf Tietze, Helmstedt, Germany

Spatial Database Systems

Design, Implementation and Project Management

By

ALBERT K.W. YEUNG
Ontario Police College, Aylmer West, Ontario, Canada

and

G. BRENT HALL
University of Waterloo, Ontario, Canada

 Springer

A C.I.P. Catalogue record for this book is available from the Library of Congress.

ISBN-10 1-4020-5393-2 (PB)
ISBN-13 978-1-4020-5393-1 (PB)
ISBN-10 1-4020-5391-6 (HB)
ISBN-13 978-1-4020-5391-7 (HB)
ISBN-10 1-4020-5392-4 (e-book)
ISBN-13 978-1-4020-5392-4 (e-book)

Published by Springer,
P.O. Box 17, 3300 AA Dordrecht, The Netherlands.

www.springer.com

Printed on acid-free paper

Dedication

Contents

Preface

The decision to write this book was motivated by a number of factors. First, although several useful textbooks on spatial databases have recently been published, this is an area of spatial information science that has lagged somewhat behind the rapid advances of the technology and the profusion of books on domain-specific applications. Second, much of the information pertaining to spatial database technologies is only available in scattered journal papers and conference proceedings, and prior to this book no single effort has been made to sift through this expansive literature and unite the key contributions in a single volume. The tasks of sourcing and coherently integrating relevant contributions is daunting for students, many of whom have a substantial number of competing demands placed on them. This book should make the task of knowledge building less daunting. Third, and perhaps most importantly, an apparent trend in many spatial information science programs is to focus, from first or second year undergraduate through to fourth year courses, on learning to work confidently and independently with increasingly complex software tools. Hence, many courses are technical in nature, and while they continue to produce technically adept students, knowledge of the broader aspects of spatial databases is often not as complete as it might be among graduates. Some programs have sought to address this by introducing courses that focus on spatial data management. However, these courses are largely unsupported by a relevant and contemporary textbook.

This book seeks to fill the void on topics that must be mastered in initiating, implementing, and managing a spatial database project. Since the mid-1990s the application-driven paradigm of GIS has gradually evolved into the database-centric paradigm that is at the heart of the approach adopted in this book. Hence, in addition to updating approaches to planning, designing, and implementing GIS as a technological infrastructure within an organisation, the book concentrates specifically on spatial databases as an institutional resource, a commodity, and a knowledge base for decision making. This provides the basis for a comprehensive and balanced discussion of recent advances in the concepts and technology that underlie the current data-based and user-centric approach to spatial information.

The approach that is adopted in this book allows a transparent integration of spatial information with mainstream information technology. Since all major database vendors currently offer spatial information capabilities and functions in their products, it is important for students of spatial databases as well as practitioners to have a sound grasp of the possibilities these systems offer. At the same time, GIS vendors have also largely re-designed and re-built their products by taking advantage of the concepts and techniques of processing spatial data within a database environment. This new approach is heavily dependent on the development and use of standards, which figure prominently in many parts of the text. Importantly, this retooling of the spatial information marketplace does not automatically assume the demise of conventional GIS. On the contrary, GIS software has, itself, moved into a new era of usage where many non-traditional users are finding new applications and, at the same time, creating new domain-specific spatial data models.

The book is intended for students of spatial information science as well as professionals already in the workplace. The bond between between these two groups of users is their common interest in the state of the art of spatial database systems. The level of the discussion assumes that readers have already completed at least one technical course in GIS, preferably two, and hence have a good understanding of the concepts of acquiring, characterising and applying spatial data, the techniques of geo-referencing and positioning, as well as domain-specific spatial analysis and spatial data modelling. The book's audience is likely to span a wide variety of academic disciplines in the humanities, the sciences and in various fields of engineering. Such heterogeneity made it a challenge to write technical content that is accessible to readers with varying technical knowledge and skills. Hence, spatial database concepts and techniques are explained using a relatively non-technical approach. Despite this, a sound basic understanding of computing and data processing is required in order to master the material covered in several of the chapters.

The content of the book is organised so that it can be used either for an intensive half year course, or for two sequential half year courses that could also be offered as a single full year course. In the first case, the ten substantive chapters (2 through 11) could be used in a half year senior undergraduate or graduate spatial database course with each chapter occupying approximately a single week in the body of a thirteen week semester. Chapters 1 and 12 serve as the book ends in a single semester course to introduce the approach at the outset and to conclude the course with a week for wrap up discussion at the end. Use of the text for a single course of this nature assumes a relatively high level of technical knowledge on the part of students, as the material included in each chapter is relatively wide-ranging and often technical in nature.

Alternatively, the text could be used to support a full-year course of 26 weeks at third or fourth year undergraduate level or two thirteen-week linked courses, that span a third and fourth year undergraduate program. In this context, Parts 1 and 2 or Chapters 1 through 6 would cover the first course with a title of Spatial Database Principles and Architecture, and Chapters 7 through 12 would cover the second course with the title of Part 3 to focus on Spatial Database Implementation and Project Management. In each of these scenarios (a full-year course, or two thirteen-week courses) each chapter would occupy two weeks or approximately six hours of lectures with examples.

To supplement and substantiate the materials presented in the book, and in keeping with the course structures noted above, an accompanying Web site (http://www.springer.com/1-4020-5391-6) has been developed that instructors and students using this book are encouraged to use. This Web site includes instructional materials such as PowerPoint slides, additional reference materials and learning resources, as well as case studies and exemplary spatial database systems in use in government, business and academic research. Spatial database systems is essentially a technical subject that can be approached most effectively by reading and practising. Hence we have also developed and posted to the Web site several project/laboratory assignments.

It is important to reiterate that this text is targeted not only at tertiary-level instruction, but also at practitioners in various professions that use spatial information technology. The text provides a highly functional conceptual framework for constructing technical solutions to address issues commonly encountered in the workplace. These include, for example, the role and characteristics of spatial metadata, the considerations and implications of the law and associated principles such as liability and intellectual property in spatial data use and distribution, the process of assessing spatial data needs within an organisation and starting a spatial database project from first principles, as well as the writing of and responding to business cases and request for proposal (RFP) documents. In addition, the adherence to principles of best practice in spatial databases as well as the use of quality control and quality assurance are essential knowledge components for spatial database sponsors, administrators, project managers and developers. For all users of the text, whether undergraduate or graduate students or practitioners, each chapter is referenced with a wide range of key and recent sources. We have not included in the text specific case studies in any detail, and have rather relied on the associated Web site to present these materials.

To conclude this Preface, we wish to thank all people who have given us encouragement and support in one way or another during the course of writing this book. Our thanks must first go to Ms. Evelien Bakker, our Editor from Springer. Her patience and perseverance with busy schedules and unavoidable delays in writing is sincerely appreciated. Dr. Yeung wishes to thank Director Rudy Gheysen and Deputy Director Bill Stephens of the Ontario Police College, for their interest in this book project. Dr. Hall wishes to note his thanks to the University of Canterbury, Christchurch, New Zealand and the Department of Information Science at the University of Otago, Dunedin, New Zealand where parts of the manuscript were written. Dr. Hall also wishes to acknowledge the camaraderie and stimulating discussions in all aspects of spatial databases and their applications provided by successive years of graduate students at the University of Waterloo.

This book has benefited from review by a number of individuals, some of whom read and commented on the manuscript at various stages throughout and others who provided encouraging comments and gave us useful suggestions to improve the presentation of materials upon completion of the first full draft. In this regard we extend our thanks to Mr. Stephen Adaran (Ontario Police College), Prof. Yvan Bédard (Laval University), Dr. Don Boyes (University of Toronto), Dr. Dongmei Chen (Queen's University at Kingston), Dr. Rob Feick (University of Waterloo), Prof. Piotr Jankowski (San Diego State University), Prof. Ian Masser, Mr. Fraser Moffatt (Canada Border Services Agency), Dr. Tony Moore (University of Otago), Prof. Nigel Waters (University of Calgary), Dr. Michael Sawada (University of Ottawa), and Mr. Lawton Tam (formerly of the Ministry of Transportation of Ontario).

We wish to reserve the final words for our families. The production of any textbook is as much a matter of time-intensive labour as the fulfilment of a perceived need in the inventory of instructional and reference materials in a given field. The act of satisfying this need requires not only experience and knowledge but also substantial support on the part of others. In this context we wish to extend our heart-felt thanks to our families, to whom this book is dedicated.

PART 1

INTRODUCTION

Chapter 1

THE CURRENT STATUS OF SPATIAL INFORMATION TECHNOLOGY

1.   INTRODUCTION

In the world of information technology, it is fair to state that the only constants are change and growth. Judging from the developments in the field of spatial information technology in the last several years, this is definitely not an overstatement. Since the inception of the idea of using the computer for cartography and mapping in the 1950s, newer hardware and software tools have evolved to enable users of spatial information to operate more easily and extend constantly the frontiers of knowledge. New concepts of and techniques for using spatial information have paralleled the advances in hardware and software, and cumulatively these have accelerated the uptake of related information technologies.

The pervasive proliferation of spatial information during the past decade contrasts sharply with the earlier era, which prevailed until the early 1990s. This era was characterised by the umbrella term of geographic information systems (GIS) as a highly specialised technology of interest primarily to professional users and researchers for specific applications. One of the major driving forces behind the recent popularisation of spatial information is the increasing availability of spatial data from government and commercial sources, distributed via the Internet through such mechanisms as spatial data depots, digital geolibraries and spatial data warehouses and clearinghouses. Another major driving force is the growing awareness of the importance of spatial information by all sectors of modern society. Policy makers in the public sector, for example, have embraced the notion that spatial information is an important requirement of good governance, a fundamental aspect of the economy, and access to this information is a civil right that enables citizens to participate in public affairs. At the same time, commercial organisations including hardware and software developers, data collectors and providers, and information service providers all see spatial information as a business opportunity bounded only by their vision. Users of spatial information are no longer limited to professionals and researchers, but include people who plan their travel itinerary using an on-line map service on the Internet, who check the weather conditions on a specialty television channel, who search for a new home in an on-line multiple listing service (MLS), and who access geographically referenced information about their communities, the environment, and national and internation affairs on the World Wide Web.

The changes in the collection, management and use of spatial information noted above would probably not have happened, at least not to the extent that is evident, without the power of spatial databases. This book charts the use of spatial information by using a holistic data-based and user-centric approach instead of the conventional application-driven and GIS-centric approach. The approach embraces database development and management, data access policies and interchange, standards and metadata, commodification and liability of spatial data services, spatial knowledge discovery and decision support, all of which are addressed in the following chapters. From the perspective of training and education, this conceptualisation implies that the traditional approach, which viewed GIS in terms of data management, cartography and spatial analysis, must be broadened to accommodate the growing realisation of spatial information as an institutional resource, a commodity, and a knowledge base for decision making. In essence, this new view of spatial information was one of the main motivations for writing this book.

This introductory chapter seeks to provide a conceptual framework for the purpose, content and structure of the book. It first overviews recent advances in the concepts and technology of the new approach to spatial information. The knowledge and skills required to design, implement and manage a spatial database are discussed. The organisation of the book is then introduced by giving a brief description of each chapter and the threads that are unwound among individual chapters.

2.   ADVANCES OF SPATIAL INFORMATION CONCEPTS AND TECHNOLOGY

The design and implementation of spatial information technology largely depends upon the trends that have characterised information technology (IT) in general. The concepts and techniques used in spatial database systems today are drawn largely from the general principles and methods of database systems that evolved in the past decade. These have brought spatial information, which until the mid-1990s was still widely perceived as a stand-alone technology, into the mainstream of IT. They have also turned project-specific GIS applications toward the multi-purpose database world of an enterprise information infrastructure that supports both day-to-day business needs as well as management decision making. It is now data and users, not applications or technology, that drive the use of spatial information in an organisation, leading to the data-based and user-centric approach to spatial information that characterises current practices in the implementation of spatial database systems.

2.1   A New Metaphor of Spatial Information

Spatial information was traditionally approached from three perspectives, namely spatial data management, cartography, and spatial analysis. These perspectives can be described as the instrumental or tool-oriented view of spatial information. Sui and Goodchild (2001) contended that such a view of spatial information is fundamentally inadequate to capture the essence of spatial information technology and its social implications. They suggested that, justifiably, spatial information should also be conceived from the perspective of *media*, where media are generally defined as the means of communicating information to the general public. The concept of media also includes mass media, which refer to the instruments of mass communication that take place in modern society, where information can reach a large number of people within a short period of time.

On the basis of the traditional views of spatial information and the proposition of Sui and Goodchild, a new conceptualisation of spatial information has evolved, as shown in Figure 1.1. At the core of this conceptualisation there are four functional aspects of spatial information, namely database systems, cartography, communication, and spatial analysis, each of which serves one or more specific but interrelated application domain denoted by the rectangular boxes in the diagram.

The spatial database component plays a central and critical role in the new metaphor of spatial information. Typical spatial information applications today are both data- and computation-intensive, and require the storage of a huge volume of traditional (alphanumeric) and non-traditional (geometric objects, images, time series) data sets. None of the data management, spatial analysis and mapping functions will work efficiently without a powerful spatial database system behind them.

Powered by its spatial indexing and data processing capabilities, a spatial database system can quickly assemble data relevant for use in a particular application and, in conjunction with spatial data analysis techniques, develop solutions for spatial decision making. Spatial information analysis for decision support can be enhanced by taking advantage of emerging techniques in *computational intelligence* that use artificial intelligence to tolerate imprecision and uncertainty in large databases in order to achieve robust and acceptable solutions (Bezdek, 1994). There are also emerging techniques, collectively known as *data mining*, which automatically scan the data held in a spatial database to discover possible relationships between data elements which, in turn, reveal phenomena that are otherwise hard or impossible to detect (Miller and Han, 2001). The database approach, therefore, is able to enhance decision support applications of spatial information tremendously when compared with traditional GIS approaches.

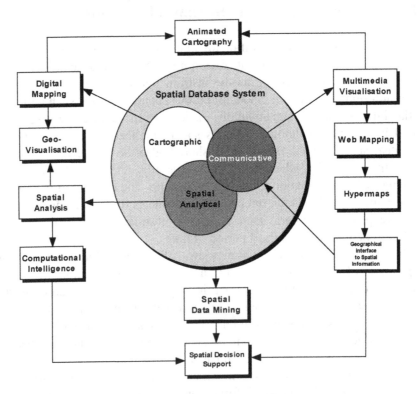

Figure 1-1. A new conceptualisation for spatial information functions

The increasing reliance of spatial database systems on the Internet and related technologies has led to many innovative methods of communicating spatial information. The focus of development in this area has gradually moved, over the last several years, from the simple dissemination of spatial data through the Internet to sophisticated interactive mapping and interoperable spatial data processing, realising the concept of "GIS as new media" proposed by Sui and Goodchild and noted above.

At present, spatial database systems are used in relatively sophisticated ways that go well beyond the simple management, display, and analysis of geographically referenced information. A typical spatial database system is enhanced by capabilities to manipulate new data types and models, complex data structures including spatial indexing, and sophisticated algorithms and operators for efficient data processing. As noted earlier, these capabilities are a far cry from the conventional file-based approaches of GIS that dominated the world of spatial data into the early 1990s. As shown in Figure 1.1, the database approach has brought the conventional and emerging functions of spatial information closer together. This approach not only articulates the use of spatial information within the conventional application domains themselves, but it also provides the necessary bridge that enables interoperable applications within mainstream information technology.

2.2   The Merging of Spatial Information with Mainstream Information Technology

The move to use a database approach for spatial information started in the early 1990s (Black, 1996). This involved considerable effort and commitment on the part of both conventional GIS and database software vendors. It also engendered many mergers, partnerships and joint development projects among companies, as well as the formation of industry consortia that brought government, commercial and research organisations together to define and work toward common goals. Today, practically all major database vendors offer spatial information capabilities and functions in their products. At the same time, GIS vendors have also largely re-designed and re-built their products by taking advantage of the concepts and techniques of processing spatial data within a database environment.

A typical spatial database system today is an ordinary commercial database with additional capabilities and functions to handle spatial data. These capabilities and functions include:

- *Spatial data types.* Spatial data are stored either as special data types following the Open Geospatial Consortium (OGC, 1999) definition of "simple features" or as a binary large object (BLOB) in a database. An OGC simple feature is defined in a database and is, therefore, handled completely within the functionality of the database itself. A BLOB, on the other hand, is a generic data type into which any binary data can be stored and, as a result, spatial data stored in this form must be indexed and manipulated using add-on software components to the spatial database.

- *Spatial indexing.* This is a mechanism to facilitate access to a spatial database by means of stored coordinates in two-dimensional space. There are many different indexing options, such as the R-tree, quadtree and B-tree, each of which has its strengths and weaknesses depending on the specific data format and application need.
- *Spatial operators.* These refer to the suite of data processing functions and processes that can be achieved through use of structured query language (SQL) to query and retrieve selected database contents, join database tables according to specific spatial and non-spatial criteria, and generate the results of processing in specific formats.
- *Spatial application routines.* These include a variety of software components for specific database application functions such as spatial data loading, versioning and long transaction control, performance tuning, database backup and replication.

There are now several database options available that offer spatial information processing capabilities and functions. Oracle was arguably the first database vendor to add spatial capabilities to its core products. In September 1995, Oracle and the Environmental Systems Research Institute (ESRI) agreed to integrate the Oracle 7 Spatial Option with two of ESRI's products, namely Spatial Database Engine (SDE) and ArcView GIS. Since then, Oracle has formed partnerships or joint-development agreements with several other GIS companies including Intergraph, MapInfo and GE Smallworld Systems. At the same time, Oracle also has perused its own internal development strategy that resulted in the creation of a structure for spatial data and the implementation of spatial operators by extending SQL to allow a wide range of spatial data processing functions within an Oracle database system.

Other key players in the database industry such as IBM, Sybase and Microsoft have all entered into partnership with GIS software vendors to develop spatial functionality, or developed their own capabilities, typically through extensions to their database products. In April 2001, IBM acquired Informix, notable for its spatial capabilities, and aggressively developed and extended the spatial capabilities of its own flagship database product called DB2. Sybase and Microsoft adopted a different approach that focused on partnering with GIS software vendors to add spatial database capabilities to their products. Sybase, for example, worked with Autometric to develop a suite of spatial database functions called Spatial Query Server (SQS) that can be implemented on a Sybase database system. Microsoft's database products, Access and SQL Server, are successfully implemented to provide database capabilities for the GIS software products of ESRI, MapInfo, Intergraph, Autodesk and several other companies.

Using the database approach to spatial information provides basic benefits such as centralising spatial and non-spatial business data in a single system for more efficient processing. It also increases the *return on investment* (ROI) by leveraging the skills of IT personnel. By merging spatial databases with mainstream IT, an information technology department can manage both spatial and non-spatial data at the same time by sharing the same resources and by administering the two types of database systems using the same security, optimisation and backup protocols and procedures. However, although the operating environment for spatial and non-spatial databases is similar, there are fundamental differences between the knowledge and skills required to work with these two different types of data and databases. This implies the requirement of regular IT staff for training in spatial information concepts and methods, and the need for spatial information staff to be familiar with the principles and techniques of operating the spatial database within the corporate database system.

2.3   Institutionalisation of Spatial Database Systems

The merging of spatial information with mainstream IT has led naturally to the institutionalisation of spatial database systems. This means that instead of using spatial information in *ad hoc* projects that have little connection between them, spatial databases are now implemented as permanent and integral components of corporate IT infrastructures with the explicit mandate to serve the long-term business goals of organisations.

An institutionalised spatial database system has the following characteristics:

- *Serveing the business goals of an organisation as a whole rather than individual departments.* The objective of implementing a spatial database is to provide the necessary information support for the day-to-day operation of an organisation as well as for executive decision or policy making. Depending on the information needs of an organisation, spatial database systems can be set up differently using a combination of system architectures. These include:

    - An operational spatial database system optimised for *on-line transaction processing* (OLTP) in support of daily business operations.
    - A *spatial data warehouse* formed by centralising the repositories of legacy spatial data structured and optimised for *on-line analytical*

processing (OLAP) in support of executive level decision or policy making.

- A *spatial datamart*, which is a small, single-subject area spatial data warehouse subset that provides decision support to information users from a specific department or business function of an organisation.

- *Senior management commitment and expectation of return on investment.* Implementing a spatial database is an expensive and time-consuming undertaking. This implies that a successful spatial database implementation project requires senior management commitment with respect to funding and the allocation of technical and human resources. At the same time, as spatial information becomes increasingly mission-critical to an organisation, the expected ROI is also high. Therefore, a spatial database system must always be implemented on the basis of a clear vision of what it can do to improve the efficiency and productivity of an organisation. Moreover a good understanding of the constraints and limitations of its functionality is required in order to avoid unrealistic expectations and minimise end user frustration.

- *Formal systems design and implementation using accepted industry practices.* Technically, implementing a spatial database system is far more sophisticated and complex than using a standalone GIS for an *ad hoc* project. Thus, the design and implementation of a spatial database system requires careful planning and must follow well established industry practices at each stage of the systems development life cycle (SDLC). This means that project management skills are at least as important as technical knowledge for successful system implementation.

- *Conformance to standards.* It is essential for a spatial database system to comply with corporate and industry standards in order to ensure interoperability with other database systems, both within and outside of an organisation. Standardisation and interoperability reduce the cost of operating and maintaining a spatial database by sharing resources, minimizing staff training, and facilitating data transmission and exchange among different systems in a distributed computing environment. Standardisation is a fundamental principle of good data management, and serves to minimise the risk of legal liabilities in the provision of spatial information services to external clients.

The institutionalisation of a spatial database system does not necessarily mean the demise of conventional GIS. Instead, spatial database systems and GIS have a relatively distinct division of work and play different but highly complementary roles, as illustrated in Figure 1.2, in the new working environment. Whereas a GIS is more suitable for *ad hoc* spatial problem-solving, a spatial database is designed more for spatial decision support and other information management functions. Spatial problem-solving is application-oriented and short-term. Once a solution to a particular spatial problem is found, the GIS will be used for another problem-solving task. Spatial decision support, however, is mission- and business-oriented and long-term. A spatial database will continue to operate as long as the business remains in operation and it will change only if the mission of the organisation changes. The differences between and the complimentary nature of GIS and spatial database systems are explained in more detail in Chapter 4.

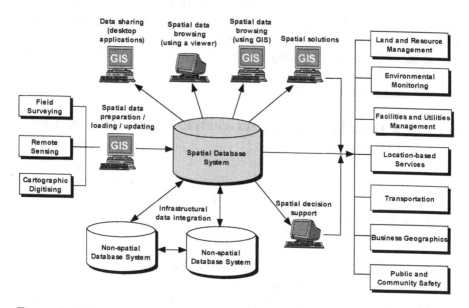

Figure 1-2. The new working relationship between spatial databases and GIS

Within the information management structure of organisations, a spatial database is a special-purpose data store. The primary functions of the spatial database are mostly concerned with the processing of data within the database system itself and the pre-processing (for example, extraction, summation, spatial joins, re-classification and so on) of data for export to application programs for further analysis and modelling. A typical spatial database system has relatively limited data capture capabilities, and has to rely on a GIS to provide data from, for example, field surveying, remote sensing and cartographic digitising for data upload.

A GIS can also be used as a convenient tool for updating the spatial database by means of its inherent graphics and database editing capabilities. A spatial database system is also typically limited in its ability to perform sophisticated spatial analysis and modelling. It has to rely on the analytical functions of a GIS and related spatial analysis software to turn its data into useful information for different application domains. In many cases, a spatial database is implemented as a data warehouse that provides the geographical referencing framework for spatial data used in GIS applications on desktop computers (see Chapter 6).

It is also important to note that a spatial database can often be used without a GIS. Although many people use desktop GIS to view spatial data, new developments in Internet technology and forms of data representation and presentation (for example, Geography Markup Language [GML] and Scaleable Vector Graphics [SVG]) have made it possible to view spatial information in a remote spatial database with the use of an Internet browser. Processes for data integration and transfer between spatial and non-spatial databases do not normally require the use of a GIS. Similarly, spatial decision support is possible without a GIS, but this is achieved with the loss of GIS-based data visualisation.

2.4   A Data-based and User-centric Approach to Spatial Information

Spatial database systems represent a data-based and user-centric approach to spatial information. Such an approach has three dimensions, namely stewardship, sharing and commodification, as shown in Figure 1.3.

These components are described as follows:

- *Stewardship* is concerned mainly with the internal use of spatial information. Some organisations, in particular those in the public sector, are mandated by regulations or legislation to collect and maintain a spatial database as part of their business functions. In such cases, the objectives and content of the database are tightly coupled with the respective business requirements and goals of the organisation. Examples of such spatial databases abound, including land title registration systems, forestry resource inventories, censuses of population and housing, inventories of highways and other public infrastructure such as water mains, electricity and gas distribution systems, and so on.
- *Sharing* is a fundamental principle of spatial database systems because this is the most effective way to minimise the cost and time required for implementing a spatial database. The primary concern of spatial data sharing is public access, which entails a delicate balance between the protection of privacy under freedom of information (FOI) legislation and the public's right to access information.

- *Commodification* is, in effect, another means of spatial data sharing but the focus here is on the commercial selling or trading of spatial data. There are two important considerations in the commercial exploitation of spatial data, namely the copyright of the original owner of the data and value-added data providers and the legal liabilities of the original data owner and value-added data providers in case of damage or loss caused by incomplete or erroneous spatial data provided to an end-user.

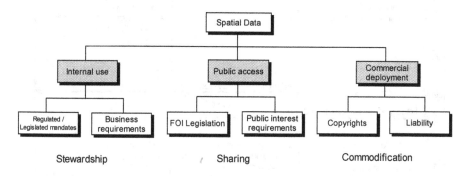

Figure 1-3. A data-based and user-centric approach to spatial information

The new and conventional approaches to spatial information are not mutually exclusive. In the new data-based approach, data take precedence over other considerations but do not replace or trivialise them. Data are now generally recognised as a valuable asset of an organisation and, as such, they should be managed and protected as carefully as possible in the same way that financial assets, facilities and human assets are managed and protected. From a corporate management perspective, therefore, a spatial database system plays a dual role of information and asset management. To implement a spatial database system from this perspective requires not only technical knowledge and technical skills but also management skill that involves dealing with people and numerous other aspects of information management. These aspects are described in the next section.

3.   KNOWLEDGE AND SKILLS FOR SPATIAL DATABASE SYSTEMS

Many organisations that are contemplating the adoption of spatial database technology are already users of GIS. To these organisations the implementation of a spatial database system is more of a migration of technology rather than the introduction of a new technology. It is necessary for existing and new GIS staff to acquire spatial database knowledge and skills in order to participate actively and productively in the transition to the new working environment. Figure 1.4 summarises a typical strategy for migrating from GIS to spatial database systems.

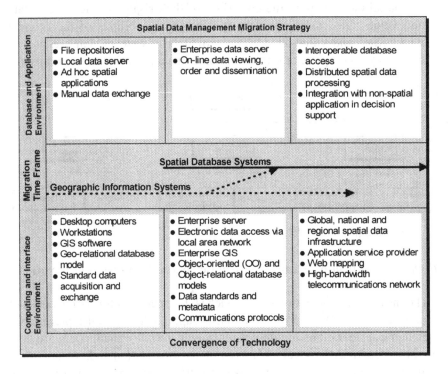

Figure 1-4. Migration strategy from GIS to a spatial database system

Two interrelated perspectives underscore the use of technology in this context, namely:

- *The database and application environment,* which changes gradually from file-based processing and *ad hoc* applications using a conventional GIS, to database-dependent processing using a spatial database system in a distributed and interoperable environment with full integration between spatial and non-spatial applications.
- *The computing and interface environment,* which specifies the convergence of progressively advanced hardware, software and communications technologies to the transition to the new database-dependent application environment.

Depending on the academic and professional backgrounds of individuals, different people require different training in order to become productive members of a spatial database project team. A fundamental requirement of everyone involved in the design and implementation of a spatial database system is to have a working knowledge of the following major areas:

- *Project management principles and skills.* The design and implementation of a spatial database requires the coordination of a wide range of tasks that include financial and human resources management, procurement of hardware, software and consulting services, database creation and loading, user training and on-going systems support and maintenance. A good understanding of the principles and skills of project management is important not only for the project manager but for all members of the project team so that everyone knows exactly his or her respective role with respect to the entire project (see Chapter 9).
- *Decision support applications of spatial databases.* Spatial decision support not only requires access to data currently residing in the database, but also access to legacy data. The growing interest in legacy data has led to the development of spatial data warehouses and data marts. In order to assist end users to use spatial databases for decision making purposes, a good understanding of the design and architecture of spatial data warehouses and data marts is important, and so is a working knowledge of the principles and techniques of decision support using a spatial database (see Chapter 11).

4. ORGANISATION AND OVERVIEW OF THIS BOOK

This book is organised around three substantive themes with a part of the book devoted to each theme, namely (i) database principles and architecture, (ii) spatial database implementation and project management, and (iii) trends of future developments.

Part 2 contains five chapters that cover the general concepts of database systems and spatial database systems with the aim to provide a solid foundation for the study of spatial database design and implementation. It starts with Chapter 2 which sets the tone for the remainder of the book by giving a comprehensive and concise introduction to the terminology, concepts and techniques of database systems as the technology stands today. Chapter 3 advances the discussion by focusing on the principles and methods of modelling in database design within the context of the systems development life cycle. Emphasis is placed on building a robust, durable and extensible database design before implementation is initiated.

Based on the foundation of the discussion in Chapters 2 and 3, Chapter 4 focuses on the representation of spatial data in a database environment. Spatial data types are defined and the concepts of geometry and topology are introduced. This chapter also explains in detail spatial operators as they are used in the conventional database environment.

Chapter 5 provides an overview of data standards and metadata, and their deployment in spatial database design and implementation. Discussion in this chapter provides the context for the study of spatial data sharing in Chapter 6. This final chapter of Part 2 explains database integration as a design methodology using emerging concepts and techniques such as ontology, data mediation, data warehousing, data marts, and database federation.

Part 3 of the book focuses on aspects of spatial database implementation and project management. It begins with Chapter 7 which considers human and non-technical aspects of spatial databases that include user education and legal issues surrounding spatial data ownership and the use of spatial information by organisations and individuals.

Chapter 8 addresses the concepts and methods of user needs assessment in spatial database implementation. The approach is founded on best practice in information system development including data orientation, user-centric design via joint application development, continuous quality assurance, iterative development, detailed documentation of project activities, focus on architecture and infrastructure, and adherence to standards. Chapter 9 focuses on the principles and best practices of project management. The concept of the project management life cycle is introduced and the skills required to shepherd a spatial database project through its five phases are explained in detail.

No discussion of spatial databases is complete without considering the emergent role of the Internet and related technologies. Chapter 10 discusses spatial database systems in the Internet environment. The discussion covers both the architecture and technical implementation of Web-enabled spatial database systems. Chapter 11 completes Part 3 by reviewing the concepts and techniques of spatial data mining and decision support, and demonstrating their use in selected application domains of spatial information.

Chapter 12, the only chapter in Part 4, reviews trends in spatial database systems design and implementation from the perspectives of technology, data, application and people. Since the only constant in this rapidly evolving field is indeed change, the trends reviewed in the final chapter conclude with a speculative summary of likely developments in spatial database systems in the foreseeable future.

To assist readers who may feel overwhelmed by the numerous technical terms that they will encounter in the text, an abbreviated Glossary is included at the end of the book. The full Glossary, which provides definitions of all terms in italics in the text, can be found in the accompanying Web site of this book (http://www.springer.com/1-4020-5391-6). In addition, the book Web site includes a rich resource of links and updated reference materials. It also contains suggestions for practical assignments and review questions that aim to reinforce the learning process. Instructors and students are encouraged to make use of the maerials in this Web site to assist in their study.

5.   REFERENCES

Bezdek, J.C. (1994) "What is Computational Intelligence?", in *Computational Intelligence Intimating Life* by Zurada, J.M., R.J. Mark II and C.J. Robinson (Eds.), New York, NY: IEEE.

Black, J.D. (1996) "Fusing RDBMS and GIS", *GIS World*, Vol. 9, No. 7.

Dolton, L.M. and Lowe, J.W. (2001) "Prospecting Spatial Database Offerings", *Geospatial Solutions*, Vol. 11, No. 10.

Limp, W.F. (2001) "From the Back Rom to the Glass Room: Spatial Database Break Computing Barriers Enterprise Wide", *GEOWorld*, Vol. 14, No. 8.

Lutz, D. (2000) "Take Advantage of That Spatial Database!", *GEOWorld*, Vol. 13, No.8.

Miller, H. and Han, J. (Eds.) (2002) *Geographic Data Mining and Knowledge Discovery*, London: Taylor & Francis.

OGC (1999) *Simple Features Specification for SQL*, Wayland, MA: Open Geospatial Consortium (formerly called the Open GIS Consortium).

Sui, D.Z. and Goodchild, M.F. (2001) "GIS as Media?", Guest Editorial, *International Journal of Geographical Information Science*, Vol. 15, No. 5, pp. 387-390.

PART 2

DATABASE PRINCIPLES AND ARCHITECTURE

Chapter 2

CONCEPTS AND ARCHITECTURE OF DATABASE SYSTEMS

1.   INTRODUCTION

The processing and analysis of spatial data are becoming increasingly dependent on the use of database management systems (DBMS) rather than conventional GIS. This approach to handling spatial data facilitates cost-effective data management using the capabilities of DBMS such as systems security, database integrity, backup and recovery, data replication and, more importantly, a close integration with mainstream business computing environments. Hence, contemporary spatial data processing is implemented more as a component of an organisation's corporate information technology (IT) infrastructure than as a traditional stand alone departmental application.

The use of DBMS has not only changed the fundamental concepts of spatial data and their management, but it has also created a demand for new skills and a new breed of spatial data managers and users, who are expected to be competent in the principles and methods of relatively complex database technologies. This chapter provides a concise, yet comprehensive introduction to the working principles of DBMS. It describes their construction in terms of hardware/software configurations, data structures and data processing operators. The discussion focuses on generic DBMS principles that serve as the theoretical and technical foundation for the study of database models and spatial databases in Chapters 3 and 4 respectively.

2.   DATABASES AND DATABASE SYSTEMS

The terminology used in the database literature is often confusing due to a lack of commonly accepted definitions. Even commonly used terms such as "data file", "data model" and "database system" are often defined differently in different database books. Hence, it is important to provide a set of commonly accepted definitions of database terms and processes for this book.

2.1   Database Terminology

Figure 2.1 depicts the relationships between the terms commonly associated with the concepts and techniques of a database. The primary function of a database is to provide timely and reliable information that supports the daily operations of an organisation. Since business operations and decision making often involve problem-solving using information obtained from a database, the business functions that the database is designed to address are commonly referred to as the *problem space* of the database. This concept is complex by nature and difficult to understand. Hence, it is necessary to start the database construction process with the development of a *data model* that is capable of describing the problem space in a comprehensible way.

A data model is a conceptual description of the database. It does not include any reference to the physical layout, or structure, of the database itself. The physical layout of the database, which describes how data are organised and stored in the database, is called the *database schema* or simply the *schema*. A schema is, in simple terms, the translation of a conceptual data model into a physical representation that can be implemented using a computer. Practically speaking, the schema is still a conceptual rather than a physical construct because it is simply a description of the database, including its tables and the relations among them. The relationships between a data model and a database schema are explained in detail in Chapter 3.

A *database engine*, which is also commonly referred to as a *database server* (for example, Microsoft Jet Database Engine and SQL Server, Oracle Database Server, and IBM DB2 Universal Server), is a collection of computer programs that manipulate the data contained in a database. It serves as the interface between the data in the database, as described by the schema and application software tools that process the data.

In database implementation projects, the development of the data model and the database schema is part of a *design process*. The actual construction of the database starts when the database administrator implements the schema by instructing the database engine to prepare the physical space where the database will reside on the computer's hard drive. This combination of physical structure and data is formally defined as a database in the context of this book. In addition to data files, the database contains a *data dictionary* that describes the contents of the database, and a set of *database integrity rules* that must be enforced in order to protect the data maintained in the database. In addition, it also contains *defined views*, which are logical extractions of specific aspects of the physical database, as well as *stored procedures*, which are blocks of *Structured Query Language* (SQL) code for defining, managing and querying data in the database.

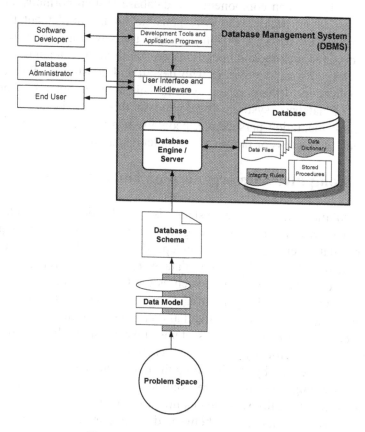

Figure 2-1. Database terminology

The term "database" does not include the *user interface* (UI), commonly referred to as the database front-end, by which users access and interact with the database. It also does not include application programs that are used to process and analyse data in the database, nor the software development tools used to create these application programs. In addition, it excludes communications software tools, known as *middleware*, that support data transmission and database processing over local area, wide area, or global communications networks. To describe a database and its associated components (the database engine, user interfaces, application programs and middleware) the terms *database management system (DBMS)*, *database system*, or *information system* are commonly used. In this book, the term "database system" is used because it is more popular than "database management system" in today's computing parlance.

Some authors, notably Date (2004) and Elmasri and Navathe (2003), expand the definition of database systems to include people as an integral component. The human component of a database system normally includes the database administrator, who is responsible for the implementation and maintenance of the database system, and end users who use the system for business decision making and operations. In addition, the human component also includes software developers who design and construct software tools for the application of the database system. This more embracing or holistic definition of database systems is adopted in this book because, from a systems implementation perspective, people typically play a crucial role in determining the success or failure of an implementation.

2.2   Computer data organisation and database

The evolution of database systems is closely tied to the changing methods of electronic data processing. In the early days of computer data processing, data were organised as independent data files. Access to data files was achieved by defining how data were stored and retrieved, using high-level programming languages such as the Common Business Oriented Language (COBOL) or Programming Language One (PL/I). To illustrate this, Figure 2.2(a) shows how individual data files were managed and used in *file processing systems*. In these systems, a particular application program was developed specifically to manage each individual data file and to generate special reports by manipulating the data contained in each data file.

File processing systems were simple in concept and their data management functionality was rather limited in practice. One of their major drawbacks was the tight coupling between data files and applications. In file-based data processing, all application programs were purpose-built for specific files. This was often a very time-consuming and expensive task. Whenever a change was required in an application due to changes in business functions, the data file had to be changed as well. Also, since different data files tended to have different structures and storage formats, it was difficult to form connections between data from different data files without extensive application programming and format conversion. As a result, it was practically impossible to represent relationships among data items in disparate files, which restricted the data from being used for an advanced level of information extraction and decision making.

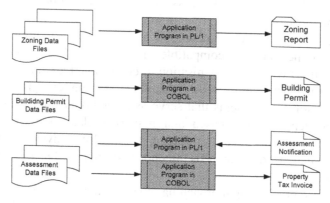

(a) File-based data processing

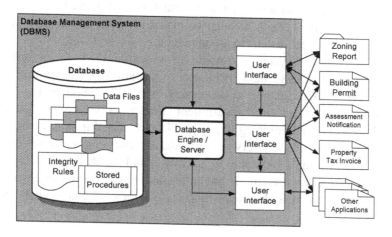

(b) Integration of data files and processing procedures in a DBMS

Figure 2-2. File and database processing

The development of modern database systems was driven by several factors. These include the desire to overcome the limitations of conventional file processing applications, and to take advantage of advances in computer technology and the changing concept of using data. In a database system as illustrated in Figure 2.2(b), the files in Figure 2.2(a) are integrated into the database. Hence, the database stores not only data, but also a description of the characteristics of the data and database integrity rules. Users can query and update the database by obtaining specific views of the database through a user interface and according to their respective security privileges.

Database processing differs from conventional file-based data processing in that it separates data from applications. This means that any changes that occur in business functions will necessitate changes in application programs only, and the database itself can remain intact as all data files are treated as an integrated whole. Since the data files are created by a single database system, all of the data are compatible. This allows data from different business functions within an organisation to be managed and manipulated in an integrated manner. Hence, today's database systems comprise complex hardware and software systems that serve as a total data library capable of managing conventional text-based and numerical data as well as raster images, vector graphics and multimedia files. More advanced database systems also include data analysis functions to support business decision making (see Chapter 11).

The database processing approach has changed not only the way data are organised in a computer, but also in the ways data are used. A database is perceived not merely as a collection of data files, but as an important asset for an organisation. As such, it must be managed in the same ways human, financial and technical resources are managed.

Modern database systems provide the essential tools for information resource or asset management in business, education and government. Up until the early 1970s, database processing was only found in the largest business corporations and government departments. Today, database processing has become an integral part of IT, and is used in practically every aspect of modern society. Every year, database processing generates billions of dollars of business revenue in software and hardware development, database construction and implementation, as well as in management and consulting services. It is the most important sector of the computer and IT industries.

2.3   Classification of database systems

Database systems can be classified in a variety of ways according to different criteria (see Table 2.1). Conventionally, they were classified according to the different data models on which they were built. These models fell into three categories that characterise the evolution of database systems, namely hierarchical, network and relational. A fourth class called object-oriented database systems emerged in the 1990s as a result of the advances in what is now commonly known as object-orientation technology. More recently, hybrid systems have been developed by taking advantage of relational and objected-oriented concepts and techniques. Such systems are commonly referred to as object-relational database systems.

Table 2-1. Classification of database systems

Classification Criteria	Categories
Data Models	o Hierarchical Systems
	o Network Systems
	o Relational Systems
	o Object-oriented Systems
	o Object-relational Systems
Primary Database Functions	o Data Storage or Inventory Systems
	o Transaction Systems
	o Decision Support Systems
Nature of Data	o Spatial Information Systems
	o Non-spatial Information Systems
Objectives of Information	o Custodial Systems and Data Warehouses
	o Project-oriented Systems
Hardware Platforms and Systems Configurations	o Distributed Systems
	o Desktop Systems

This classification of database systems is, however, somewhat misleading as the five categories represent more the different stages of the evolution of database technology than distinct typological classes. Hierarchical and network database systems, which were used on mainframe and mini-computers in the 1960s and 1970s, are now of historical interest only in the database world (Ullman and Widom, 1997).

The concept of the relational database system was first proposed by Codd (1970). The relational system sought to overcome the database rigidity of hierarchical and network systems, and to insulate database users from the physical implementation of the system. The central building block of a relational database system is the *table*, which represents a collection of data pertaining to a particular business area of an organisation. Advances in relational database concepts and techniques quickly established the relational class of database systems as the *de facto* standard in the database world.

The introduction of object-oriented programming concepts and techniques in the 1990s led to the development of object-oriented database systems. In object-oriented database concepts, the problem space is perceived as being composed of *data objects*, which are individually identified as *observable entities* that are grouped into *classes* by the similarity of their respective properties. Today, practically all database systems are based on relational or object-oriented design principles, or a combination of both. The characteristics of relational, object-oriented and object-relational database systems are examined in detail in Chapter 3.

There are several other ways of classifying database systems. One of these is to establish different types according to primary database functions, such as data storage, transaction processing, and decision support. Data storage or inventory systems are designed basically for the management of large volumes of relatively static data. Natural resource inventories and land information systems are typical examples of this class of database system. Transaction processing systems, such as banking and airline seat reservation systems, are designed to serve business functions that require frequent changes in the database. These systems are driven by very powerful computer processors for instantaneous access to and modification of the database contents, and are capable of serving many users concurrently. Decision support systems are designed to help end users make logical decisions on semi-structured and structured decision problems. The domain and architectures of decision support systems vary widely ranging from relatively simple systems that serve the operational needs of a particular function such as wildlife management or land use permit processing, to highly complex management information systems (MIS) that provide summary information from all of the data files in a corporate database for senior decision making.

Another way of classifying database systems is to use the characteristics of the data in the database as the principal criterion of classification. Database systems classified in this way can be labelled either as spatial or non-spatial in terms of their contents. The former includes GIS database systems designed to process location-based data pertaining to land resources, the natural environment, as well as human activities associated with resources and the environment. The latter includes a wide variety of systems in domains such as banking, accounting, human resources management and customer records systems. Historically, spatial and non-spatial database systems were developed independently of one another. However, in recent years, these two types of systems have become increasingly integrated into a single corporate or enterprise database, sharing the same data and computing resources. As a result, although spatial database systems are still perceived as a special type of information system the conventional boundary between spatial and non-spatial database systems is rapidly disappearing.

It is also possible to classify database systems into *custodial* and *project-oriented systems* according to the objectives of an implementation. Custodial systems are large-scale, domain-specific data stores set up to serve the long-term information needs of an organisation or a community of users. National, provincial, and regional natural resource inventories and topographic mapping systems are typical examples of custodial database systems. Many custodial database systems are now constructed on various interpretations of the concepts of data warehousing. In this context, a *data warehouse* is different from a conventional database in that it is not only a data repository and an information management tool, but it also represents a new approach of managing and thinking about data (see Chapter 6).

Project-oriented database systems, as the name implies, are systems created to serve a particular function or event that lasts for a relatively short period of time. Once the objectives of the function or event are achieved, the database will cease to operate. Typical examples include those set up for university-based research projects and environmental assessment exercises by government agencies associated with particular land development proposals. Unlike custodial systems, these project-oriented systems are *ad hoc* applications and have a comparatively short operation life span.

It should be noted that the above database system types are not mutually exclusive of one another. Database systems are generally relatively complex systems whose characteristics transcend the boundaries of different classification schemes. For example, a spatial database system can be designed as a data repository, a transactional system, or a spatial decision support system and can contain equal aspects of all three functions. It can be constructed using a derivative of the relational data model such as geo-relational data model, the object-oriented, or object-relational data models. At the same time, depending on the purpose of the system, it can be implemented as a custodial system or a project-oriented system.

3. DATABASE OPERATIONS

Database systems require several processes and procedures to operate in an articulate manner. The working principles of database systems are explained in the following sections by examining the characteristics and requirements of the major data processing operations in a typical database.

3.1 Database Storage and Manipulation

Contemporary database systems are characterised by the large volumes of data they are designed and required to handle. Banking and retailing systems, for example, often store terabytes (1,000 gigabytes, or 10^{12} bytes) of data recording details of every transaction or sale over a long period of time. Remote sensing databases that store satellite imagery may require several petabytes (1,000 terabytes, or 10^{15} bytes) to store the images acquired in a single year.

In order to accommodate this huge volume of data, database systems store data in arrays of disks, called *secondary storage devices*. When data are required by a query, they are read into the Central Processing Unit (CPU), which is called the *primary storage*. Some large database systems require the use of *tertiary storage devices* that have terabyte storage capacities. Unlike secondary storage devices, tertiary storage devices cannot be accessed directly by the reading device of the computer. When data in tertiary storage are required, the storage medium (for example, a digital video disk [DVD]) is retrieved and loaded into the data reader, and the data are transferred onto the secondary storage device. The typical access time for data in secondary storage devices is 10 to 20 milliseconds, whereas it commonly takes several seconds to access data on tertiary storage devices.

Both secondary and tertiary storage devices are divided into *disk blocks*. These are regions of contiguous storage locations containing 4,000 to 16,000 bytes of data. Data are transmitted between secondary or tertiary storage devices and the main memory by disk blocks, not by data files (see Section 5.2). Database systems control the storage of and access to data in the storage devices by means of the database engine, as noted Section 2.1. There are two pieces of data manipulation software in the database engine:

- The *buffer manager* handles the main memory by allocating the data read from secondary storage devices to a specific page in the main memory. It re-uses the page by allocating it to new data coming into the main memory if the existing data are no longer required.
- The *file manager* keeps track of the location of data files and their relationships with the disk blocks. When a specific data file is required, the file manager locates the disk block(s) in the secondary storage device containing the file, and sends them to the buffer manager which then allocates the block(s) of data to a main memory page.

The use of the database engine separates the users from the physical implementation of the database, and provides them with the processes that are needed to access the database contents. The separation of applications from databases is fundamental to all contemporary database design principles. It allows different groups of users to access different parts of the database, each using his/her own interface that is designed to do exactly what is required of the data. All application programs accessing the actual data files are coordinated and controlled centrally by the database engine, and the rules of database integrity and validity can be enforced relatively easily, as explained in the next section.

3.2   Database Security and Integrity Constraints

Closely related to database storage and manipulation are the concepts and methods of *database security* and *integrity constraints* that are designed to protect the data in the database from being corrupted, compromised, or destroyed.

In the context of database management, security refers to all rules and measures that are enforced to protect the database against unauthorised use, and/or modification or destruction of database contents. Modern database systems typically provide two forms of data security:

- *Discretionary security*, also known as privileges, controls the ability of users to access specific data files, records or fields in a particular mode (for example read-only, read-write).
- *Mandatory security*, classifies users and data into different security levels, and then implements the appropriate security measures to allow users within a particular security level to access the data of the level that they are allowed to access.

Although database integrity constraints are also designed to protect the database, they are distinct concepts and different techniques are used for their implementation. Instead of protecting the data from unauthorised access and physical destruction, integrity constraints are applied in database systems to protect the intangible "value" of the data by safeguarding their accuracy, correctness and validity. Database integrity constraints are enforced by applying certain rules, called *business rules*, which govern both the structure and the use of the data in the database. Since the accuracy, correctness and validity of the data must be maintained from the time they are created, it is necessary to apply integrity constraints even *before* the data are populated into the database. Therefore, integrity constraints are always identified during the data modelling process and, as a result, they form an integral part of the database schema and must be enforced during the data collection process.

Business rules are application-specific. This implies that different database systems tend to have different, and possibly unique, integrity constraints. There are, however, constraints that are common to all databases. These include:

- *Domain constraints*, which specify the types of data values such as numeric, character or string, Boolean, date and time, and user-defined.
- *Key and relationship constraints*, which govern the use of entities as primary, secondary and foreign keys in data tables.
- *Semantic integrity constraints*, which are written rules stating what is allowed and not allowed in both data structure and data management.

The concepts of identifying and developing integrity rules and constraints are discussed in detail in the context of their application to data modelling in Chapter 3.

3.3   Database Query

In database processing, a query is a question or task a user asks of a database. Queries are handled by a database tool called the *query manager*. The function of the query manager is to turn a user's high level database access or manipulation command in SQL (usually pronounced as either "sequel" or as the individual letters spelled out) into a sequence of operations on the stored data (see Section 3.7). In many database systems, the query manager is capable of performing *query optimisation*. This is the process of selecting a *query plan* that will answer the query in the most efficient way. The most commonly used method of query optimization is to use an index created for the database, which allows the system to find the required data quickly and directly without going through the entire database structure (see Section 5.3).

A database query is made up of one or more operators supported by a database system and, as a result, its format and functionality are governed by the database model of the database. The relational database model, for example, uses eight operators in queries to manipulate the contents of tables in a database (Codd, 1970). These relational operators are explained as follows and in Figure 2.3:

- *SELECT.* This operator is used to query the rows of a table. It lists either all of the row values or only those which match specific criteria (Figure 2.3a).
- *PROJECT.* This operator is used to query the columns of a table. It generates a subset of columns from a table, removing duplicate values from the result (Figure 2.3b).
- *JOIN.* This operator horizontally combines (that is, concatenates) data from one row of a table with rows from another table using the relationship between particular columns in the two tables (Figure 2.3c). A JOIN is called an equi-JOIN if it is based on equality of column values. It is called a natural-JOIN when redundant columns are removed.
- *PRODUCT.* The product of two tables, which is also called the Cartesian Product, is obtained by concatenating every row in one table with every row in another table (Figure 2.3d).
- *UNION.* This operator generates a new table by appending rows from one table with those of another table (Figure 2.3e). Tables used in the UNION operation must be compatible in terms of the number of columns and data types. Duplicate rows are eliminated during the UNION operation.

- *INTERSECT.* This operator generates a table consisting of all rows appearing in both of two tables (Figure 2.3f). Tables used in the INTERSECT operation must be compatible in terms of the number of columns and data types.
- *DIFFERENCE.* This operator generates a table consisting of all rows that appear in the first but not the second of two tables (Figure 2.3g). Tables used in the DIFFERENCE operation must be compatible in terms of the number of columns and data types. As with arithmetic, the order of subtraction is important. Therefore, Table R – Table S is not the same as Table S – Table R, as illustrated in Figure 2.3g.
- *DIVIDE.* This operator takes a binary (that is, two-column) table and one unary (that is, one-column) table, and creates a new table consisting of all values of one column of the binary table that match, in the other column all values in the unary table (Figure 2.3h).

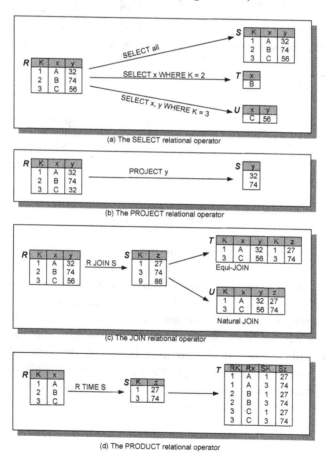

Figure 2-3. Relational operators

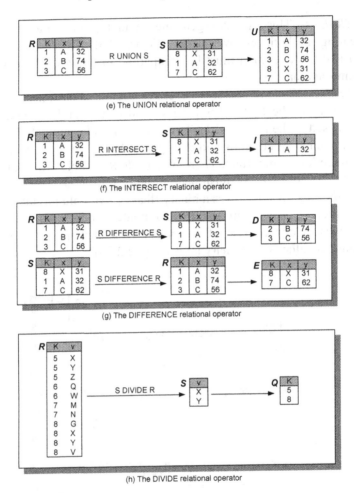

Figure 2-3. Relational operators (continued)

In theory, the outcome of a query using any of these operators is a new relational table as shown in Figure 2.3. In practice, however, most database systems generate data listings that are more "readable" than data tables by having customised formats complete with column headers and other types of explanatory information. It should also be noted that not all relational database systems are capable of supporting every one of the above operators. However, virtually all systems support the most commonly used and important operators, including SELECT, PROJECT, JOIN, UNION and INTERSECT.

3.4   Database Transactions

Database queries retrieve the values of specific data items stored in the database. They do not change the original values. In contrast, when values are changed, a *transaction* is said to have taken place. A database transaction is a more complicated process than a database query because of the need to handle possible conflicts caused by concurrent transactions (that is, when two or more users access the database system and attempt to change the same values at the same time). In order to avoid potential conflicts between concurrent users, database transactions are designed according to the following principles:

- *Atomicity*, which means that either all of a transaction is executed or none of it is (that is, a transaction can never be completed only partially).
- *Consistency preservation*, which means that data in a database remain, before and after the transaction, in a "consistent state" as specified by the database schema and other constraints and integrity rules imposed on the database.
- *Isolation*, which requires the results of simultaneous transactions to be independent of each other.
- *Durability* or *permanency*, which means that after the completion of a transaction, its results can and always will be traced even if the system fails or crashes.

Each and every transaction in database processing is strictly controlled by the *transaction manager* of the database engine that enforces the above principles by following the sequence of operations shown in Figure 2.4. In addition, four transaction control mechanisms are also used. These are as follows:

- *Concurrent control*, which "locks" the data items involved in the transaction, thus isolating its results from affecting other transactions, and at the same time avoiding them being affected by the results of other transactions.
- *Logging the transactions*, which keeps track of all changes made to the database in a "redo" log, and backs up the log on a secondary storage device so that the log data will survive a power failure that may cause loss of data in the main memory.
- *Transaction commitment*, which prevents any changes to the database unless the transaction is ready to complete, and the changes are logged.

- *Rollback,* which allows the database to "undo" an incomplete transaction process and return to its original consistent state if and when the system fails before the transaction is complete.

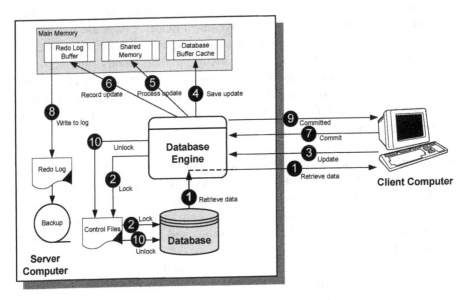

Figure 2-4. Steps in a database transaction

3.5   Database Backup and Recovery

The methods of logging, transaction commitment and rollback described in the previous section are designed to handle relatively minor disruptions to data processing that prevent a transaction from completing. These data recovery techniques work well as long as the transaction log is not lost due to disruptions such as power failure or improper shut down of the database.

However, if a more catastrophic failure such as a disk crash occurs, then more sophisticated procedures are required to restore the database to its consistent state prior to the failure. The standard database management procedure in these cases is to use the *backup and recovery manager* of the database system to copy periodically the entire database and the transaction log on to an external storage medium such as magnetic tape, CD-ROM and DVD. In the case of critical data, a second copy of the backup is often made and stored in a secure site away from the place where the database system is housed, for example, in another building.

The frequency of backup can be daily or weekly, depending on the nature of the database applications and the importance of the data. Some database systems allow *hot backup* which takes place while the database system is in operation. A hot backup can be used to recover the database to the point in time of the system failure. In systems where only *cold backup* is allowed, the database must be shut down during the backup process. Database recovery from a cold backup will only restore the database to the time when the last backup was made.

In database management, it is common practice to back up the transaction log more frequently than the entire database because the size of the log is substantially smaller than the database. This approach actually allows the database to be recovered not only to the date of the system failure, but to the date of the backup of the transaction log. After a system crash, the database backup tape is first loaded into the computer to reconstruct the database to the date of the latest database backup. The backup tape of the transaction log is then used to re-create all the transactions that it stores to the latest date of the transaction log backup.

3.6 Database Replication and Synchronisation

Database replication is the process of making a copy, or *replica*, of a database onto one or more additional computers that may be located at different sites. The computers that the database is replicated to are called *replication servers*. Many organisations require the use of database replicas in distributed database systems to support their business needs:

- To improve system performance by allowing users in satellite offices to use a local replica of the database rather than to communicate directly to the central database server.
- To improve database availability by ensuring that data always remain accessible when the database server is shut down, for example, for systems maintenance.

Ideally, database replication should be "transparent" to the user, meaning that they can use the replica as if it were the master database. In practice, however, transparent replication is not easy to implement because it is relatively difficult to keep the master database and the replicas synchronised. A database replication environment requires the master database server as well as all replication servers to be running in order to "commit" a transaction on all the systems simultaneously. The transaction will fail if any one of the systems is unavailable at the particular instance of the transaction commitment. Various methods have been proposed to overcome this problem but none of these has appeared to satisfy completely the principles of database transaction as noted in Section 3.5 (Date, 2004).

3.7   Structured Query Language (SQL)

In database processing the user uses SQL to interact with the computer. Originally designed and developed to retrieve data from relational databases, SQL is now the standard language for querying as well as managing both relational and object-relational databases. Since its adoption by the American National Standard Institute (ANSI) as a standard database language in 1986, SQL was formally accepted as an international standard by the International Organization for Standardisation (ISO) and the International Electrotechnical Commission (IEC). It has also been adopted as a Federal Information Processing Standard (FIPS) of the United States Federal Government.

SQL is a non-procedural computer language because it does not have IF statements for testing conditions, nor does it contain WHILE, FOR, GOTO and CASE statements for program structuring and flow control. It is best described as a *database sub-language* used specifically to create statements for database query and management by means of approximately two hundred reserved words. These statements resemble sentences in the English language. They include "noise" words that do not add meaning to the statements but make them read more naturally.

SQL statements can be used in one of the following five ways:

- *Interactive processing*, through a command-line user interface (for example, "sqlplus" of Oracle, "db2i" of DB2, and the open source "psql" of PostgreSQL and "mysql" of MySQL).
- *Embedded in a high-level computer language*, such as COBOL, FORTRAN, PL/1, C++ and Java, to provide the necessary facilities for application programs written in a high-level language to query a database.
- Using a *call level interface* (CLI), which consists of routines implemented at the operating system level (as opposed to the application program level in the case of embedded SQL noted above). A CLI can be invoked to process SQL statements from programming languages, such as C++ and Visual Basic (VB), at execution time to connect to a database, extract data from data tables, obtain data processing status information, and present the results of a database query.
- Using two standard Java *application programming interface* (API) protocols, namely Java Database Connectivity (JDBC) and embedded SQL for Java (SQLJ), to create Java applications and applets that are able to extract data from a database.
- In stand-alone application program modules in the form of *stored procedures* (blocks of code that perform one or more data processing

actions), *functions* (blocks of code that take parameters to be invoked in a procedure), or *packages* (a collection of stored procedures and functions grouped together under one module name). These program modules are created by adding programming structure, flow control and subroutines to the conventional SQL processing environment. A good example is Oracle's Procedural Language Extensions to SQL (PL/SQL).

SQL statements specify what tasks must be done, but not how they are done, in database operations. To query a database, for example, it is not necessary for a user to know exactly how the database is structured or the complex procedures that take place to obtain the required results. When the database engine receives SQL statements in one of the above forms, its query manager transforms the statements into a sequence of internal operations that optimise data retrieval and present the query results as specified. Therefore, SQL is relatively easy to learn and use. Figure 2.5 illustrates how SQL is used in typical database operations that include:

- *Database query*, which allows a user to retrieve data stored in a database (Figure 2.5a).
- *Data definition*, which helps a user to organise data and establish their relationships in the database (Figure 2.5b).
- *Data manipulation*, which allows a user to modify the contents of a database by inserting new data, removing old data and changing the values of existing data (Figure 2.5c).
- *Database connection and access control*, which enforce database security measures by enabling/disabling a user's privilege to access or modify a database or a part of it (Figure 2.5d).
- *Data sharing*, which coordinates data replication (Figure 2.5e) and controls concurrent access to a database.
- *Data integrity*, which defines integrity constraints that protect a database from corruption due to inconsistent data input, database update or systems failure (Figure 2.5f).

Under the auspices of the various national and international standardisation organisations noted above, SQL is continuously evolving in order to take advantage of advancing computing technologies and to respond to changing user needs. Development is currently underway to enhance SQL into a computationally complete language. Enhancements in the new SQL standard, called SQL3, which were published by ANSI and OSI in 1992, include user-defined data types, multimedia objects, distributed data processing support and other facilities normally associated with object-oriented data management. The enhancements of SQL with special reference to spatial data are explained in Chapter 4.

SQL Function / Example	Explanation
(a) Data retrieval SELECT parcel_id, area FROM lu_2002 WHERE lu_code = 'agr'	These SQL commands retrieve the identification numbers and areas of parcels whose land use code is 'agr' (agricultural) from data table lu_2002

(b) Data definition
 	
CREATE TABLE lu_2002
        (parcel_id   VARCHAR2(8)   PRIMARY KEY,
        lu_code   VARCHAR2(3)   NOT NULL,
        area   NUM(8,2)   NOT NULL,
        survey_date   DATE DEFAULT SYSDATE)
	The SQL command **CREATE TABLE** specifies the structure and data types of the data table lu_2002. In the example, **VARCHARS, NUM** and **DATE** specify the data types of the four columns of the table; the numbers in the brackets specify the number of characters and digits string and numeric data types respectively. The value of survey_date will be set to the system date automatically if no date value is supplied duirng data entry. Note how the constraints **PRIMARY KEY** and **NOT NULL** are specified.

(c) Data manipulation INSERT INTO lu_code VALUES ('12322', 'res', 4500.00, '06-may-2002')	These SQL statements add data into the table lu_2002. Character strings are put inside quotation marks but numerical values are not. Since a date value is given, the default system date as specified in the data structure in the previous example will be not used.
(d) Database connection and access control CONNECT system/passwordxxx CREATE ROLE db_maintenance GRANT INSERT, UPDATE ON lu_2002 TO db_maintenance GRANT db_maintenance TO john.young	These SQL commands connect to the system, create a database maintenance role, and give this role to a user called john.young

(e) Data sharing
 	
CREATE TRIGGER copy_data
AFTER INSERT ON lu_2002
FOR EACH ROW
BEGIN
        INSERT INTO lu_2002_copy@fes.uwaterloo.ca
        VALUES (:new.parcel_id, :new.lu_code, :new.area,
        :new.survey_date);
END;
/
	This sequence of SQL commands copy newly input data to a data table at a remote site

(f) Data integrity
 	
CREATE TABLE lu_2002
        (parcel_id   ARCHAR2(8)   PRIMARY KEY,
        lu_code   VARCHAR2(3)   NOT NULL,
        area   NUM(8,2)   NOT NULL,
        survey_date   DATE DEFAULT SYSDATE,
CONSTRAINT UNIQUE (parcel_id),
CONSTRAINT area_chk CHECK (area > 0.00)).
	This example shows how data input constraints are imposed to ensure that the parcel_id is a unique number for each land parcel and that the area is not a negative value

Figure 2-5. Examples of database operations using SQL

4. HARDWARE AND SOFTWARE ARCHITECTURE

Database systems can be implemented on different computer platforms ranging from a single desktop personal computer to a cluster of computers and workstations in a networked environment. Since it is practically impossible to describe all types of system configurations, the discussion is limited to central and distributed architectures using the client/server mode of processing on which most database systems in academic, commercial and government organisations are implemented.

4.1 Centralised and Distributed Database Architecture

Early generations of database systems were developed during the mainframe era when a single mainframe computer served hundreds of terminals. Mainframe-based database systems contained the database, managed database operations and stored applications all in one computer. This centralised database architecture dominated relational database systems when they were first introduced.

Figure 2.6 illustrates the configuration of a centralised database system in which all processing is done in a single computer, and all data are stored in the same computer's secondary memory. A centralised database system usually runs under a time-sharing and multitasking operating environment, which allows several processors to run concurrently on a single CPU. Users access the database via "dumb" terminals connected to it locally or over a telephone line. A *dumb terminal*, as its name implies, is used only as a means to query the database and display the results. It does not contain any application programs or data retrieved from the database. All database input/output (I/O), concurrent control and applications are done on the host computer.

The advent of microcomputers and advances in communications technologies in the 1990s rapidly changed the hardware configuration of database systems. Most database systems today are constructed on the concept of distributed processing which separates database operations from data storage functions. Figure 2.7 shows the configuration of a distributed database system that typically includes the following components:

- Computers, operating system and database system software that form individual *sites*, also called *nodes*, of the system.
- *Network cards* and *software* that are installed in the computers to enable them to interact and exchange data with one another regardless of their respective configurations.

- *Communications network protocols* (for example, the Transmission Control Protocol/Internet Protocol [TCP/IP]), that carry data from one computer to another.
- A *data processor*, also known as a *data manager*, which is the database engine that stores and retrieves data located at a particular site.
- The *transaction processor*, which is the software in each computer that requests data from its own and other databases. The transaction processor, also known as the *transaction manager* or *application processor*, controls all database transactions (for example, inserting new data, removing old data and modifying values of existing data) and related database operations (for example, concurrent control, transaction commitment and rollback, and data replication).

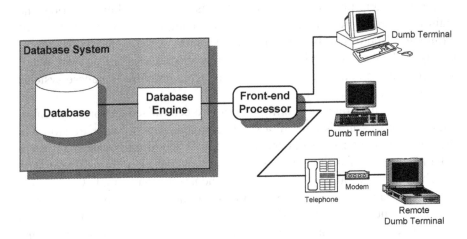

Figure 2-6. Configurations of a typical centralised database system

A distributed database system not only allows the physical separation of processes and data by storing and managing them in different computers in a communications network, it also permits the partitioning of a particular process or data file into smaller units residing at different sites. This approach to the division of work in a distributed database system seeks to optimise and share the resources of its constituent sites. It is made possible by enforcing a number of transparency features in database architecture and operation, including:

- *Distribution transparency*, which allows a user to access any database within the system without the need to know where and how the data are stored.

- *Performance transparency*, which is designed to make a distributed database system work and behave as if it were a single centralised database system.
- *Transaction transparency*, which allows a user to update the databases at different sites.
- *Heterogeneity transparency*, which enables the integration of different database systems (for example, relational, object-oriented and object-relational) under a common schema.

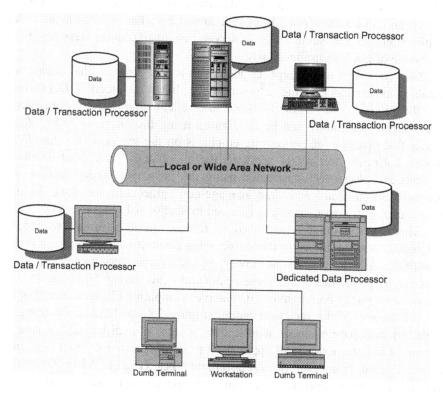

Figure 2-7. Configuration of a typical distributed database system

A distributed database architecture, therefore, is defined not only by the geographical distribution of computers located at different places, but also by the physical and logical distribution of database operations and data storage in these computers. Distributed database systems are now mostly implemented using the client/server computing model as explained in the next section.

4.2   Client/server Computing

Client/server computing is the method of running different processes on separate computers by sharing their resources. The idea of splitting the data processing environment along client/server lines was developed when various forms of Local Area Network (LAN) solutions were created in the late 1980s (Orfali et al., 1999). In the client/server computing environment, the computer that requests a service from another computer is called the *client,* and the computer that provides the service is called the *server.* A client can request the service from several servers when running a particular application. At the same time, a server can also provide services to a multitude of clients simultaneously.

The client/server concept is increasingly adopted in the design of database systems as they move toward a distributed architecture. Depending on the division of work between the client and server computers, client/server computing can be configured using the *fat server, thin client model* that places more processing emphasis on the server, or the *fat client model* hat does most of the work on the client-side. Centralised database systems are obviously constructed using the fat server model because all database applications and data management processes are done in the mainframe or minicomputer, as explained in Section 4.1 (Figure 2.8a).

When microcomputers are used to replace dumb terminals to access databases, some of the application processing functions can be done on these machines. As the processing power of microcomputers increases, it is possible to move more and more application processing functions to the client side, and the paradigm of client/server computing changes accordingly toward the use of the fat client model (Figure 2.8b and 2.8c). By using a high-end microcomputer or workstation, a site in a distributed database system is able to perform all application processing using its own data and data retrieved from one or more remote sites (Figure 2.8d). Database systems in which applications running on one computer rely on the data processing and/or database management services of another computer are said to have a *two-tier client/server architecture* (Figure 2.8a to 2.8d).

The *three-tier client/server architecture* is an extension of the two-tier model. In this architecture, the functions of each tier are dependent on individual implementations. However, the tiers are typically configured as follows (see Figure 2.8e):

- A *client,* used for interaction with the database.
- An *application server,* where application programs are stored and executed.
- A *database server,* used for storage and retrieval of data.

In a three-tier architecture, the data processing tasks are moved from the client to the application server tier. This shift of workload away from the client means that there is no longer any need for using costly high-end computers as client machines. The three-tier architecture is better known as a thin-client configuration. Web-based database systems on the Internet and Intranets are invariably constructed on the three-tier architecture in which the browser supplies everything required for assessing both the database and the application servers. This topic is discussed further in the context of spatially enabled Internet-based applications in Chapter 10.

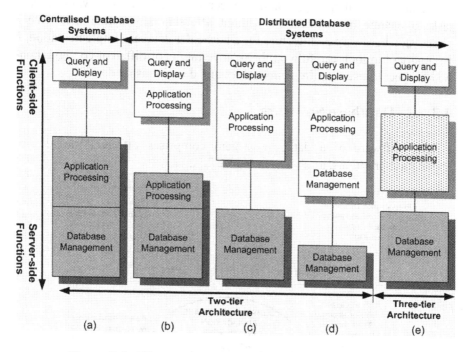

Figure 2-8. Client-server computing for database systems

A three-tier architecture has many advantages over a two-tier one. Typically, in the three-tier case the application server is more powerful than the client(s), and the performance of the system improves as a result. Performance is further improved by reducing the amount of network traffic between the database server and the client(s). There are also considerable cost savings in operating the systems in the long run. On the one hand, the cost of setting up the database system is lower as there is no need to buy resource-intensive computers to use as clients. On the other hand, because the applications reside on the application server, most of the upgrades are required only on one or a small number of computers. Further, the maintenance of one application server is generally easier and less costly than maintaining many client computers.

However, it should be noted that the three-tier architecture is not a panacea for all database systems. A two-tier system in a well-configured network, with controlled client machines and well-tuned applications, is just as cost-effective as a three-tier system to operate and maintain. Further, supporting a three-tier architecture is probably more technically challenging than supporting a two-tier one. For example, communication in a three-tier system is more difficult to design and support than that the two-tier case. Since the usage of the applications is centralised, it is not always an easy task to change an application without affecting any single user. When choosing a hardware architecture for a particular database implementation, it is important to look carefully at the relative merits of different configurations with respect to user requirements before any decision is made.

4.3   Database Software

The software of a database system comprises different modules of application programs, as shown in Figure 2.9.

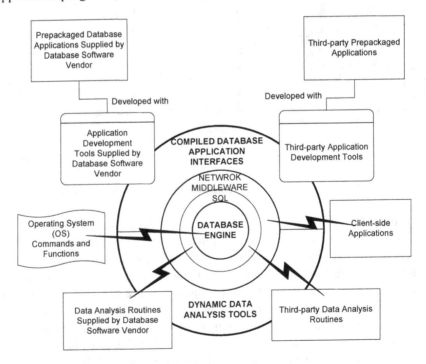

Figure 2-9. Software layers of a database system

Central to the software system is the database engine which, as noted in Sections 2.1 and 3.1, is responsible for storing, retrieving and manipulating data in the database, and managing background processes, such as data conversion, transaction logging and memory management, that are associated with the storing and manipulation of data. The functions of the database engine are dependent on the operating system of the host computer. This is why Microsoft Windows-based database systems work differently from their UNIX- or Linux-based counterparts.

Earlier generations of database systems required COBOL or PL/1 programmers to code mechanisms for managing data storage and retrieval routines that interacted directly with the files in the database. Modern database systems handle these tasks using SQL or programs written in SQL extensions (for example, Oracle's PL/SQL) that allow users to request the data that they need. SQL extensions are also used by database software vendors and third-party consultants to develop "front-end" applications that can run on the database server, the application server or the client machine. These applications include data analysis modules, form generators, report generators and graphics generators.

For database systems that use a client/server architecture, client and server machines are connected in a network and communicate with one another using database connectivity middleware (for example, Microsoft *ActiveX Data Objects* (ADO), *Open Database Connectivity* (ODBC), *Common Object Request Broker Architecture* (CORBA), and *Java Database Connectivity* (JDBC)). A *middleware tool* is typically made up of an API on the client machine to invoke a request, and communication software tools residing on both the client and server to facilitate the transmission of the request over the network and the resulting responses.

The proliferation of database technology has resulted in the growth of a support and consulting industry for custom software tools and packages for different application domains including finance, human resources, goods inventory, and health care among others. Database users can now purchase software solutions from database software vendors or third-party software developers to minimise start-up cost and time.

4.4   Web-based Database Architecture

The advent of the Internet and the proliferation of the World Wide Web (Web) protocol have had significant impacts on the development of database systems. By using a Web browser as an interface, the Internet enables a user to access data located anywhere in the world almost in real time. Given the advantages of the Internet as a means for efficient information dissemination and communication, many organisations have implemented their database systems using a *universal access architecture* based on Internet standards. Such an architecture allows users within an organisation (in an *Intranet*), as well as users outside of the organisation (in an *Extranet*), to access its databases dynamically. A database that is connected to the Internet and is also made accessible to internal and external users within an organisation is commonly called a *Web-based database*. The role of the Web in spatial database implementation is discussed in detail in Chapter 10 and discussion here is limited to its implications for database systems.

The architecture of a Web-based database can be implemented in different ways, depending on the hardware and software configurations being considered. Figure 2.10 shows the architecture of a typical Web-based database and the flow of information when a user uses a Web browser to query a remote database dynamically.

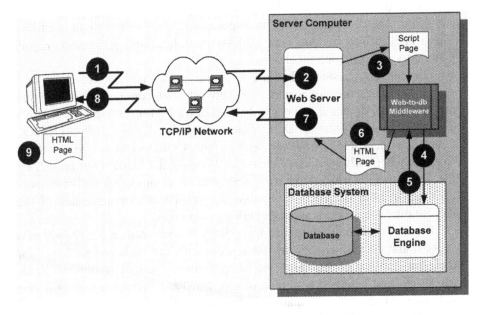

Figure 2-10. Architecture of a typical Web-based database (Note: the numbers in the circles show the sequence of flow of information in a database query)

The architecture of such a Web-based database is a three-tier client/server configuration as explained in Section 4.2. It is made up of the following components:

• A client computer where a user submits a request to the database by means of a *Hypertext Markup Language* (HTML)-formatted page that is sent over the Internet to the Web server using the *Hypertext Transfer Protocol* (HTTP).

- A Web server, which is equipped with a program called a *server-side extension* (also known as *Web-to-database middleware*) that understands, validates and processes database queries using either the *Common Gateway Interface* (CGI) or the API protocols.

The database server, which receives queries from the Web server (which may reside in the same or a different computer), then dynamically generates an HTML-formatted page that includes the data retrieved from the database, and sends it back to the Web server. The client computer in the architecture in Figure 4.2 is a thin client, as defined in Section 4.2. It is possible to enhance the functionality of the Web browser on the client computer by adding *client-side extensions*. Such extensions are available in various forms, including:

- *Plug-ins*, which are external application routines that are automatically invoked by the Web browser when needed.
- *Java*, an object-oriented language that can be used to develop applications to run on top of the Web browser.
- *JavaScript*, a scripting language that can be used to develop applications for interacting with the Web server that are embedded in a Web page.
- *ActiveX*, Microsoft's alternative to Java, which extends the functionality of the Web browser by adding "controls" to a Web page.
- *VBScript*, another Microsoft product derived from Visual Basic to develop Web page-embedded applications for interacting with the Web server in a similar fashion to JavaScript.

Web-based databases are now so common that all database administrators and developers are expected to have a good understanding of the principles and methods of these database systems.

5. DATA STRUCTURE

The content of a database includes not only the *application data* that the user collects and stores in the database, but also a data dictionary and the metadata about the data in the database (see Chapter 5). In addition, it includes data that are generated during the operation of the database, such as the transaction log and other database control files that keep track of all data files and the database processes using them. As system-generated data files are database-specific and have little to do with database users other than the database administrator, the following sections focus only on the structure of the application data.

5.1   Logical Data Structure

When discussing data structure in a database, it is important to distinguish between *logical* and *physical data structures*. The logical data structure is a conceptual configuration of how data are organised in a database for optimal performance and ease of administration. The logical data structure is used in logical data modelling that aims to find the most effective distribution of the data and data files inside the database (see Chapter 3).

The concept of a *tablespace* in Oracle's Optimal Flexible Architecture, for example, is a logical data structure. Each tablespace in an Oracle database must consist of one or more physical data structures called *datafiles*. When a data table is created, a user can specify in which tablespace this table is stored. The system will automatically find space for it in one of the datafiles of the tablespace.

Figure 2.11 shows the relationship between tablespace, datafiles and tables in an Oracle database. There are two tablespaces in this database, TBLSPC-1 and TBLSPC-2. When a new table d01.dbf is created, the user can specify its tablespace, and the system will place it in one of the datafiles where space is available.

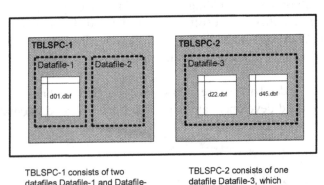

TBLSPC-1 consists of two datafiles Datafile-1 and Datafile-2, and the table d01.dbf is allocated to Datafile-1

TBLSPC-2 consists of one datafile Datafile-3, which houses two tables d22.dbf and d45.dbf

Figure 2-11. Relationship between tablespace, data files and tables in Oracle's optimal flexible architecture (OFA) logical database structure

A tablespace, as a logical data structure, is a conceptual view of the physical storage of the Oracle tables. It helps the database designer and database administrator to organise logically data in the database. Using a logical data structure relieves database designers and application programmers of the burden of managing the physical storage of data because they only need to know the logical tablespace names, and leave the actual storage and management of the physical database structure to be handled by the database system itself.

5.2   Physical Data Structure

The physical data structure is the actual organisation and placement of data files in the database. It describes the physical data files that make up the database. The physical structure of data files in a particular database is dependent on the model of its database system (see Chapter 3). For example, a data file in the relational database model is a collection of rows, or records of data pertaining to an *entity* or *entity class* (for example, provinces of Canada). Each row in a table represents an *instance* or *occurrence* of the entity (for example, Ontario) and is made up of one or more columns pertaining to the *attributes* or *characteristics* of the entity (for example, name, capital city, population, area, annual average rainfall, and average lowest and highest temperature).

Values of the attributes are stored in a table in a certain *data type*. The data types describe and limit the type of data that can be stored in a column as well as limit the database operations that can be applied to the data. Different database systems support different data types but generally all data types can be classified into four basic categories:

- *Character* or *string data type*, including fixed and variable-length characters, and long character type that is capable of holding up to 2GB of text-based data.
- *Numeric data type*, including integer and floating-point numbers of different precisions.
- *Date data type* for storing date and time data.
- Other data types including BLOB and user-defined data called *abstract data types* (ADT).

An important concept in a data type is the *NULL* value. This is a fundamental feature of the relational database. By definition, NULL represents the lack of a value. When a column in a table is declared NOT NULL in database creation, it means that the column must contain valid entries. If no value is assigned to a NOT NULL column during database creation or data processing, an error will be returned by the system.

In the design and administration of database systems, it is important to understand the difference between *data structure*, which is concerned with tables of rows and columns, and *memory structure*, which is concerned with the actual storage and handling of data within the computer's memory. The user collects and stores data in tables, but the system stores and handles data in database blocks. The size of one block varies from one database system to another. It can also be adjusted by the database administrator. When data are moved between secondary and main memory during database processing, they are transferred by blocks rather than tables. This means that if a database query requires a table occupying 2.3 blocks of space in secondary memory, all the three database blocks used by the table will be sent to the main memory, not just the 2.3 blocks of memory space that contain the required data.

5.3   Database Indexing

Indexing is a fundamental database concept and technique. An index is an element of data structure that is used to speed up access to a specific part of the database. It is based on the same concept of using the index of a book to look up information quickly. Many indexing methods have been proposed for tables in relational databases (Date, 2004). However, the *B-tree index* (where B stands for "balanced") is by far the most commonly used form of index (Figure 2.12). This particular indexing paradigm is the default of many commercial database systems including Oracle and IBM DB2.

When a user issues the command to index a table, the database system adds a column to the table that contains a row identification for each row of the table, and creates a separate index file. The B-tree index is composed of one or more levels of branch blocks and a single level of leaf blocks. The branch blocks contain information about the range of values contained in the next level of branch blocks. The number of branch levels between the root or header block and the leaf blocks is called the *depth* of the index. The leaf blocks contain the actual index values and the row identifications for the associated row.

During data processing, if an application asks for the attributes pertaining to a particular instance in the table (for example, FID 1501), the system will scan the index file instead of the table itself for the row identification of the instance of interest and, once it is found (i.e. ROWID 0010 in Figure 2.13), this row identification is used to retrieve the attributes of its associated record from the table.

The B-tree index structure does not contain many blocks at the higher level of the branch blocks. Therefore, it takes relatively few I/O operations to reach the leaf blocks quickly. Since all leaf blocks are at the same depth in the index, all retrievals require the same amount of I/O to get to the required row in the table, thus evening out the performance of the index in the search.

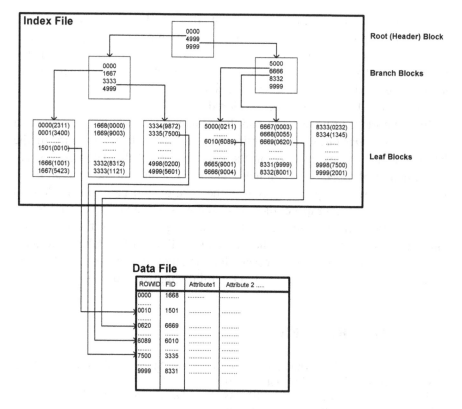

Figure 2-12. A B-tree index

6.   SUMMARY

This chapter presented the concepts and architecture of database systems. A *database* was defined as a collection of structured data, together with the data dictionary that describes the data, the data integrity rules that are enforced to protect the data, as well as stored procedures that enable users to access the database. A *database system* was defined as a data processing system comprising the database and its database engine, user interfaces, application programs and communication software. Database systems are also commonly called a *database management system (DBMS)* or an *information system*. The advent of database systems represents a new approach to data processing and management. Database systems are now found in practically every aspect of modern society.

The working principles of database systems were explained by examining the various operations by which these systems store and process data, as well as procedures and constraints that are developed to protect the database from physical destruction and degradation of the quality of its data. The *client/server systems architecture* was explained along with the construction of database systems using two-tier and three-tier client/server architectures. Also the characteristics and the configuration of a typical Web-based database system were explored.

It is important to understand and distinguish the concepts of *logical* and *physical data structures*, and identify the impacts that the structure of the database and its indexing have on the performance of the database system. The structure of *tables* in relational database systems and the method of *B-tree* indexing were explained, as was the relationship between the logical data structure of a tablespace, the physical data structure of database blocks, and the data storage structure of the operating system of the computer. Knowledge of the general concepts and methods of database systems discussed in this chapter is necessary for the study of database models and spatial database systems discussed in the following two chapters.

7.   REFERENCES

Date, C.J. (2004) *An Introduction to Database Systems*, 8th ed., Reading, MA: Addison-Wesley Publishing Co.

Dodge, G. and Gorman, T. (2000) *Essential Oracle 8i Data Warehousing*, New York, NY: John Wiley and Sons, Inc.

Elmasri, R. and Navathe, S.B. (2003) *Fundamentals of Database Systems*, 4th ed., Boston, MA: Addison-Wesley Publishing Co.

Haas, L. and Lin, E. (2002) "IBM Federated Database Technology", San Jose, CA: IBM.

Lopez, X.R. and Gopalan, A. (2001) "Managing Long Transactions Using Standard DBMS Technology", in *Proceedings* of Annual Conference, Geospatial Information and Technology Association (GITA), Sydney, Australia.

Maguire, D. and Grisé, S. (2001) "Data Models for Object-component Technology", in *Proceedings* of Annual Conference, Geospatial Information and Technology Association (GITA), Sydney, Australia.

Orfali, R., Harkey, D. and Edwards, J. (1999) *The Essential Client/Server Survival Guide*, 3rd ed., New York: John Wiley & Sons.

Ullman, J.D. and Widom, J. (1997) *A First Course in Database Systems*, Upper Saddle River, NJ: Prentice Hall.

Chapter 3

DATABASE MODELS AND DATA MODELLING

1. INTRODUCTION

Development of database systems always starts with a modelling phase that seeks to capture the user's requirements and turn them into technical specifications for implementation. Using a model to assist in the design process is a standard practice in architecture and engineering. Architects and engineers rely on models to validate design concepts, evaluate alternatives, estimate costs, analyse potential risks, develop and refine design specifications, and visualise the possible outcomes of implementing the design. In the design of database systems, the designer uses a model for exactly the same reasons before the actual system is built.

As database systems get larger and more complex, and user requirements become more sophisticated, data modelling as a design paradigm becomes correspondingly more important. Constructing a database system without a proper model is analogous to building a complex engineering structure without a blueprint. No database implementation project should be allowed to proceed without a model that is acceptable to the database sponsor, administrator and developer.

This chapter examines the principles and methods of modelling in database design within the context of the *systems development life cycle* (SDLC). The subject matter is approached first by defining key terms and concepts relating to database models and data modelling. The characteristics of several commonly used database models are then examined for conceptual, logical and physical database design. Finally, advances in the principles and methods of documenting data models and data modelling are discussed.

2.   DEFINITIONS AND CONCEPTS

The ability of a database system to satisfy its intended objectives depends on its design using various models developed at different stages of the database development process. The following sections explain the terminology relating to database models, how these models are developed and how they are used in the implementation of database systems.

2.1   Definition of a Database Model

In the context of database design, a model is considered to be a collection of concepts, language and graphics that are used to describe the data structure and data processing operations in a database. The focus of a model is the description of a database, rather than the methods used to build it. In other words, a database model is only a blueprint for the construction of a database. It describes what is to be included in a database but not how the database is constructed.

To use a common analogy, a database model in database design is like an architectural plan in construction design. An architectural plan describes the components of a building, the relationships among its structural parts, and the building codes and regulations that must be followed in its construction and intended use. An architectural plan does not typically specify the actual methods or procedures of constructing the building as these are left to the building contractor to decide. The function of an architectural plan is to serve as the means of communication between the developer (who stipulates the requirements and intended use of the building), the architect (who translates the developer's requirements into an architectural design), the construction contractor (who constructs the building), and the individual(s) who will occupy the building.

A *database model*, similarly, describes the design of a database but not the ways of constructing it. Just like an architectural plan, a database model also serves as the means of communication between the database sponsor (who commissions the development of the database and is normally assumed to represent the user community at large), the database designer (also commonly known as the data modeller, who designs the database), the database developer (who implements the database by assembling the data, setting up the specified data structure, and loading the data into the database), and the intended end users.

As noted above, a database model is constructed using three building blocks, namely concepts, language and graphics. The word "concept" in the context of data modelling has a special and unique meaning. It refers not only to abstract ideas, but also to tangible and intangible real world features/phenomena that are relevant to the information needs of database users. Since the identification of these features/phenomena in data modelling is largely a mental process, particularly at the initial stage, the word "concept" is an appropriate term to use in this context. As the data modelling process evolves, an identified concept becomes an "entity" in the language of the relational database model or an "object" in the language of an object-oriented database model. Within a database, an identified concept is commonly referred to as a "data object" and its presence or occurrence in the database is called an "instance".

The concepts in a database model are typically developed using a variety of data abstraction techniques and are documented by means of a definitive set of linguistic syntax and graphical notation (Figure 3.1).

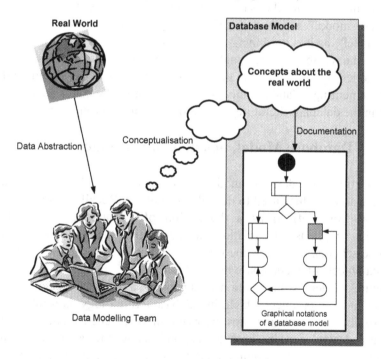

Figure 3-1. The process of data modelling

By using different levels and techniques of data abstraction to conceptualise the real world, and by representing the resulting concepts by means of different linguistic syntax and graphical notation, different types of database models were developed over the past twenty or so years (for example, the hierarchical, network, relational, and object-oriented database models noted in Chapter 2). Further, as database design is an evolutionary process during which the user requirements are progressively refined until they are turned into implementation specifications, different database models describe a database at different stages of the design process (for example, conceptual, logical and physical database models). The characteristics of different database models and the roles they play at the different stages of database design are described in Section 2.3.

Up to this point, the terms "model" and "database model" have been used in the discussion rather than "data model". In the database literature, many authors tend to use the term "data model" (for example, Elmasri and Navathe, 2003; Ullman and Widom, 1997). However, there are other authors who prefer to use "model" instead (for example, Batini et al., 1992; Date, 2004; Orfali et al., 1996). There are also authors who choose to use "database model" (for example, Rob and Coronel, 2002). Although all the three terms refer to the same thing in practice, the terms "model" or "database model" are used in this book to avoid confusion with the term "data model", as it is used in association with the various types of spatial data such as the vector, raster, spaghetti and arc-node data models. The term "data modelling" is retained to describe the processes of developing, refining and documenting a database model since it has a well-established and unique usage in the database literature and within the computer industry.

2.2   Database Model, Schema and Instance

In Section 2.1 a database model was defined as a collection of concepts, language and graphics used to describe a database. Recall that in Chapter 2 a "schema" was defined as a description of a database. The following discussion explains the relationship between a database model and a database schema, as well as the relationship between a database schema and the data objects that a database model describes.

Figure 3.2 depicts the relationships between a database model, a database schema and a data object which, in database terminology, is formally called an *instance*. The relationships between model, schema and instance in database design can be best understood using the analogy of human languages. In database design, a model provides the methods and tools for a high-level description of the real world, in much the same way as a human language. A database model, just like a language, contains the necessary vocabulary and the linguistic/graphical rules that allow a database designer to describe a database in a way that is comprehensible to other people. In this context, the phrase "high-level description" means a description that is general and universal and that can be used to describe any aspect of the real world. The English language, for example, can be used to describe scientific activities as well as artistic works. By the same token, a database model can be used to describe a scientific database, a business database, or a spatial database.

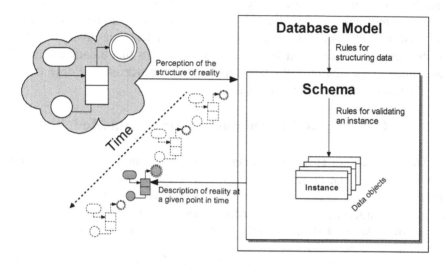

Figure 3-2. Relationship between a database model, a database schema and an instance of a data object

A *schema* in database design is a representation of a specific portion of the real world, built using a database model. More formally, a schema is a static, time-invariant collection of linguistic and graphical representations that describe the data structure of a database and database processing operations. Again, using the analogy of human language, a schema is equivalent to the grammar of a language, which provides rules to validate the "correctness" of written or spoken expressions. A schema is also considered as being equivalent to the collection of technical terms and graphical notation used by a particular profession such as architecture, engineering, accounting or medicine. A schema is said to be static and time-invariant because, just like the grammatical rules in a language and the technical terms used in a profession, it contains established procedures and standardised terms that are relatively stable over time. Recall in Chapter 2, it was noted that the data structure of a database always remains the same regardless of a change in applications. This implies that a schema must also be static and time-invariant.

An *instance* is an occurrence of a data object in a database. More formally, an instance is a dynamic, time-variant collection of data that conform to the data structure specified by a database schema. A schema can contain multiple instances, each of which corresponds to a state of the database at a particular point in time. An instance is said to be dynamic and time-variant because, although the occurrence of a data object can remain untouched in a database over a long period of time, the value of a data object can be changed, sometimes very frequently and rapidly, as a result of database transaction operations. The dynamic, time-variant nature of a database instance is exactly the opposite of the static, time-invariant nature of the database schema that specifies it. This implies that a database model and schema describe only the structure and occurrence of data objects, but not the values of data objects in a database.

2.3   Conceptual, Logical and Physical Data Modelling

Database design is an evolutionary process. It starts with a *conceptual database model* that represents the real world at a high level of abstraction. This means that the description of the database is entirely conceptual and completely independent of any hardware and software considerations. This phase of data modelling is aptly called *conceptual data modelling* or *conceptual database design*.

In essence, conceptual data modelling is a mental process where the database designer or data modeller abstracts the characteristics and properties of real world features that are relevant to the purposes of a database and, equally importantly, excludes those that are not. There are three types of data abstraction in conceptual data modelling (Batini et al., 1992):

- *Classification abstraction*, which is used for defining one concept as a *class* of real world features characterised by common properties. In a spatial database, for example, the concept "land parcel" is a class whose members are all land parcels (Smith's land parcel, Brown's land parcel, land parcels owned by the Federal Government of Canada, land parcels owned by the Province of Ontario, and so on). In a spatial database, a land parcel is characterised by a set of properties and is represented by thematic shading that distinguishes it from other polygon-based classes such as "body of water", "zoning boundary", "land cover type", and so on.
- *Aggregation abstraction*, which defines a new class from one or more sets of other classes that represent its component parts. Using the same spatial database example, the class of "residential subdivision" can be abstracted by aggregating the classes of "land parcel", "right of way", "road", "easement", "land use zone", and so on.

- *Generalisation abstraction*, which defines a set-to-subset relationship between the elements of two or more classes. For example, the class "administrative boundary" in a spatial database may be a generalisation of the classes "provincial boundary", "county boundary", "city boundary" and "ward boundary". Likewise, the class "residential land use" is a generalisation that includes the classes "low-density residential land use", "medium-density residential land use" and "high-density residential land use".

The end product of conceptual data modelling is a conceptual database model. As noted, this database model is a hardware- and software-independent, high-level abstraction of the real world. A conceptual model of a database can be described by diagrams, verbal descriptions or a combination of both. Figure 3.3 is an example of a graphical conceptual schema of a land parcel registration system using the *Entity-Relationship (E/R)* method of conceptual data modelling.

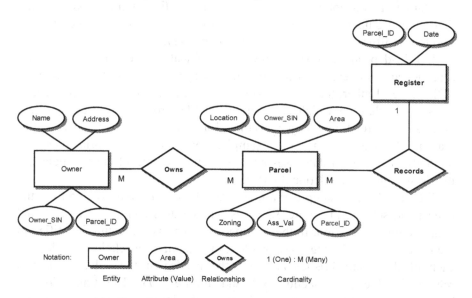

Figure 3-3. An Entity-relationship (ER) diagram

This schema depicts the real world objects to be stored in the database (that is, OWNER, PARCEL, and REGISTER) and their respective characteristics (which in the case of OWNER, for example, include the parcel owner's name, address, social insurance number [SIN], and the identification number of the parcel [Parcel_ID] that this person owns). It also describes how one object is related to another in the database (for example, an OWNER "owns" a PARCEL) and any constraints that may exist in the relationships (for example, a REGISTER can record many PARCELS but a PARCEL can be recorded in only one REGISTER). The principles and method of conceptual data modelling are explained in Section 3.1 where the terms entity, attribute, relationship and cardinality are defined in the context they are used in the diagram.

A conceptual schema as illustrated in Figure 3.3 describes the content and structure of a database by means of a commonly accepted set of graphical notation and vocabulary. The schema focuses solely on what the database will be like from a data perspective without considering any computer hardware and software requirements. In order to turn such a high-level abstraction into database implementation specifications, it is necessary to take into account the specific software and hardware requirements of a particular DBMS during the subsequent phases of data modelling. A database schema that includes software considerations in its description of a data structure is called a *logical schema*. Such a schema is developed by translating a conceptual schema according to the linguistic syntax and diagrammatic notation of a selected DBMS. A logical schema, therefore, is said to be software-dependent or DBMS-dependent.

DBMS-dependence does not mean that logical schemas are dependent on a particular commercial DBMS product such as Oracle, DB2 or Microsoft SQL Server. Rather, the term DBMS is used generally to refer to the various database models on which commercial DBMS products are designed and built, such as relational, object-relational and object-oriented. This implies that a relational schema, for example, conforms to the principles and methods of relational databases and, as such, is applicable to any database that is managed by a relational DBMS.

Figure 3.4 is an example of a logical database schema translated from the conceptual schema in Figure 3.3. This schema uses conventional relational database model notation. In the schema, each real world feature identified in the conceptual modelling phase becomes the name of a relational table, and each of the characteristics becomes the header of the columns in this relational table (see Section 3.2 and Figure 3.7). The translation or "mapping" from the conceptual schema to the logical schema is relatively straightforward. However, other factors that need to be taken into account during this process, including steps to refine and re-confirm the description of user requirements and data structure, identification of primary, secondary and foreign keys (PK, SK, and FK respectively in the Figure 3.4, also see Section 3.1 for definitions), as well as the proper documentation of the schema itself. All activities involved in the generation of a logical schema are collectively referred to as logical data modelling or logical database design. The techniques of logical data modelling using various database models are explained in Sections 3.2 through 3.4.

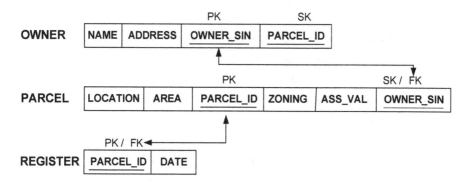

Notation: PK - Primary key; SK - Secondary key; FK - Foreign key

Figure 3-4. A relational logical schema developed from the conceptual schema in Figure 3-3

A logical schema describes a data structure with specific reference to a particular database model, but it is not yet usable for computer-based implementation. This is because hardware and software requirements must be taken into consideration in order to implement a data structure. The software, or DBMS, requirements in logical data modelling were noted earlier. Hardware requirements in data modelling include the computer and systems architecture of the database system, the physical location of data files, as well as the specific allocation of storage space to data objects in the respective data files. Such detailed specifications of the data structure of a database are collectively called the physical schema.

The process of translating or mapping a logical schema to a physical schema is commonly referred to as *physical data modelling* or *physical database design*. Physical modelling is a more complex and technical process than logical modelling because it requires competency in using both the DBMS and the hardware system used to install the database. Since this process is both hardware- and DBMS-dependent, it is practically impossible to describe a physical schema in generic terms. Figure 3.5 depicts how the "PARCEL" entity in the logical schema in Figure 3.4 will appear when it is mapped to a physical schema to be implemented in a typical relational database. It must be understood that in practice, even the simplest physical schema will be many times more complex than what is shown in this example.

PARCEL

Definition: A taxable unit of land within city limits.
Feature type: Polygon
Implemented as layer: Cadastral_fabric
Business table: LAND_PARCEL

ATTRIBUTE DEFINITION

Name	Type	Size	Optional	Unique	Indexed	Key
LOCATION	Char	100	M	Y	N	
AREA	Num	10.2	M	N	N	
PARCEL_ID	Char	15	M	Y	Y	P
ZONING	Char	5	O	Y	N	
ASS_VAL	Num	5.2	M	N	N	
OWNER_SIN	Char	9	M	Y	Y	S / F

ATTRIBUTE DESCRIPTION

LOCATION	Municipal address of parcel, including Street Number, Street Name, Street Type, County Name, Province Name, Postal Code
AREA	Size of parcel, as obtained from survey plan, in sq. m. to 2 decimal places
PARCEL_ID	Unique identification number assigned by Property Assessor's Department for a PARCEL, used as priamry key
ZONING	Zoning code, assigned by planning department (Refer to Appendix D for Zoning Codes)
ASS_VAL	Assessed value as determined by the Property Assessor's Department
OWNER_SIN	Parcel owner's social insurance number, used as secondary key and foreign key for OWNER table.

DATA LOAD / STORAGE

Initial volumn: 10000   Growth: 10% per year
Space: 25 Blocks, 50206 bytes   Initial allocation: 600k   Next: 60k

Figure 3-5. A portion of the physical model developed from the logical schema in Figure 3-4

2.4 The Importance of Database Models and Data Modelling

A database model, just like its counterparts in architectural and engineering projects, provides the means for the database designer to consolidate user requirements, test design concepts, compare alternatives and visualise what the database will be like when it is completed. A database model provides the rules and tools to document design decisions and the final implementation specifications. In this regard, it serves as an indispensable communications tool to facilitate the interaction between the database sponsor, designer, developer and end users.

As noted in Chapter 2, it is now the rule rather than the exception to build database systems using a distributed architecture and the client/server model of computing. As a result, the database design process has become much more complex and sophisticated, both technically and intellectually, than ever before. Through data modelling, the database designer is able to overcome human limitations to comprehend and solve the real world problems that a database is designed to address. Data modelling allows the designer to break a complex problem into smaller and more manageable parts, and work with each of these individually in a coordinated manner. It is important to stress that database models and data modelling should not be limited only to large and complex database implementation projects.

All database projects, large or small, can benefit from the use of a database model and data modelling to a certain extent. Data modelling requires both time and financial commitments and resources. It sometimes represents a fairly large capital cost that must be invested at the initial stage of a database project. However, the benefits resulting from reducing risks and shorter turn-around time in systems development will sooner or later prove that data modelling is a price that is worth paying. All developers of database systems, including spatial database systems, should have a good understanding of the principles and methods of database models and data modelling, and commit themselves to the practice of data modelling in their database projects.

3.   COMMON DATABASE MODELS

The following discussion consolidates the understanding of database models by examining the concepts of four commonly used models, namely the entity-relationship model, the relational model, the object-oriented model, the object-relational model and their respective uses in database design and development.

3.1   Entity-relationship (E/R) Model

An Entity-relationship (E/R) model is a conceptual database model that describes, at a high-level of abstraction, the nature of an organisation's data and how these data are used. The process of creating an E/R model is the first stage of the data modelling process in database design.

As noted at the beginning of this chapter, it is necessary to start data modelling with a high-level of abstraction because at this initial stage the database sponsor, designer and user all tend to view the data and their use from different perspectives. With this plurality of perspectives, communications between stakeholders in the problem domain (that is, sponsors and end users of the database) and those in the technical domain (that is, database designers, developers and administrator) are as important as the design of the database itself. Thus, the primary objective of an E/R model is to serve both as a means of communication between interested parties and as a record of design decisions. As such, the E/R model is characterised by the use of diagrams to express and describe its concepts. Because of its heavy reliance on the use of diagrams, the end product of E/R modelling is more commonly referred to as an *E/R diagram* rather than as an E/R schema (Figure 3.6).

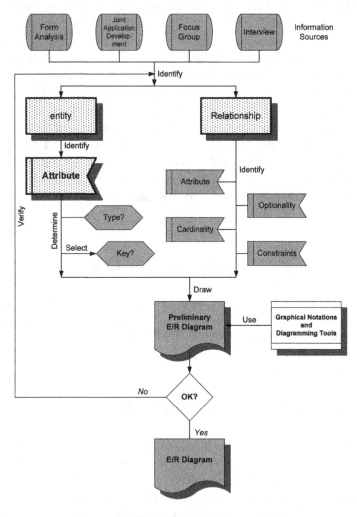

Figure 3-6. Workflow of ER modelling

In essence, the workflow of E/R modelling is a top-down approach to data modelling. Its objective is to identify the entities, the relationships between them, and the attributes that the database users require. During the modelling process, the nature and properties are determined for each of the entities, relationships and attributes identified. Information necessary for E/R modelling can be obtained from a variety of sources, including interviews, focus group meetings, joint application development (JAD) sessions and analysis of existing business forms (see Chapters 8 and 9).

The central concept of the E/R model is an *entity*, which is also called a *data object* or simply an *object*. An entity is a real world feature or phenomenon that has an independent existence (for example, people, buildings, cars, highways, and so on) but it can also represent an abstract concept (for example, temperature, land value, contour lines, and so on). Entities that share common properties are collectively called an *entity type* or *entity class*. An entity has a unique name, which is always singular and is used for the entity type and entity class as well (that is, there is no distinction between the name of an entity and its entity type). A typical database normally contains many different entities.

A *relationship* in an E/R model is an association between entities. It must be uniquely identifiable, and is given a name that describes its function (for example, belongs_to, managed_by, has, and so on). In the E/R diagram in Figure 3.3, "owns" is the name of the relationship between OWNER and PARCEL, and "records" is the name of the relationship between PARCEL and REGISTER. Besides the name, a relationship in an E/R model also has the following four properties:

- *Cardinality*, which denotes the number of occurrences of the entities participating in a relationship (for example, in Figure 3.3, an OWNER can own more than one PARCEL and a PARCEL can be owned by more than one OWNER, the relationship to REGISTER therefore has a cardinality of "many-to-many").
- *Optionality*, which denotes whether the relationship is optional or mandatory for either or both participating entities (for example, in Figure 3.3, since no OWNER will exist without registering a PARCEL, and no PARCEL will exist without being registered to an OWNER, the relationship is not optional).
- *Constraints*, which are business rules governing a relationship (for example, no one under 18 years of age is allowed to register a PARCEL).

Another important component of an E/R model is the *attribute*. By definition, an attribute is a particular characteristic or property of an entity or a relationship. An attribute in an E/R model can be classified in different ways, including:

- *Simple or composite attributes.* A simple attribute is one that cannot be subdivided (for example, the social insurance number of an OWNER) whereas a composite attribute is one that can be divided to yield smaller components (for example, the name of OWNER, which can be subdivided into a first name component, a middle name component and a surname component).
- *Single-valued and multi-valued attributes.* A single-valued attribute is one for which each occurrence has only one value. Most attributes are single-valued for a particular entity but multi-valued attributes are not uncommon (for example, a building where the ground floor is used for retail business and the upper floors are used for residential purposes).
- *Derived attributes.* A derived attribute is one that is computed from another attribute (for example, the tax payable for a property is computed by multiplying the assessed value of the property by a mill rate).
- *Keys.* An attribute can be used as a "key" or index to search a database for the required information. A key can be a primary key (PK), which is "the" key field used to search a particular occurrence of an attribute in a relational table. It can also be used as a secondary key (SK) or foreign key (FK) whose function is to supplement or help a primary key in the search. When a column appears in more than one table, its appearance usually represents a relationship between data in the two tables. In the logical schema in Figure 3.4, for example, the inclusion of the owner's social insurance number in both the OWNER and PARCEL relational tables is deliberate because it helps to match parcel owners with the parcels that they own. In this case, OWNER_SIN in the OWNER table is a primary key, and OWNER_SIN in the PARCEL table is a SK.

The concept and process of E/R modelling look relatively simple and straightforward. In practice, however, it is rather demanding with respect to the time and resource commitments of the database sponsor, designer and representatives of the user community. It also requires the database designer to have considerable skills in facilitating team meetings and developing E/R diagrams from disparate sources of largely unstructured information. Since there is no standardised notation for an E/R model, traditionally the database designer is free to use one of several sets of graphical notations such as Chen's, Crow's Feet, Rein85 and IDEFIX (Rob and Coronel, 2002). There is a growing trend toward the use of Unified Modelling Language (UML), which is introduced in Section 4.5.

3.2 Relational Model

In database design, the relational model is an "implementation" database model because, unlike the E/R model that is DBMS-independent, it is used to describe how a database will be implemented using a particular DBMS. In the relational model, data are logically structured within *tables* that are formally called *relations*. The idea of storing data using a logical table structure looks relatively simple. However, behind this logical structure is a sound mathematical foundation based principally on set theory and predicate logic (Codd, 1970 and 1990).

As noted above, the central concept of the relational model is a table or relation (Figure 3.7).

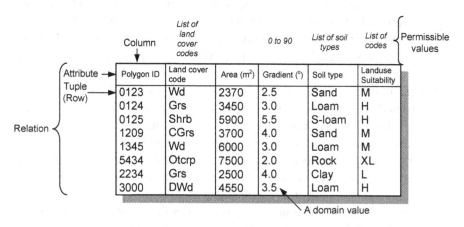

Figure 3-7. Features of a relational table

In essence, a table is a data file representing an entity type in a relational database. It has the following characteristics:

- *A unique name* that distinguishes it from other tables in the database.
- A *column* of a table represents one of its attributes. It is also called a *field*. A table has as many columns as the number of its attributes. The order of the columns in a table is unimportant.
- One or more the columns serve as the key of a table. Depending on its function, a key can be a primary key, a secondary key or a foreign key (see Section 3.1). If a table has a primary key and a secondary key, the two keys are collectively called a *composite key*.
- A *row* of a table represents an instance or occurrence of an entity. It is also called a *tuple* or *a record*. In a particular table, each row is distinct,

which means that no two rows are identical in terms of their cell values. The order of the rows in a table is unimportant.

- Each cell of a table contains exactly one value. If the value of a particular cell is not known or not available temporarily, a special value called "null" is assigned to it. Using null is the way to deal with missing or incomplete data in a relational table. A null value is not the same as the numerical zero, or as blanks embedded in a character string. It indicates the absence of a value, rather than the value of a cell.

- The type of values stored in the cells of a particular column of a table is called a *domain*. The term "domain" has a special meaning in the context of database systems.

In a relational database, a domain is the smallest unit of data representation. In this context, "smallest" means that the value of a domain is not divisible into smaller components. Composite attributes and multi-valued attributes, therefore, are not legitimate domains. The values of a domain can be in one of several *data types*, including character, numeric, and logical among others. In many cases, the values of a domain in a table are restricted to a particular list of names (for example, the categories of a land-use classification scheme) or a range of numerical numbers (for example, the ages of people or the heights of trees). Such a list of names or range of numerical numbers is called the *permissible values* of a domain. Since the permissible values limit what values can be stored in a table, they are called the *domain constraint* of the database.

Using a domain constraint is a very effective way to maintain the integrity of a database. For example, if a value of 200 appears in the "age" column of a table of demographic data, it is easy to establish that there must be something wrong because it is an impossible value. This ability of a domain constraint allows the database administrator to be alerted to potential anomalies or irregularities in the database, and it can be usefully employed to build automated quality control mechanisms in data entry.

Besides the domain constraint, there are two other constraints that a relational table must conform to in order to maintain database integrity. One of these is the *entity constraint* that applies to the primary key of a table. Since the primary key is used to search a table for a particular row or record, any column chosen as a primary key cannot be null. This characteristic of the primary key is used as a means to stop a user from inserting a new row into a table with a null in the primary key column.

Another important constraint is called the *referential constraint*, which applies to the secondary or foreign key. This constraint does not allow the insertion into a table of a new row with a particular value unless that value has already existed in another table where it is a foreign key. The reason behind this is simple. If the insertion of such a row were permitted, it would not be possible to relate two tables by using the primary key in one table and the foreign key in another because there is no corresponding foreign key value for the primary key value. As a result, the integrity of the database, not the tables, will be severely compromised.

In addition to the integrity constraints, the operations of relational tables are also subject to constraints imposed by *business rules*. Simply put, a business rule is a condition that the use of a relational table must satisfy. Using the example of a land parcel database, in most countries a parcel can only be registered when a certified survey plan is supplied with the application for registration, the land transfer documents are verified, a fee is paid, and so on. All these are business rules governing the process of registering a land parcel that involves the insertion of new rows to the relational tables concerned.

The integrity constraints and business rules govern basically the use of a relational table. For a table to be useful in a relational database, it must also conform to another set of rules that govern its structure. These rules are called *normal forms* which are all concerned with the relationships among the columns of a relational table. There are several normal forms but for most database applications, it is only necessary for a table to satisfy three normal forms, namely first normal form (1NF), second normal form (2NF) and third normal form (3NF). Whether a particular table is in 1NF, 2NF or 3NF depends on the results of testing it against the rules that define these normal forms (Table 3.1). The process of testing a table to determine its normal form is called *normalisation*.

Table 3-1. First, second and third normal forms

Normal Forms	Rules
First (1NF)	○ There are no repeating attributes in the table (that is, no two columns are allowed to store identical attributes, for example, the land use status of a parcel at different points in time)
Second (2NF)	○ The table is in 1NF, and
	○ All non-key attributes are functionally dependent on the primary key
Third (3NF)	○ The table is in 1NF and 2NF, as well as
	○ There is no transitive dependency of attributes on the primary key ("transitive" in this context means indirect)

Figure 3.8 depicts the workflow of developing the schema of a relational model from an E/R diagram according to the integrity constraints, business rules, and normalisation rules as explained above. This process, which is called logical modelling, as defined in Section 2.3, starts with a careful re-examination of the E/R diagram. The focus is on the formal definition of entities and, if necessary, to decompose an entity into more specific entities.

This is then followed by a similar re-examination and refinement of the relationships. Once the entities and relationships are stable, keys are identified, and the logical schema is constructed. The final stage of the modelling process is to normalise the table structure to ensure compliance with the various normal forms.

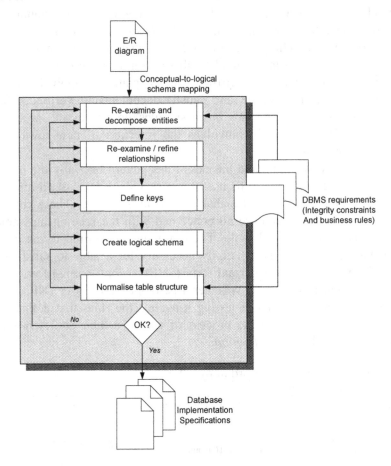

Figure 3-8. Workflow of developing a logical schema

During the modelling process, references have to be made from time to time to DBMS requirements such as the integrity constraints and business rules noted earlier that are relevant to the objectives of the database. In this context, the different tasks shown in Figure 3.8 are sequential as well interative. This means that it is always possible to revisit previous stages of work if new ideas or proposals are found at a later stage. The outcome of logical modelling is a set of DBMS-dependent implementation specifications called *logical schemas*.

3.3   Object-oriented (OO) Model

User requirements in the early days of database systems were limited largely to needs that could be satisfied by text-based applications. Current databases are required to store and process not only text, but also graphics, video, sound, maps, and different combinations of these in a multimedia environment. The rapid changes and advances in user requirements prompted database vendors and researchers to look for new solutions, which eventually led to the development of *object-oriented database systems* (OODB).

The concepts and methods of OODB were drawn largely from object-oriented (OO) programming, which was developed in response to the inadequacy of conventional programming languages to handle increasingly complex application requirements. In this regard, OODB are distinguished from conventional databases in that they are not based on a formally and universally accepted database model. Instead, they are based on a set of programming principles and a particular method of modelling the real world that is generally referred to as *object-orientation* or *object-oriented technology*.

The best starting point to learn about object-orientation is to understand what an "object" is. In the principles of object-orientation, an object is a *conceptually autonomous* data item in the computer that represents a real world *entity* with the ability to act upon itself and to interact with other objects (Rob and Coronel, 2002). An object, just like an entity in the E/R model and a relation (that is, a table) in the relational model, is a conceptual or abstract construct in object-oriented parlance. An object is said to be "autonomous" in the sense that it is independently and uniquely identifiable, and is not subject to the influence of any other objects. The word "entity" is a classic example of the lack of standardised terminology in the database world. In OO terminology, an "entity" is the equivalent of a "feature" in relational database terminology. Note also that some authors use the word "object" instead of the word "feature" to refer to real world things, such as people, or ideas in relational database modelling. If that is the case, the meaning and usage of the two words are exactly the opposite in relational and OO terminology.

Another important property of an object is that it is able to "act upon itself" and "interact with other objects". This means that an object represents not only the occurrence of an identifiable real world entity, but also the tasks it performs, including those it performs on others as well as those it performs on itself. Each object has several important components and characteristics, including:

- A *name*, which is assigned by the database designer.
- A *unique identity*, formally called an *object ID* (OID), which is assigned by the system when the object is created. An OID cannot be changed under any circumstance for obvious reasons.
- *Attributes*, formally called *instance variables,* which can be stored as any legitimate data type such as string, integer, real, and so on. The attributes form the data structure of an object.
- *Object state*, which is the set of values for an object's attribute at a given point in time.
- A *base data type*, which includes conventional data types, such as string, real, integer, and so on. Base data types use a predefined set of arithmetic operators such as addition, subtraction, and multiplication.
- An *abstract data type* (ADT), which, unlike the base data type, has user-defined operations called *methods* that govern their behaviour and use.
- A *method*, which is a computer program that performs a specific operation on the object's data. A method is also called a *service*. An object can have multiple methods. Collectively, these methods represent the *behaviour* of an object.
- A *message*, which is used to invoke a method by specifying a receiving object, the name of the method, and any required parameters. When invoking a method, the sender of the message accesses an object's actions, not its internal structure. This hiding of an object's internal details, which is known as *encapsulation*, is designed to protect the integrity of the state of the object.
- A *type,* which is the specification of an interface that an object will support. An object implements a type if it provides the interface described by the type. All objects of the same type can interact through the same type. An object can implement multiple types at the same time.
- *Control and business rules*, which govern the use of an object (that is, its behaviour).

A collection of similar objects with shared structure (attributes) and behaviour (methods) is called a *feature class* or simply a *class*. In essence, a class is the equivalent of an entity class and table in the E/R and relational database models respectively. Classes are the major building blocks of an OODB. Thus, it is important to understand the way classes are organised.

In an OODB, all classes are organised into a *class hierarchy* which resembles an inverted tree, as shown in Figure 3.9. In a class hierarchy, the classes at different levels have a superclass-subclass relationship. This hierarchical class structure makes possible the use of the concept of *inheritance*, which means that a class at a certain level is able to inherit the data structure and behaviour of the class above it (that is, its superclass).

The method of inheritance allows a particular class to use the method of its superclass. However, it is possible to override the method of a superclass at the subclass level. This concept is simply called *overriding*. It is also possible for different objects to respond to the same message in different ways by using the same name for a method defined in different classes in the class hierarchy. This concept, which is called *polymorphism*, is a very important feature of OO systems because it allows individual objects to respond to the same message according to their respective characteristics.

The object-oriented database model is one that is constructed using the concepts of an object and an object class, together with the related OO characteristics described above. Hence, an object as defined is a representation not only of a real world entity, but also the operations that it performs as well as those that are performed upon it. Such a concept distinguishes the OO model from conventional database models such as the relational model in one very important way. Instead of focusing solely on the description of data, the OO model describes data, database operations and processes all within a single object. From the perspective of database design, the OO model is therefore a more complete and meaningful description of a database than the E/R and relational models.

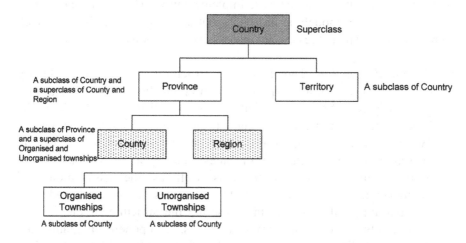

Figure 3-9. The concept of class hierarchy in OO

Up until the late 1980s, an OO model meant different things to different people because there were no standards in place to define such a model. Many OO models and data modelling methods were proposed with relatively distinct features and methodologies (Graham, 2000). In 1989, an industry-sponsored standardisation organisation known as the *Object Management Group* (OMG) was formed, with its headquarters in Needham,

Massachusetts. The objective of the OMG is to establish industry guidelines and detailed object management specifications that serve as a common framework for OO application development across different hardware platforms and operating systems (OMG, 2003).

The OMG produces and maintains vendor-independent standards and specifications for OO models, systems and databases. These include the *Object Management Architecture* (OMA), which is a set of standards to facilitate interoperation of objects across different systems (Orfali et al., 1996), and *Unified Modelling Language* (UML), which is a diagrammatic language for modelling, designing and visualising OO systems (Booch et al., 2000; see also Section 4.5).

OMG guidelines and specifications are widely accepted by the computer industry as *de facto* standards in systems design and implementation. Most database software vendors are now offering products that comply with the *Common Object Request Broker Architecture* (CORBA), a standard of the OMA for client/server-based database systems. At the same time, UML has also become the modelling language of choice not only for OO modelling, but also increasingly for relational modelling as well (RSC, 2000).

Figure 3.10 depicts the method of OO modelling as proposed by Booch et al. (2002), and its use in database implementation. This particular method is made up of three modelling processes that are grouped together and work with each other. The processes can be described individually as follows:

- *Structural modelling.* The objective of structural modelling is to identify all the "things" that are important to an application. These things form the *vocabulary* of the system that is to be modelled. For example, in the land parcel example, things that are important include land parcel, owner, land registrar, application form, and so on. Structural modelling also aims at determining the attributes and operations associated with each of the things identified, as well as the association and interdependencies between them. The outcomes of structural modelling are four types of diagrams that describe the structure of the system, namely object diagrams, class diagrams, component diagrams and deployment diagrams. In essence, therefore, structural modelling is a conceptual modelling process similar to E/R modelling, as explained earlier in this chapter.
- *Behavioural modelling.* The objective of behavioural modelling is to identify the dynamic aspects of the system. These include the roles played by each of the objects or object classes identified in structural modelling, the interactions among them, as well as flows of control involving branching, looping, recursion and concurrency. In other words, behavioural modelling is concerned with the methods and messages

associated with objects. The outcomes of behavioural modelling include a variety of behavioural diagrams such as activity diagrams, sequence diagrams, interaction diagrams and collaboration diagrams. In many ways, behavioural modelling resembles the logical modelling process in relational data modelling, and the behavioural diagrams are logical schemas of the system

- *Architectural modelling.* The objective of architectural modelling is to model the implementation aspects of the system. This is normally done by dividing the system into physical parts, called *components*, which conform to and realise a set of interfaces. The subjects of architectural modelling are the individual components of a system, not the system itself. Architectural modelling covers a wide spectrum of modelling tasks, including the modelling of user interfaces, data files and tables, executables and code libraries, and the mapping between the logical and physical aspects of the system implementation. The outcomes of architectural modelling include a series of component and deployment diagrams that can be used to construct a system and a database.

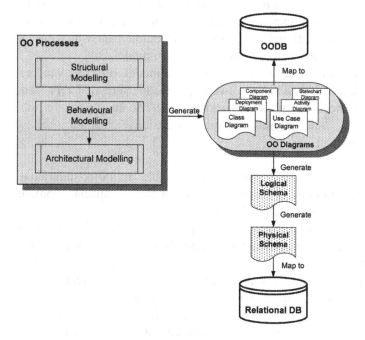

Figure 3-10. Workflow of OO modelling

OO modelling is distinct from conventional data modelling methodologies by virtue of its focus on application processes rather than data structures. If the target DBMS is object-oriented, it is relatively straightforward to map application processes to the data structure of a particular OODBMS. However, if the target DMBS is relational or object-relational, then the mapping is not as straightforward. To overcome the logical-to-physical mapping problems in these cases, the conventional modelling techniques of logical and physical modelling are often used as intermediate steps between data modelling and database implementation.

3.4   Object-relational Model

Object-relational databases were developed to overcome the limitations of relational systems to handle complex data required by new database applications in multimedia data management, engineering design, medical imaging, and scientific simulation and visualisation among others. Vendors of database software extended the capabilities of traditional relational systems by introducing many of the concepts of OO systems discussed in the previous section into the relational model. These include, for example, object storage, user-defined data types, inheritance, encapsulation of methods with data structures, and so on. A database that is constructed using a relational database model with object-oriented extensions is said to be an object-relational database, and the model on which the database is built is called an object-relational database model.

To accommodate the object-relational model, different vendors extended their relational database software products using different concepts and different mechanisms (Rennhackkamp, 1977). In generic terms, however, practically all object-oriented extensions include the following characteristics:

- *User-defined data types*, which enable an object-relational database system to manage complex data types that encapsulate complex internal data structures and attributes.
- *User-defined functions*, which define the methods by which applications can create, manipulate and access the data stored as a user-defined data type.
- *User-defined access* methods, which define the techniques by which user-defined data types are accessed and indexed.
- *An extensible optimiser*, which helps the DBMS determine the best way to access data stored as user-defined data types by using an appropriate user-defined function and user-defined access method.

Object-relational databases have inherited the robust transaction management capabilities of relational systems on the one hand, and the flexibility of data storage and access of object-oriented systems on the other hand. Practically all leading database software vendors, notably IBM and Oracle, have extended their conventional relational software products to include some object-oriented functionality. The introduction of the object-relational model has had particularly significant impacts on spatial data management because spatial data are complex data types by nature. Chapter 4 explains in detail the concepts and methods of using the object-relational model to develop spatial database models.

4. PRINCIPLES AND TECHNIQUES OF DATA MODELLING

Data modelling as a database design paradigm can be carried out in different ways in different database projects depending on the model of the database system to be implemented. Data modelling for a relational database, for example, is different from data modelling for an object-oriented database because, as noted above, the logical modelling phase of the process is DBMS-dependent. Similarly, data modelling for a spatial database is different from data modelling for a non-spatial database because the data structures of these two types of databases are different from one another (see Chapter 2). However, although the detailed methods of data modelling vary between one project and another, the principles and techniques used to develop these methods are more or less the same at a high-level of generalisation.

4.1 The Four Principles of Data Modelling

Booch et al. (1999) propose four principles of data modelling that are excellent starting points to understand this process. These include:

- *The choice of a model has a profound influence on how a problem is approached and how a solution is formed.* Database models are created to represent the real world and different models use different concepts, languages and diagrams to describe a particular aspect of reality. Hence, it is important for a database designer to choose a model carefully by always taking into account the complexity of the real world problem to be solved, the nature of the application of the data, the technical environment, the corporate policies governing the use of information technology in an organisation, and the expected performance standard of the database.

- *Every model may be expressed at different levels of precision.* Although a model can be used to represent the real world to a high level of detail, it is not necessary for the model to be used to the fullest extent of its descriptive capacities. The most appropriate model is one that allows the database designer to describe the real world at a chosen level of detail, according to the needs of the user of the model.
- *The best models are connected to reality.* Although a model aims to simplify the real world for ease of understanding, it is essential that the simplification does not mask any important details or characteristics that the model is supposed to represent. When a database model is not capable of depicting the real world accurately, it is important to know where the connection between the model and the real world is weak, and how the weaknesses of the database model can be addressed, for example, by using another database model.
- *No single model is sufficient and every non-trivial system is best approached through a small set of nearly independent models.* It is sometimes not possible to represent a database by means of a single model. A typical example is a distributed or federated database (see Chapter 6) made up of a complex assemblage of a variety of databases. In cases where multiple models are used, it is important to make sure that these models are logically interrelated but can be built and used separately.

4.2   The Systems and Database Development Life Cycle

A systems development life cycle (SDLC) is a generic description of the process of developing a database system in six phases, namely planning, analysis, design, building, implementation and maintenance (Figure 3.11 and Table 3.2). In practice, the six phases of the SDLC are interactive as well as sequential. This means that although the analysis phase in Figure 3.11 is listed "after" the planning phase, results of detailed analysis during the database development process may have some significant impacts on the original findings of the planning phase. This will consequently lead to modifications of the database implementation plan as it was proposed. Similarly, during the design phase, the database designer may find that the information obtained from the analysis phase is not sufficiently detailed for a particular aspect of the database and, as a result, the database will need to be refined.

This approach to systems development is adopted because the development of a database system is a very complex process, both intellectually and technically. It is practically impossible for one phase to be completed before another phase can start. Instead, it is more realistic to build in a certain degree of flexibility in the SDLC to accommodate changes in the outcomes of one phase that are necessitated by findings in another phase.

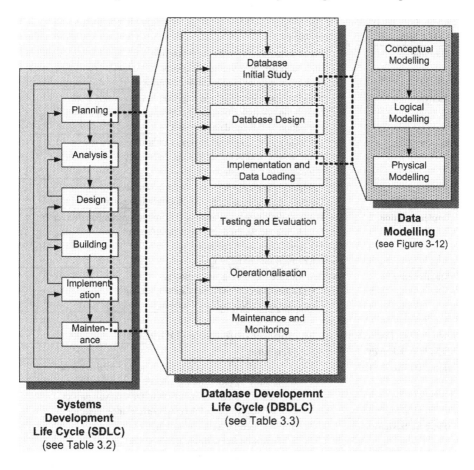

Figure 3-11. The systems development life cycle (SDLC)

Clearly, since a database is the central component of a database system, its design, building and implementation are also subject to the SDLC. A *database development life cycle* (DBDLC) is a generic description of the processes of developing a database in six phases, namely database initial study, database design, implementation and data loading, testing and evaluation, operationalisation, and maintenance and monitoring (Figure 3.12 and Table 3.3).

Just like the phases of a SDLC, the phases of a DBDLC are also sequential as well as interactive. Although the six phases of the life cycle are carried out more or less sequentially, as shown in Figure 3.12, it is always possible to change the outcomes of one phase if they are found to be impractical or deficient at a later phase of the database development process.

Table 3-2. Activities of the systems development life cycle (SDLC)

SDLC Phases	Activities
Planning	o Initial understanding of business functions
	o Initial assessment of user requirements
	o Feasibility study to implement the database
Analysis	o Systematic assessment of user requirements
	o Evaluation of existing business practices and operations
	o Evaluation of existing data resources
Design	o Development of hardware/software architecture
	o Development of systems performance standards
	o Development of data structure
Build	o Application programming (on a development computer)
	o Database programming (on a development computer)
Implementation	o Hardware/software installation of the production computer/server
	o Data loading to the production computer/server
	o Systems testing and fine tuning
	o User education and training
Maintenance	o Performance monitoring and evaluation
	o Regular maintenance including database backup
	o Continuing user education and training

Table 3-3. The database development life cycle (DBDLC)

DBDLC Phases	Activities
Initial Database Study	o Analysis of business functions and information needs
	o Identifying problems and constraints
	o Defining goals and objectives of the database project
	o Defining scope and database performance standards
Database Design	o Conceptual database modelling
	o Hardware and software (DBMS) selection
	o Logical database modelling
	o Physical database modelling
Implementation and data load	o Hardware and software (DBMS) installation
	o Creating data structure
	o Data loading, including any data conversion
Testing and evaluation	o Testing and fine-tuning database
	o Testing and fine-tuning application programs
Operationalisation	o Putting the database into production mode
	o User education and training
Maintenance and monitoring	o Regular maintenance of hardware and software, including change management in hardware and software upgrades
	o Database backup and replication
	o Continuing user education and training

The task of data modelling transcends the three phases of the DBDLC shown in Figures 3.11 and in detail in Figure 3.12. Conceptual data modelling always starts after the initial database study has identified the problems that a database system is intended to solve. It is used to formalise the findings of the initial study by describing the design of the database in its conceptual schema. The major part of data modelling is done in the design phase.

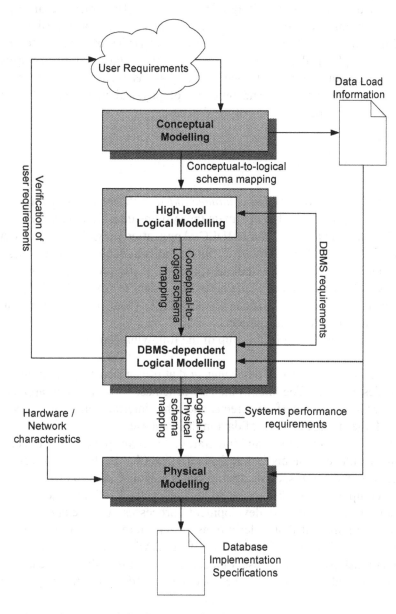

Figure 3-12. Conceptual, logical and physical modelling in the context of the database development life cycle (DBDLC)

The first step in database design is to map the conceptual schema to a high-level logical schema. This schema is then refined by including additional detail such as the identification of keys, partitioning of entities, determining permissible values of the attributes, and so on. The outcome of logical modelling is a DBMS-dependent schema that contains sufficient detail for the physical modelling of the database. The objective of physical data modelling, which is often called physical database design, is to determine the data storage and access characteristics of a database. These include the physical location and partitioning of the data files, hardware devices that support data storage and access, as well as the performance standard of the database. The outcome of physical modelling is a set of implementation specifications of the database.

4.3   Case Tools

CASE is an acronym for *computer-aided software engineering*. It is a structured approach used to automate the systems development activities in a SDLC and, by extension, in a DBDLC. Application programs used in CASE are collectively called *CASE tools*. Since the construction of databases is a very time-consuming and tedious process, there are many compelling reasons to use automated methods in order to expedite design and implementation activities. As a result, CASE tools play an increasingly significant role in database projects.

A typical CASE tool is made up of the following components:

- *A systems development environment* that uses a rich set of drawing tools to describe and document database schemas, flows of data, application processes, user interface screens and other interface elements associated with the storage and use of data in the database.
- *A repository* that stores and integrates all systems development decisions and results of systems and design activities (that is, the database schema, data flow diagrams, application process diagrams, and so on).
- A complete *data dictionary* which keeps track of all objects created during the systems development processes (for example, entity descriptions, attribute definitions, data store, and screen interface formats). The data dictionary of a CASE tool also records the relationships among these objects and the rules that are capable of automatically checking the accuracy and consistency of these relationships.

There are three broad classes of CASE tools according to their functionality in relation to the SDLC:

- *Front-end* CASE tools, which provide support for the planning, analysis and design phases.
- *Back-end* CASE tools, which provide support for the building and implementation phases.
- *Cross life cycle* CASE *tools*, which, as their name implies, support all the activities across the entire SDLC, including functions that support project management and documentation in some products.

Obviously, data modelling is concerned primarily with front-end CASE tools. This class of tools enables a database designer to create a model of the database to be constructed. These tools also allow the database designer to check the quality of the model by performing consistency and validity checks. By using a front-end CASE tool, a database designer is able to develop a database model quickly and detect any design errors before they propagate into the building phase or, worse still, the implementation phase.

A new generation of CASE tools, called *analysis and design engines*, use a rule-based architecture that enables the tool to be customised for analysis and design methods. By using these advanced CASE tools, a database designer can tailor the modelling process to the specific needs or development environment of a database project. CASE tools support the analysis and design phases by building the necessary diagrammatic notation, formulating the rules governing the semantics of the symbology, and creating an interface to serve as the development platform.

Despite the usefulness of CASE tools in data modelling, it must be noted that data modelling is more an intellectual than a technical exercise. No matter how powerful a CASE tool may be, a database designer using a CASE approach must be as well-versed in the concepts and methods of data modelling as a designer who does data modelling manually. In the hands of a poor database designer, a CASE tool can at best produce good-looking database diagrams which will not necessarily convert into good databases.

4.4   User-centred Database Design

Traditionally, systems development has been technology-centred and application driven, focussing on hardware performance and software reliability. Therefore, it was an effective means of managing large and complex systems development tasks involving close interaction between hardware limitations and application programs. The central concern in systems development today is not hardware performance, but rather systems usability or, as it is often called, user friendliness. The new generation of systems is designed to be efficient and easy to use in order to improve the user's overall experience and productivity. As a result of the change in design focus and objectives, the conventional technology-centred concepts in systems development are being gradually abandoned in favour of what have come to be known as *user-centred design* (UCD) methodologies.

The principles and methods of UCD are not new. For decades, UCD was applied successfully in the design of products and services in various industrial sectors. In the context of systems development, UCD is a comprehensive development environment driven by (a) clearly specified, task-oriented business objectives, and (b) recognition of user needs, preferences and constraints. Information collected using UCD analysis is scientifically applied in the design, testing, and implementation of computer-based systems (Vrendenberg et al., 2001).

Although UCD applies generally to the SDLC as a whole, it is particularly useful as an approach in the analysis and design phases. From the data modelling perspective, the merits of UCD include, but are not limited to, the following:

- *Providing a well-structured framework compatible with the concepts of the SDLC*. The UCD process begins with the identification and definition of the business functions of an organisation which are then translated progressively into implementation specifications that are readily usable for creating and testing the proposed system. The primary driving forces behind USD are user requirements and the systems solutions to meet them. Technical considerations play a part at a certain stage of the design process, but they are never a driving force in and of themselves.
- *User participation in the modelling process*. Design concepts in UCD are derived from a comprehensive business function analysis carried out by professional systems designers together with users from all profile types. A collaborative approach to business function analysis allows the development of a database model from the user's perspective, while at the same time conforming to basic systems development concepts and philosophies validated by the database designer.
- *A high-level of user-designer interaction during the modelling process*. UCD is characterised by a strong emphasis on user response in the design decision-making process. The aim is to ensure that most if not all of the user's needs and limitations are accounted for at the data modelling stage, thus avoiding needless expenses that might be incurred in correcting design errors in the implementation phase of SDLC.

- *An iterative approach to data modelling.* UCD is a highly iterative approach which encourages design changes in response to user input. As their experience grows with increasing involvement in the modelling process over time, users may develop different views of the systems that will more realistically represent the real world. Therefore, the ability to revisit previous design decisions and make necessary changes is essential in data modelling. To this end, UCD is compatible with the concepts of the SDLC discussed above.

UCD is now a universally accepted systems development methodology in the computer industry. Its use is spearheaded by industry leaders such as IBM and Microsoft. Since the focus of UCD is the usability of computer systems (which includes non-discriminatory access to information resources), it forms the cornerstone of government and corporate policies and regulations on the use of information technology. In this regard, UCD should and must be the data modelling standard in database projects.

4.5   Data Modelling Documentation

The previous subsections discussed the principles and techniques of data modelling used to create practical and realistic representations of the real world. In order to serve the other function of a database model (that is, as a communications tool among the database sponsor, designer, and developer), representations of the real world obtained during the data modelling process must be accurately and clearly documented. This means that capturing user requirements is only part of the objectives of data modelling. No data modelling task is considered complete without proper documentation. Since documentation is the only tangible outcome of data modelling, the quality of a data modelling task can only be as good as its supporting documentation.

Batini et al. (1992) suggested that for a conceptual model to represent the real world, it must have four qualities:

- *Expressiveness*, which means that it has a large variety of concepts, including linguistic syntax and diagrammatic notation as well as constraints, for a comprehensive representation of the real world.
- *Simplicity*, which means that schemas built using a particular conceptual model must be easily understandable to the database sponsor, designer, developer, and other interested parties. It should be noted that simplicity and expressiveness are conflicting requirements, and a good conceptual model must be able to balance the requirements of these two qualities.

- *Minimalism,* which is achieved if every concept presented in a conceptual model has a distinct and unambiguous meaning with respect to every other concept.
- *Formality,* which refers to the formal specification of data and their structure. Formality requires that all concepts of the model have a unique, precise and well-defined interpretation.

Although the above qualities were originally proposed for conceptual models, they can in fact be conveniently used more generally as a guideline for the development of documentation standards for logical and physical database models as well.

Many methods and tools exist for the purposes of documenting database models. However, UML has now emerged as a *de facto* industry standard for visualising, specifying, constructing and documenting the artefacts of computer systems, including database systems (Booch et al., 1999). It is not within the scope of this chapter to present fully the characteristics and use of UML. However, because of its increasing importance in the computer industry, in the discussion below a brief introduction to UML is provided for data modelling with special reference to its documentation capabilities.

UML was formally proposed in 1994 to provide a standard notation for object-oriented data modelling. It is now maintained by the Revision Task Force (RTF) of the OMG. Although UML was created by fusing the concepts of Booch et al., Object Modelling Techniques (OMT) and Object-oriented Software Engineering (OOSE), it is a non-proprietary standard that is open to all users. Despite its origin as a modelling language for object-oriented systems, it is now commonly used for modelling relational systems as well (RSC, 2000). Because of the very rich set of linguistic and diagrammatic notation available in UML, some database designers find it helpful to model a system first in UML, and then map the resulting model onto an operational database model for implementation (Ambler, 2000; Fussell, 1997).

The documenting function of UML can be best summarised by its *metamodel.* In essence, this is a set of definitions that describe in fairly precise syntax the underlying meaning of each *element* used in visual modelling and the relationships among all elements. The UML metamodel has a four-layer architecture (Figure 3.13a) that includes the user object layer, the model layer, the metamodel (2M) layer and the metametamodel (3M) layer.

The user object layer is the bottom layer of the UML metamodel architecture. This layer contains *object diagrams* populated with the facts from the problem space of the database (that is, the aspects of the real world represented by the database model) (Figure 3.13b). Using the land parcel database from earlier in this chapter, "Parcel" is an object diagram that contains a line item for "parcelId" with a value "REG1234567890KK", an "address" with a value of "1185 THORNLEY STREET, LONDON, ON", an "area" with a value of "10000.00" and a "zoning" code of "RES-3". Object diagrams are built using the rules defined by the next layer above it, the model layer.

UML Layer	Description	Example
Metametamodel	Defines the language for speficying metamodels	Meta-class, Meta-attribute Meta-operation, etc.
Metamodel	Defines the language for specifying models	Class, Attribute, Operation
Model	Defines the language for describing subject domains	Parcel, Owner, Registers
User object	Defines specific subject domain information	Parcel (id, location, area, zoning) Owner (name, address, SIN) Registers (sin, pid, date)

(a) The Four-layer Metamodel Architecture of UML

(b) An object diagram of PARCEL

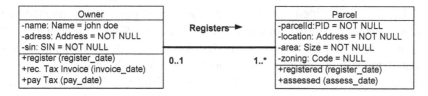

(c) A Class Diagram

Figure 3-13. UML layers and features

The model layer explains the *classes* that describe *the subject domain objects*. In the above example, "Owner", "Parcel", and "Register" are subject domain classes. The model layer describes what each of the subject domain objects looks like, the attributes it contains, the operations it can perform, and so on (Figure 3.13c). These class descriptions conform to the rules specified in the next layer above, the metamodel (2M) layer. This layer defines a class as a concept having attributes, operations and associations or relationships. It defines an attribute as having a name, a data type, a default value, and constraints. These class and attribute definitions in turn conform to the specifications of the metametamodel (3M).

The metametamodel (3M) layer is the top layer of the UML metamodel architecture. It defines the concepts of meta-class, meta-attribute and meta-operations. These are all abstract definitions that collectively form UML and which serve as templates for constructing a wide variety of concepts.

Documenting data modelling is a tedious and time-consuming process. However, it is important to realise the value of good documentation to the quality of any data modelling task and the consequences of poor documentation to a database project. Experience has shown that a database designer who is not willing to commit the time to create proper documentation at the data modelling stage will often unwittingly create unforeseen problems at subsequent stages of database implementation and use.

5. SUMMARY

This chapter approached the study of database models and data modelling in general terms. The objective was to provide an account of database models and data modelling methodologies as they stand today. Understanding of the material presented in this chapter is important for the study of spatial databases in the rest of the book, in particular Chapter 4 on spatial data, Chapter 8 on user needs assessment and multi-user data modelling, and Chapter 9 on spatial database project management.

The chapter defined key terms in data modelling, and explained the characteristics of several commonly used data modelling methodologies in database design, including E/R modelling, relational modelling, object-oriented modelling and object-relational modelling. The principles of data modelling were also explained in the context of the SDLC and DBDLC. In addition, the newest developments were explained in data modelling techniques including CASE and documentation tools.

The importance of data modelling can never be emphasised enough in database projects. Data modelling takes time and is a major commitment, both financially and technically, in all database projects. However, the risk of jumping right into the design and development of a database without first undergoing a rigorous data modelling process can produce unforeseen and unanticipated problems after the fact. In any database project, therefore, it is important not to start the first line of code before all stakeholders are satisfied that they have a good data model in hand.

With the prerequisities for data modelling now covered, attention turns in the next chapter to discussing spatial data and spatial database systems.

6.   REFERENCES

Ambler, S.W. (2000) "Mapping objects to relational databases", IBM DeveloperWorks: Components Overview Library paper. (http://www-4.ibm.com/software/developer/library/mapping-to-rdb/index.html)

Batini, C., Ceri, S. and Navathe, S.B. (1992) *Conceptual Database Design: An Entity-Relationship Approach*, Redwood City, CA: The Benjamin/Cummings Publishing Company, Inc.

Booch, G., Rumbaugh, J. and Jacobson, I. (1999) *The Unified Modelling Language User Guide*, Reading, MA: Addison-Wesley Longman, Inc.

Chen, P. (1976) "The Entity Relationship Model: Toward a Unified View of Data", *Communications of the Association of Computing Machinery*, Vol. 1, No. 1, pp. 9-36.

Codd, E.F. (1970) "A relational model for large data banks", *Communications of the Association of Computing Machinery*, Vol. 13, No. 6, pp. 377-387.

Codd, F.E. (1990) *The Relational Model for Database Management*, Reading, MA: Addison-Wesley Publishing Co.

Cooper, R. (1996) *Object Databases: An ODMG Approach*, Boston, MA: Thomson Learning.

Date, C.J. (2004) *An Introduction to Database Systems*, 8th ed., Reading, MA: Addison-Wesley Publishing Co.

Elmasri, R. and Navathe, S.B. (2003) *Fundamentals of Database Systems*, 4th ed., Boston, MA: Addison-Wesley Publishing Co.

Filev, A., Loton, T., McNeish, K., Schoellmann, B., Slaer, J. and Wu, C.G. (2002) *Professional UML with Visual Studio .NET*, Birmingham, UK: Wrox Press Ltd.

Fussell, M.L. (1997) *Foundations of Object-Relation Mapping*, Sunnydale, CA: ChiMu Corp.

Graham, I. (2000) *Object Oriented Methods*, 3rd ed., Boston, MA: Addison-Wesley.

Halpin, T. (2001) *Information Modelling and Relational Databases: From Conceptual Analysis to Logical Design*, San Francisco, CA: Morgan Kaufmann Publisher.

OMG (Data Management Group) (2003) Web site at http://www.omg.org/

Orfali, R., Harkey, D. and Edwards, J. (1996) *The Essential Client/Server Survival Guide*, 2nd ed., New York, NY: John Wiley & Son, Inc.

Pressman, R.S. (2005) *Software Engineering: A Practitioner's Approach*, 6th ed., New York, NY: McGraw-Hill, Inc.

Rennhackkamp, M. (1997) "Extending Relational DBMSs", *DBMS*, Dec-1997.

Rob, P. and Coronel, C. (2002) *Database Systems: Design, Implementation and Management*, 5th ed., Boston, MA: Thomson Learning.

RSC (Rational Software Corporation) (2000) "UML and Data Modelling", a Rational Software white paper, Cupertino, CA: Rational Software Corporation.

Ullman, J.D. and Widom, J. (1997) *A First Course in Database Systems*, Upper Saddle River, NJ: Prentice-Hall, Inc.

Vredenberg, K., Isensee, S. and Righi, C. (2001) *User-centered Design: An Interactive Approach*, Upper Saddle River, NJ: Prentice-Hall, Inc.

Chapter 4

SPATIAL DATA AND SPATIAL DATABASE SYSTEMS

1. INTRODUCTION

Chapters 2 and 3 discussed the concepts and techniques of database systems and data modelling in general. This chapter turns attention to spatial database systems. Spatial database systems are different from conventional database systems in two important respects. First is the requirement to store complex data types such as points, lines and polygons. Second is the functionality needed to process such complex data types using spatial operators that are considerably more sophisticated than conventional database operators for the processing of alpha-numeric data types. Accordingly, this chapter first focuses on the representation of spatial data in a database environment. Attention is then turned to the characteristics of database systems that are designed specifically to manage and process these types of data.

The chapter is organised as follows. Section 2 defines different types of spatial data and introduces the concept of "geometry" as a means of representing spatial data. The concept of "topology" in the representation of spatial relationships is described in Section 3. This section also discusses geo-relational and DBMS approaches to representing complex spatial data types in a spatial database. Section 4 explains the concepts and techniques of spatial database systems with particular reference to spatial data operators and spatial database processing. Finally, Section 5 summarises the material presented in this chapter in the context of systems design and implementation discussed in Chapters 2 and 3.

2. DEFINITION AND CLASSIFICATION OF SPATIAL DATA

Spatial data are also commonly called *geographically referenced data* and *geospatial data*. In the following discussion, the characteristics of spatial data are explained from a database perspective, which is somewhat different from the conventional definition that stems from cartographic or mapping perspectives.

2.1 Spatial Data and Pseudo-spatial Data

Spatial data are data that can be displayed, manipulated and analysed by means of a spatial attribute that denotes a location on or near the surface of the Earth. This spatial attribute is normally provided in the form of coordinate pairs that allow the position and shape of a particular spatial feature to be measured and represented graphically. Spatial data have two important properties:

- They reference to a *geographic space*, which means that the data are registered to an accepted geographic coordinate system spanning some area of the Earth's surface, so that data from different sources can be cross-referenced and integrated spatially.
- They are represented at a variety of geographic scales and when spatial data are recorded at relatively small scales, representing large areas on or about the Earth's surface, they must be generalised and symbolised.

Spatial data are collected and stored in two fundamental forms called *vector* and *raster* (Figure 4.1). The basic unit of spatial data in the vector form is the *geographic object*, which is an identifiable discrete real world feature or phenomenon represented by a point, a line or a polygon. In a spatial database, vector data can be stored as part of the topographic base whose function is to provide the spatial referencing framework for data collection and analysis. Data in a topographic base data set include geodetic and survey control points as well as all features that are commonly found on a typical topographic map such as roads, rivers, urbanised areas and natural vegetation features.

Application vector data can describe the state of the natural environment (for example, environmentally protected areas, forest resource inventories, land cover, and air and water qualities) as well as data about human activities on and utilisation of the land and its resources (for example, land use, land subdivision and registration, and transportation and public utility networks).

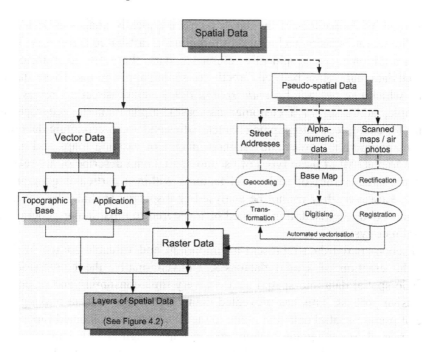

Figure 4-1. Types of spatial data

In practice, topographic base data and application data are often distinguished by their functions rather than by their types because many spatial data sets serve both purposes. For example, when a street network is used as the geographical reference for the collection, analysis and display of traffic accident data, it is part of a topographic base. However, when it is used to record the flow of traffic or the routing of vehicles, it becomes an integral part of a specific application data set used by transportation planners and engineers.

In the raster form of geographic data, the basic spatial unit most commonly takes the form of a square grid cell, or a *pixel* (which is the short form for a "picture element"), nestled within a tessellation or grid of equally sized pixels. This basic spatial unit serves simultaneously the functions of a data store (that is, it contains the thematic attribute pertaining to the space it represents) and geographical referencing (that is, the location of a cell in a raster grid implicitly denotes its location in the real world). The size of a single pixel or spatial unit is known as its *resolution* and this defines how well a raster data set can represent features in the real world. A small size or resolution gives a better spatial representation of real world phenomena, but it also results in a much larger data file. The resolution of a raster data set is sometimes chosen by the database developer or administrator, for example when raster data are acquired by image scanning. However, it is also often governed by the sources of the data which are typically images collected by satellite remote sensing techniques or through aircraft-based aerial survey.

In addition to geographically referenced data, there are other forms of spatial data that cannot be used directly for spatial applications. These spatial data, which can be called *pseudo-spatial data*, include street addresses (for example, addresses in a customer database), alpha-numeric geographical data (for example, demographic characteristics and socio-economic data of a census enumeration area) and scanned images of existing maps and aerial photographs. All of these types of pseudo-spatial data describe or are related to the characteristics of real world features within a particular geographic space. However, they cannot be analysed or displayed spatially before they are geocoded, digitised, transformed or registered to a particular geographic coordinate system.

Pseudo-spatial data represent an important and valuable source of data for the creation of spatial databases. Conventionally, the conversion of pseudo-spatial data into spatial data is a very time-consuming and resource-intensive process. This has prevented many organisations from making full use of pseudo-spatial data that relate to their business or mandated functions. However, data conversion technologies are now significantly more mature than in the earlier days of spatial data base construction. Practically all GIS software packages nowadays are equipped with built-in functions for geocoding, map digitising and transformation, and image rectification and registration to convert pseudo-spatial data into fully functional data layers of a spatial database.

2.2   A Functional Perspective of Spatial Data

In a spatial database environment it is logical to classify data according to their respective functions in database management and application as shown in Figure 4.2. There are four categories of spatial data in a functional classification, namely:

- *Base map data layers*. These include the geodetic (survey) control network that forms the spatial reference framework for all data in the database, and various types of topographic base data that are used to provide the geographical referencing required for the collection, analysis and display of application and business solutions data that are described below.
- *Framework data layers*. These include three interrelated layers for the geographical referencing of human activities on the land, namely (a) a parcel layer that provides the framework for land development and land administration applications; (b) a facilities layer that forms the basis for

facilities management in public utilities and resources management; and (c) an address layer which is used to support various land and resource applications requiring the use of a civic or postal address.

- *Application data layers*. These include a multitude of spatial data sets collected and used for different database applications in land and resource management using the base map and framework data layers as the basis of geographical referencing.
- *Business solutions layers*. These are collections of spatial data layers, including framework and application data layers and their derivatives, together with relevant and related non-spatial data, that are assembled to support the operations and decision making functions of departments or divisions within an organisation.

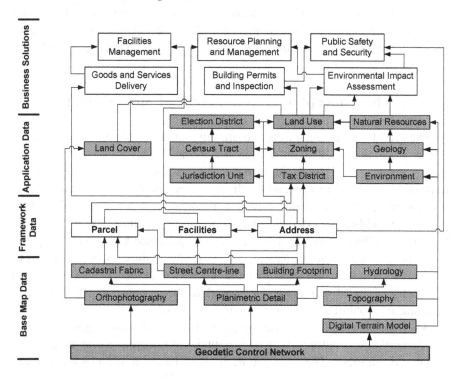

Figure 4-2. A functional classification of spatial data

In the above classification, the focus is the roles or functions that each data layer plays in a spatial database system. In spatial database applications, the roles of a particular layer are the most important factors governing its specifications and characteristics such as scale, accuracy, classification, timeliness, and so on (see Chapter 6). The type of data in a spatial data layer is of secondary importance because the conventional differences between vector and raster data are no longer as relevant in today's spatial database technology as they were as recently as half a decade ago. As shown in Section 3.5, both types of spatial data can now be managed and used in a totally integrated manner.

3.   SPATIAL DATA STRUCTURE AND DATABASE MODELS

There are two key aspects of spatial data, namely the representation of data and their relationships. In the following discussion, the concept of "geometry" is introduced as a means of representing spatial data, and the concept of "topology" is introduced as a way of representing spatial relations. The characteristics of the geo-relational database model used in conventional GIS are examined in the context of a database model in a database environment.

3.1   The Concept of a "Geometry" of Spatial Data

The field of *geometry* is commonly understood as a branch of mathematics that deals with the properties and relations of points, lines, angles, surfaces and solids in zero and higher dimensional spaces. In the context of spatial data processing, the word geometry assumed an entirely new meaning when the Open Geospatial Consortium (OGC) formalised its use in the publication of their *OpenGIS Simple Feature Specification for SQL* (OGC, 1999). In this document, the OGC proposed a hierarchy of spatial data types, called the *geometry object model*, which allows spatial features to be represented in a database. In the geometry object model, the word "geometry" is used to represent a spatial feature as an "object" having at least one attribute of a *geometric type* in a database.

As a root class of the hierarchy geometry object model, geometry is a *non-instantiable* construct (that is, it contains no occurrence or instance of real world features). The base geometry class has four sub-classes, namely point, curve, surface and geometry collection. These geometric sub-classes are *instantiable* constructs (that is, they contain occurrences of instances of real world features) and contain a definitive set of processes, called *methods*, that are used to test their respective geometric properties, define their spatial relationships, and support their use in spatial analysis (see Section 4.3).

There are many geometric types in the geometry object model, as shown in Figure 4.3. These geometric types are graphical primitives that are used to construct a geometry for objects using one or more graphical primitives. A lamp post, a highway segment, and a land parcel are examples of geometries constructed using a single graphical primitive. These geometries are called *simple geometries*. On the other hand, a group of islands that is treated as a single geographical or political entity, such as New Zealand, the Cook Islands, and Fiji, are constructed using multiple graphical primitives. These geometries are called *complex geometries*.

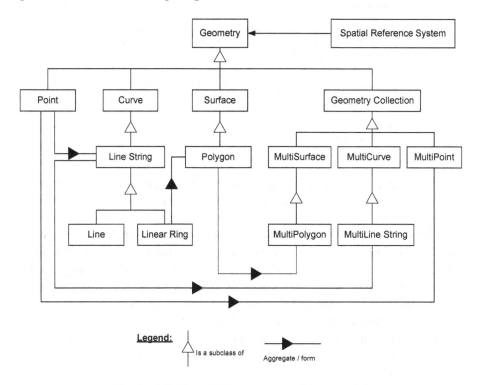

Figure 4-3. The OGC geometry object model

Since graphical primitives are spatially referenced to a particular coordinate system, the position and orientation of a geometry is always known and fixed in a particular geographical space. This spatial reference to a coordinate system helps to relate one geometry to another geographically, no matter what graphical primitives they were constructed from. In addition to its location, a geometry has other attributes that describe its characteristics such as name, classification and dimension. In a spatial database, geometries that share the same attributes collectively form a *layer* (also called *feature class*) and, consequently, are stored in the same table. Figure 4.4 shows the relationship between "geometry" and other common elements used to construct and represent spatial features in a database system.

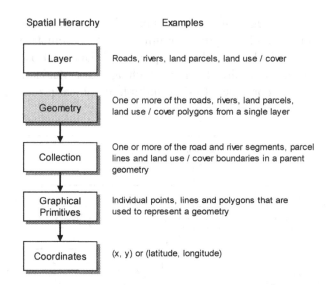

Figure 4-4. The concept of "geometry" and its relationship with other elements as representations of spatial features

The OGC definition of geometry is now widely adopted in the spatial database industry. Many database software vendors have implemented different geometries in their products (see Section 4.2.1). By defining one or more columns of a relational table as *abstract data types*, a user can store geometries in a conventional database and manage them in much the same way alpha-numeric data are managed. Database systems that are enabled to store and manage geometries as special data types are called *spatial database systems* (see Section 4.1 for a formal definition of a spatial database system).

3.2   The Concept of Topology and Topological Data Structures

Topology is a field of mathematics that studies the properties of geometric figures that remain unchanged when the shape of the figure is twisted, stretched, shrunk or otherwise distorted without breaking (West et al., 1982). In other words, it is the study of geometry on a rubber sheet and is concerned with the relationships among and between geometric figures, rather than their rigid coordinates. When topology is applied to a spatial data structure, it is typically defined as the spatial relationships between neighbouring real world features, including their *adjacency*, *connectivity* and *containment* (Lo and Yeung, 2006). A topological data structure, therefore, is one in which the inherent spatial relationships between real world features are explicitly stored.

The concept of topological data structures assumes that spatial features occur on a two-dimensional plane and, therefore, relationships in spatial data can be defined by three primitives, including *nodes* (0-cells), *edges* (1-cell), and *polygons* (2-cells) on a linear graph (Corbett, 1979). In a spatial data set, a *one-dimensional* or *network topology* is built from node and edge primitives. When topology is built for closed polygon objects, the resulting spatial data set is said to have *two-dimensional* or *planar topology*.

In a topological data structure, a node (also called a *point*) is a spatial feature of 0 dimensions where one or more edges (also called *arcs, chains,* and *segments* or *lines*) connect to form a *topological junction*. An edge is a spatial feature of 1 dimension that is formed by a directed, non-branching sequence of non-intersecting line segments bounded by a 'from' and a 'to' node. A polygon (also called a *topological ring, face* or *area*) is a 2-dimension spatial feature that is closed on itself by a sequence of connected and directed edges. The bounding edges of a polygon reference the polygons to their left and right through the 'from' and 'to' nodes that define the direction of the edge. If no polygon exists to the left or the right of a particular edge, then the "empty" space is assigned as the *world* (also called a *universe polygon*) in order to complete the adjacency relationships (that is, to ensure that there is always a "polygon" to the left or right of a bounding edge of a particular polygon).

Arc-node topology is enforced on a spatial data set by requiring the insertion of nodes at all line intersections and at line ends. Figure 4.5 illustrates the concept of topological arc-node enforcement on a pair of neighbouring polygons A and B. In Figure 4.5a, the polygons are represented by the sequence of coordinates that are used to draw or plot them graphically. A spatial data set represented in this way is said to have a *"full polygon"* data structure. A full polygon data structure is also more commonly called a *cartographic data structure* because digital map data in the world of computer-assisted cartography are typically stored in this manner (see Section 3.3).

When topology is enforced on the same two polygons, the polygon boundaries in Figure 4.5b are broken at intersecting points a and d to form separate edges 1, 2 and 3. At the same time, spatial relationships are built and explicitly stored in two topology tables. The network topology table stores the spatial relationship of connectivity of the edges defined by the from-nodes (F_node) and to-nodes (T_node). On the other hand, the polygon topology table stores the spatial relationship of adjacency of the polygons as defined by the left-polygons (L_poly) and right-polygons (R_poly) of the edges. Recall how the idea of the world or universe polygon noted above is applied to complete the adjacency relationships in the topology table in Figure 4.5b. A spatial data set represented with explicitly stored spatial relationships has a topological data structure.

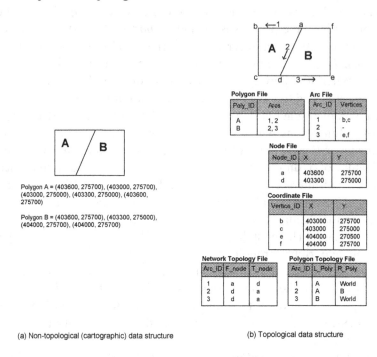

(a) Non-topological (cartographic) data structure   (b) Topological data structure

Figure 4-5. Topological enforcement and topological data structure

The theory of spatial topological data structures was intensively researched in the formative years of GIS in the 1960s and 1970s. Over the intervening years several topological data structures have been proposed and implemented (Peuquet, 1984; Theobald, 2001). The geo-relational data model of an Arc/Info *coverage* is probably the best known topological data structure to many GIS users (Morehouse, 1989; see also Section 3.1 below). A coverage is essentially a map layer that stores topological relationships among adjacent objects in what are called an Arc Attribute Table (AAT) for edges or line objects and a Polygon Attribute Table (PAT) for polygon objects. Connected edges/arcs/lines are linked through nodes at intersections and the relationship of connectivity is stored in an Arc-node table. GIS software packages enforce the planar topology of a coverage when it is created and updated with use of special polygon and arc building commands.

The need to automate the detection and correction of object digitising and editing errors and artefacts is one of the driving forces behind the adoption of a topological data structure. There are other advantages to using such a data structure. These include, for example, the reduction of storage requirements because boundaries shared by adjacent polygons are stored once only. The stored spatial relationships of adjacency, connectivity and containment also enable sophisticated spatial analysis and applications to be carried out relatively easily because their existence practically eliminates the need to develop programming routines for the calculation of these spatial relationships every time a spatial data set is used. In addition, the topological data structure as described above and its associated spatial relationships serves as a basis from which to develop more sophisticated and potentially more analytically powerful spatial data models, such as Molennar's hyper-attributed dual graph model of planar graph-based topology (Molennar, 1998) and its extension to a potentially more powerful *primal-dual multi-valued vector map* (Roberts et al., 2006).

As a correctly topologically structured spatial data set is free of digitising and editing errors and artefacts, a cartographic product generated from it will be not only aesthetically more pleasing, but it will also be geometrically more correct than one created from a non-topological spatial data set. It is the latter criterion that lends the topological spatial data structure its utility in real world applications.

3.3   Non-topological Data Structure

The introduction of the non-proprietary or "open" (which allows other software vendors and application developers to use it freely) *shapefile* data model with the release of ArcView 2.0 by the Environmental Systems Research Institute, Inc. (ESRI) in the early 1990s successfully brought the use of a "full-polygon" data structure to the spatial data world. A full-polygon data structure, which stores polygons individually as discrete entities much in the same way as a cartographic data structure did before the early 1980s, as noted above, is a non-topological data structure because it does not explicitly store spatial relationships.

The success of ArcView in the use of a non-topological data structure, notwithstanding its constraints for specific types of analyses that require information on topology, prompted many spatial data users to question seriously the wisdom of requiring all spatial data sets to be topologically structured (Theobald, 2001). There are indeed numerous applications of spatial data that do not require the use of topological relationships. The processing power of computers today also means that the computation of spatial relationships "on-the-fly", that is not requiring use of specific commands, in spatial analysis is no longer a concern for many spatial data users. This reality has practically eliminated or trivialised the initial driving forces behind the use of a complex topological data structure for routine data display and query.

However, as noted by Strand (1998), spatial data in the shapefile format have several constraints that stop them from being used for a number of important spatial applications. These include, for example, limited cartographic rendering capability, incompatibility with relational database management principles and techniques, absence of topology for efficient spatial (especially logistics-based networking) analysis, and lack of support for transferring metadata that describe spatial data and their characteristics. In this regard, the use of non-topologically structured spatial data should be perceived more as a complement to, rather than a substitution for, the conventional approach of using topologically structured spatial data. Both topological and non-topological data have their place in the spatial data arena because different applications and different users have different requirements for the use of topological relationships.

3.4   The Geo-relational Model

Spatial data were conventionally stored in the proprietary structure of a particular GIS using what is commonly called the *geo-relational data model* (Morehouse, 1985 and 1989). In this model, spatial data are abstracted into a series of independently defined layers. Each of these layers represents a selected set of associated spatial features such as roads, soil types, land cover, land parcel and drainage. Clearly, all layers must occupy all or at least of some spatially coincident area on, below or even above the earth's surface to be modelled within the same geographic space.

These layers are depicted in Figure 4.6, where the spatial features that represent each layer are classified and stored separately according to the forms of the basic graphical primitives or elements that represent them. With this model, spatial features in point form are stored separately from those represented by lines, which are in turn stored separately from those represented by polygons. Spatial features represented by the same basic graphical primitive, such as lines, are classified and stored separately, again according to their feature or entity type. For example, roads and streams which are both represented by lines, are stored in separate layers because they are of different feature types. While this representation may be seen as duplicating the same basic feature types across many different layers, it allows behaviours and characteristics that are specific to each layer to be coded into the features. For example, while all streams that intersect within a watershed normally flow into one another, the same is not necessarily true for the roads that comprise a transportation network. Hence, separation of line features into thematically distinct layers makes application of these differences easier to manage.

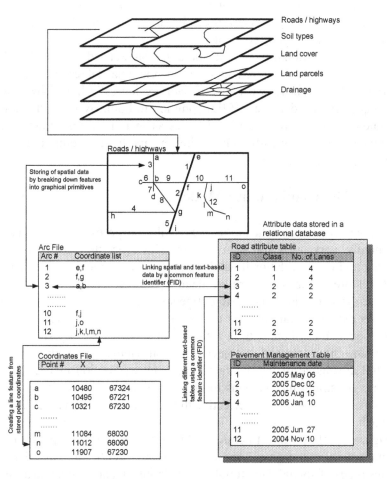

Figure 4-6. The geo-relational model

The specific graphical data structures used and their storage are typically proprietary to a given software product and, consequently, they vary from one GIS to another. Generally speaking, however, most GIS use some form of the *arc-node topological model* described in Section 3.2 that breaks down all features in a line and polygon layer into independently identifiable arcs at their intersections. These arcs are stored in separate tables together with the coordinates of the nodes that form their end points, as well as the coordinates of all other points, known as *vertices*, that form the individual arcs. During data processing, spatial features are reconstructed from stored coordinates retrieved by using information stored in the arc and polygon tables. In Figure

4.6, for example, the road segment with feature identifier (FID) 12 can be reconstructed using its coordinate list in the Arc Table to obtain the coordinates of the points j, k, l, m and n stored in the Coordinate Table. Very complex line and polygon features can be stored and reconstructed using this relatively simple arc-node concept.

In the geo-relational model, attribute data associated with spatial data in each of the layers are stored in separate relational tables (also known as *attribute tables*). These two types of data are logically linked by means of the unique feature identifiers (FID) common to both. By using the common attribute data in different relational tables as search indices or keys, data in these tables can be logically joined with one another as required during data processing and spatial analysis.

3.5   The Geodatabase Model

The geo-relational data model served the purposes of spatial data user community very well for many years. However, the advent of object-oriented and object-relational database systems has gradually eliminated the need to store spatial and attribute data separately. This new generation of database systems allows a user to define spatial data as specific abstract data types, as explained earlier, thus making it possible to store spatial data and their associated attribute data in a single database system. By storing spatial data in an object-oriented or object-relational database, it is also possible to take full advantage of the available indexing, transaction management and database constraint mechanisms to maintain the integrity of the spatial data and to make them more cost effective than in the conventional geo-relational approach.

Figure 4.7 shows the structure of this new generation of spatial database systems, such as the *geodatabase* of ESRI's ArcGIS software, that stores various types of spatial data, topology, attribute data and metadata all using a single database system. Such systems can function at multiple levels of complexity from a 'personal' geodatabase to much more complex 'enterprise' databases with a change in the database software platform. In the new generation of systems, spatial data that share the same attributes (that is, data of the same feature class) are stored in a single table. In this table there are two sets of fields, namely predefined fields and custom fields. The predefined fields include the feature identifier (FID), the geometry which describes the shape of the feature, and a geometry-tracking field that records the area of the feature.

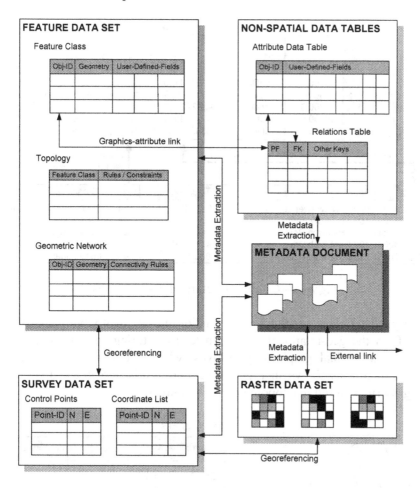

Figure 4-7. Structure of a spatial database using a DBMS for the storage of spatial data and topological relationships

Predefined fields are managed by the database system and cannot be modified by a user through any given database application. Custom fields, on the other hand, are implemented by the database administrator. They contain attribute data associated with spatial data in the predefined fields. The number of custom fields is determined by the database administrator during the process of database creation. The values of custom fields can be updated during transaction processing, and custom fields with unique values can be used as primary keys to relate to other relational or lookup tables in the database, as illustrated in Figure 4.8.

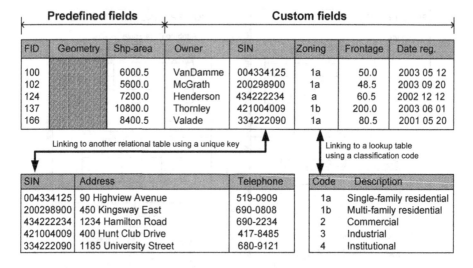

FID	Geometry	Shp-area	Owner	SIN	Zoning	Frontage	Date reg.
100		6000.5	VanDamme	004334125	1a	50.0	2003 05 12
102		5600.0	McGrath	200298900	1a	48.5	2003 09 20
124		7200.0	Henderson	434222234	a	60.5	2002 12 12
137		10800.0	Thornley	421004009	1b	200.0	2003 06 01
166		8400.5	Valade	334222090	1a	80.5	2001 05 20

Predefined fields — Custom fields

Linking to another relational table using a unique key

Linking to a lookup table using a classification code

SIN	Address	Telephone
004334125	90 Highview Avenue	519-0909
200298900	450 Kingsway East	690-0808
434222234	1234 Hamilton Road	690-2234
421004009	400 Hunt Club Drive	417-8485
334222090	1185 University Street	680-9121

Code	Description
1a	Single-family residential
1b	Multi-family residential
2	Commercial
3	Industrial
4	Institutional

Figure 4-8. Table structure of a geodatabase

A geodatabase stores topological relationships explicitly but in a different way from the geo-relational model. Instead of storing spatial relationships between individual graphical primitives, which requires a very complex table structure as shown in Figure 4.6, topology is implemented by using integrity rules stored in a topology table (Figure 4.9).

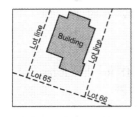

Topology File

Feature Class	Rule	Feature Class
Lot_lines	Must not have dangles	
Lots	Must not overlap	
Owner_parcel	Must be closed	
Lot_lines	Must be covered by	Lots
Buildings	Must be covered by	Owner-parcel
Buildings	Must be covered by	Lots
Buildings	Must not overlap	Lot_lines
Lots	Must be formed by	Lot-lines
Lot_lines	Must not overlap	Buildings

Figure 4-9. Storing topological relationships using an integrity rule

Such integrity rules can be applied to a particular feature class in a single table to enforce topological relationships such as adjacency (for example, removing gaps between parcel polygons), connectivity (for example, ensuring proper linking between different segments of a highway), and containment (for example, determining islands in a lake). They can also be applied to different feature classes in different tables to maintain the integrity of coincident geometry between spatial features of different feature classes (for example, ensuring the matching between political boundaries in one layer and rivers used as part of the boundaries in another layer).

4. SPATIAL DATABASE SYSTEMS

Spatial database systems are treated in this book as a special type of database system that is enabled specifically to manage and process spatial data. The following discussion explains how the database concepts and methods that were described in previous sections are applied and enhanced to enable a database system with spatial data.

4.1 Definition and Classification of Spatial Database Systems

Güting (1994) defined spatial database systems as a class of database systems that have the following three characteristics:

- A spatial database system is a database system.
- It offers *spatial data types* (SDT) in its data model and query language.
- It supports spatial data types in its implementation, providing at least *spatial indexing* and efficient algorithms for spatial joins.

The first characteristic of spatial database systems emphasises the fact that these systems are fully-fledged database systems capable of performing all standard data modelling and query tasks, but with additional functionality to perform tasks specific to spatial data. The notion of spatial database systems as a "spatial" extension of ordinary database systems implicitly stresses the need for the integrated processing of geographically referenced data and text-based data that characterise corporate data processing environments. This ability to handle and relate spatial and non-spatial data simultaneously distinguishes spatial database systems from other forms of database systems that also use geographically-referenced or location-based data, such as computer assisted design/computer aided manufacturing, cartographic systems, and remote sensing image processing systems.

SDT in the second characteristic above refers to the same points, lines and polygons discussed earlier that represent geometric entities or objects in space, as well as their relationships (for example, i is adjacent to j), properties (for example, area(a) > 1,000 km^2), and operations (for example, feature object m overlays object n). A spatial database system as defined by the first characteristic is not able to perform its intended functions without data collected as appropriate SDT. SDT, therefore, are an indispensable parameter in the definition of spatial database systems (see Section 6.2a).

It was noted in Chapter 2 that modern database systems are characterised by their ability to handle very large amounts of data, and that these systems achieve database access efficiency by means of different indexing techniques. These capabilities are also required characteristics of spatial database systems. However, since data in spatial databases are represented and referenced in two-dimensional and three-dimensional space, data in spatial databases must be able to be indexed spatially. This allows data in spatial databases to be accessed and analysed by location (using the geographic coordinates that all features are registered to) and the topological relationships that define the position of features in space relative to each other (adjacency, containment and connectivity). Together, location and topology allow various methods of overlaying and combining spatial features on map layers, collectively called *spatial joins*, to be used for analysis purposes (see Section 4.3).

In his definition of spatial database systems, Güting (1994) considered spatial databases only as the underlying technology for GIS. Hence, spatial database systems are by no means fully operational GIS and may only facilitate a small subset of the functions that GIS provide. As explained in Chapter 6, many spatial databases are now developed as *spatial data warehouses* that aim more at supplying GIS users with timely and relevant spatial data, rather than using the spatial data they contain for a particular application by or within the spatial database itself. At the same time, GIS are now increasingly implemented using external data resources obtainable from government agencies and commercial spatial data suppliers, rather than by in-house digitising and database creation. This trend in spatial database development and deployment implies that the trend in spatial databases and GIS is increasingly to serve relatively distinct but interdependent purposes in the data processing functions of an organisation.

Figure 4.10 illustrates the evolution of spatial data processing concepts and techniques since the early 1970s when the first generation GIS were developed. Until the mid-1990s, spatial data processing was characterised by a dependence on GIS software constructed using proprietary API and proprietary data formats (Figure 4.10a). The need to handle large volumes of spatial data and advances in database technology in the mid-1990s together led to the increasing use of RDBMS for GIS applications. However, relational databases remained largely as an attribute data store, which communicated with GIS by means of conventional SQL statements. Practically all data processing was still performed within the GIS (Figure 4.10b).

(a) Data file-based spatial data processing using a GIS before the mid-1990s

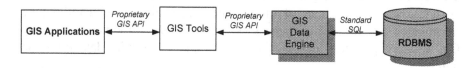

(b) DBMS-based spatial data processing using a GIS in the late 1990s

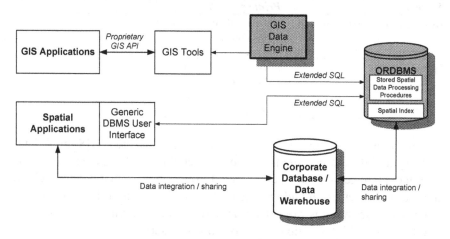

(c) Today's spatial data processing environment

Figure 4-10. Evolution of spatial data processing

The advent of object-relational database management systems (ORDBMS) in the late 1990s made possible the use of generic DBMS functionality to process and apply spatial data for end use applications. By using extended SQL capable of handling spatial data, an ORDBMS can be used to store not only attribute data but spatial data as well (see Chapter 3). Further, an ORDBMS can be relatively easily connected to other database systems in a local or global communication network, making it an integral part of the corporate computing architecture of organisations in business, government and education (Figure 4.10c).

In contemporary spatial data processing environments the division of work between spatial database systems and GIS is quite clear (Table 4.1). By taking advantage of advances in database technology, spatial database systems are implemented to store and manage large volumes of geographically referenced or location-based data. These systems index the data spatially for efficient query and retrieval, and provide the mechanisms and procedures required to protect the data from physical destruction and loss or degradation of logical integrity. However, spatial database systems are generally functionally weak in data collection and editing, spatial data analysis, and generation of maps and other forms of cartographic information products. These processes and functions are better handled by GIS that are primarily designed to collect and use spatial data for these purposes.

Table 4-1. The division of work between spatial database systems and GIS

Systems	Primary Tasks
Geographic Information Systems	o Data Collection and Editing
	o Data Analysis
	o Generation of Maps and Cartographic Information Products
Spatial Database Systems	o Data Storage and Management
	o Spatial Indexing
	o Data Security and Integrity
	o Spatial Data Query

This book distinguishes between application-oriented GIS and data-centric spatial database systems as two distinct classes of systems. The focus is on the management of spatial data through a systematic approach to database implementation and project management. Hence, it is assumed that spatial database systems are set up as part of an infrastructure intended to serve the general information needs of an organisation, with or without the use of a specific GIS.

4.2   Characteristics of Spatial Database Systems

Spatial database systems allow the management and processing of geographically referenced data. The following discussion explains their specific features relative to the conceptual definition provided in the previous section. The features discussed include spatial data types, spatial database indexing, spatial data integrity constraints and transactions in spatial data processing.

4.2.1   Spatial Data Types (user-defined or abstract data types)

Conventionally, database systems were designed to manage and process alpha-numeric data represented by character strings, numerical values, date and Boolean or logical expressions (see Chapter 2). There was no provision in conventional relational database systems to support the storage and processing of spatial data represented by points, lines, polygons and surfaces. Object-oriented and object-relational databases allow users to define ADT that describe complex column structures in a database. Several database software vendors have made use of this capability to define spatial data types (SDT) that enable their products for managing and processing spatial data.

The OGC geometry object model noted in Section 3.1 provides a conceptual standards-based framework for various SDT. Different software vendors implement the concept in different ways. Oracle Spatial, for example, has nine SDT called *geometric primitive types* including points and point clusters, line strings, polygons, arc strings, arc polygons, compound polygons, compound line string, circle, and optimised rectangles (Figure 4.11a). IBM's DB2 Spatial Extender, on the other hand, uses the term *geometry type* to describe its SDT, which are organised in a hierarchy of subtypes of geometry, including points, multi-points, line strings, multi-line strings, polygons, multi-polygons and ellipses (Figure 4.11b). While this terminology differs, the basic underlying data models and the types of functions and operations that they facilitate in software tend to be fairly consistent.

In ESRI's ArcGIS geodatabase approach, spatial features are represented by four types of geometry that describe their shapes, namely point, multipoint, polyline and polygon (Figure 4.11c). Point geometry has a single x, y or x, y, z co-ordinate value that defines its location in space. Multipoint geometry has a number of x, y or x, y, z co-ordinate values that serve the same function. A polyline geometry comprises one or more paths made up of one of four supported segment types, namely line, circular arc, elliptical arc, and Bezier curve. Polyline geometry can be optionally associated with a z value (for elevation) or an m value (for distance measurement). Polygon geometry has one or more rings which, by definition, form a connected, closed and non-intersecting set of segments. Just like polyline geometry, polygon geometry can also be optionally associated with a z value (for elevation or some other extrusion from the plane).

In addition to vector spatial data of points, lines, polygons and surfaces, it is also possible to define raster spatial data (for example, geo-referenced digital aerial photographs, scanned maps, remote sensing images and continuous surfaces such as a digital elevation model) as ADT in object-oriented and object-relational databases. As a result, vector and raster data can now be managed and processed within a single spatial database system. While the focus in this book is on vector spatial data, it is important to remember that raster data can also be incorporated as a particular SDT in most contemporary spatial database systems.

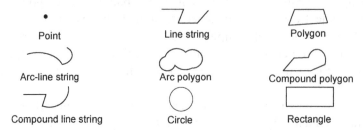

(a) Geometry types used in the object-oriented model of Oracle Spatial

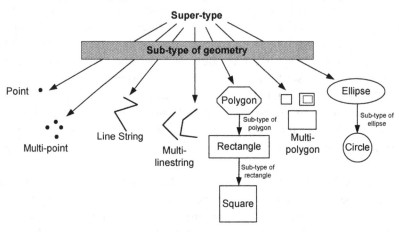

(b) Geometry types and sub-types of DB2 Spatial Extender

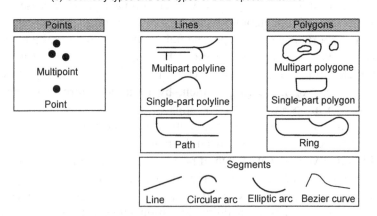

(c) Feature geometry of ArcGIS Geodatabase

Figure 4-11. Spatial data types used by Oracle Spatial, IBM DB2 Spatial Extender and ESRI Geodatabases (Source: Oracle, 2002; Davis, 1998; and Zeiler, 1999)

4.2.2 Spatial Data Indexing and Access Method

Spatial indexing serves the same purpose as the table indexing that was explained in Chapter 2, namely to expedite access to and return of data to a user from a database. Spatial indexing is much more complicated than table indexing because it deals with a two-dimensional space rather than a more straightforward linear array of text-based data in tables.

A fundamental concept of spatial indexing is the use of approximation whereby the spatial access process gradually narrows its search area until the required database objects are found. Numerous spatial indexing methods are being used for spatial database systems (Brinkhoff et al., 1994). One of the more commonly adopted spatial indexing methods is the *R-tree* (where R stands for "region") shown in Figure 4.12.

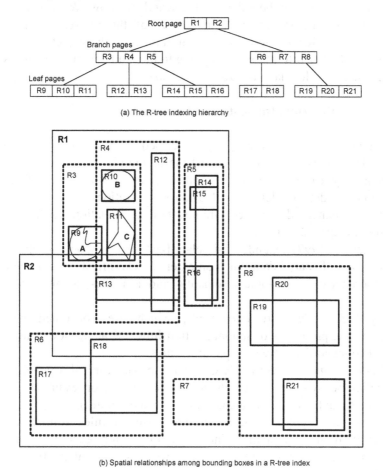

(a) The R-tree indexing hierarchy

(b) Spatial relationships among bounding boxes in a R-tree index

Figure 4-12. The R-tree index

This is a multi-level tree that stores a set of rectangles in each node, just like the B-tree discussed in Chapter 2. In an R-tree spatial index, individual data objects are represented by their corresponding *minimum bounding rectangles* (MBR) at the leaf page level. As shown in Figure 4.12, for example, objects A, B and C are represented by MBR R9, R10, and R11 respectively. The R-tree index stores the reference numbers of the MBRs and the coordinates of their four corners in an index file. When the database system receives a request for a spatial search, for example to find all the land parcels passing through a user-defined window on the computer screen, the system will scan the stored coordinates of the MBR of the root pages to determine which MBR at the branch level fall within the search window.

Once the MBR at the branch level are identified, the system then scans these rectangles to identify the MBR of objects at the leaf level to select those falling within the boundary of the search window. When these MBRs are identified, the system will make use of the reference information between the MBR and the identification number of the object to access the attribute tables for information pertaining to the land parcels such as owner's name, assessed value, land use zoning and so on.

4.2.3   Spatial Data Integrity and Constraints

Integrity constraints are best described as business rules that are applied in database systems to protect the intangible "value" of the data by ensuring their accuracy, correctness and validity. Three classes of integrity constraints in data modelling and database operations were introduced in Chapter 2, namely domain constraints, key and relation constraints, and semantic integrity constraints.

Cockcroft (1997) extended this classification to encompass the special requirements of spatial data. Such an extended classification is based on the distinction between topological, semantic, and user rules, as follows:

*   *Topological integrity constraints*, which are concerned with the geometric properties of spatial relationships (i.e. adjacency, containment and connectivity) between spatial features.
*   *Semantic integrity constraints*, which are database rules governing the spatial behaviours of objects in the database (for example, no land parcels can be located in areas shown as water bodies).
*   *User-defined constraints*, which are business rules similar to those identified in non-spatial data modelling (for example, no wood harvesting is allowed in a 200-metre buffer zone along the shore of lakes).

Cockcroft (1998) further proposed that each of the above three classes of integrity be applied to both data in the consistent state of the database and data in transaction processing, thus resulting in the following six classes of spatial data integrity constraints:

- *Static topological integrity constraints*, for example all polygons must be closed.
- *Transition topological integrity constraints*, for example if the boundary of a polygon is modified, both the polygon itself and all the conjugate polygons must be updated simultaneously.
- *Static semantic integrity constraints*, for example the area of a land parcel must not be negative.
- *Transition semantic integrity constraints*, for example after the subdivision of a land parcel, the sum of the areas of the subdivided units must be equal to the area of the original parcel.
- *Static user-defined integrity constraints*, for example rivers and streams wider than 2 metres must be stored as polygon features.
- *Transition user-defined integrity constraints*, for example after an application for re-zoning of a land parcel has been approved, the land use status of the parcel concerned must be updated within two working days.

Integrity constraints are captured during the conceptual data modelling process and must be enforced throughout the entire life cycle of systems development (see Chapter 3). They must be also properly documented and stored as an integral component of the metadata of the database (see Chapter 5).

4.2.4   Long Transaction Management

As noted in Chapter 2, database transactions are very complex processes. They involve not merely the movement of data in and out of the database, but also the recording of each and every step of the transaction process so the database can restored back to a consistent state if required. Data processing in banking, retailing and airline seat reservations is characterised by a large number of short transactions that usually last no more than a second or two. Concepts and techniques of concurrent control and data locking as described in Chapter 2 are well-defined and capable of handling a large number of short-duration transactions.

Spatial data processing, on the other hand, differs from conventional data processing in many ways, but transaction management is probably one of the most significant differences between the two. Transactions in spatial data processing, for example updating the road fabric as the result of an annual pavement maintenance program, commonly take several days or even several weeks to complete, while many users may need to access the data at the same time. If the same concept and techniques of concurrent control for ordinary database processing are used, then the availability of the database will be greatly reduced. Thus, the ability to handle long transactions typically found in spatial data processing is a prerequisite for using database technology for spatial data processing.

Different database software vendors use different solutions to address the problem of long transactions in spatial data processing. Oracle, for example, uses the techniques of *workspace management* to handle long transactions (Lopez and Gopalan, 2001). Workspace management refers to the ability of the database to hold different versions of the same record or row in one or more database workspaces. Users of the database can then change these versions independently. The unit of versioning in Oracle is a database table. A table in an Oracle database can be version-enabled, which means that all rows in the table can support multiple versions of data. The versioning mechanism is handled by a database system component called the *workspace manger* and is not visible to the database end users. After an Oracle table is version-enabled, users automatically see the correct version of the record they are interested in. In this way, the workspace creates a virtual environment that one or more users can share resources to make changes to the data in the database.

4.3   Spatial Data Processing

Spatial queries are more complex than conventional database queries because they are concerned with two- or three-dimensional data and the graphical presentation of query results. The following discussion considers the fundamentals of spatial queries and explains how they are implemented in spatial database systems using extended SQL.

4.3.1   Classification of Spatial Operators

A spatial query is formulated using one or more operators, including predicates to express spatial relationships, on the content of a spatial database. A knowledge of the the characteristics of different types of spatial operators is required to form acceptable spatial queries.

There are different methods of classifying spatial operators. One method is to classify these operators as *unary* and *binary* operators (Egenhofer, 1994). These two types of spatial operators are applied, respectively, to obtain information about the properties of a single geometry (for example, attributes including location, area, length and volume) and the relationships between two geometries (for example, distance between, direction, adjacency, connection, and containment).

Another method is to classify spatial operators as topological, projective and metric operators (Clementini and Di Felice, 1997). In this particular classification, topological operators are those which use topological relationships (for example, touch, in, intersect, cross, overlap and disjoin) to obtain the properties of a particular geometry. Projective operators are those which represent predicates about the concavity/convexity of a geometry and other relationships (for example, the *convex hull* which is defined as the containment of a set of points by a boundary formed by the smallest number of points). Finally, metric operators are those which represent predicates about the measurements of or relationships between geometries such as distance and direction.

The OGC has developed a comprehensive classification of spatial operators, as shown in Table 4.2 (OGC, 1999).

Table 4-2. OGC spatial operators defined on the class geometry

Classes	Operators	Operator Functions
Basic Operators	Spatial Reference	Returns the reference system of the geometry
	Envelope	Returns the minimum bounding rectangle of the geometry
	Export	Converts the geometry into a different representation
	IsEmpty	Tests if the geometry is the empty set or not
	IsSimple	Returns TRUE if the geometry is simple
	Boundary	Returns the boundary of the geometry
Topological Operators	Equal	Tests if the geometries are spatially equal
	Disjoint	Tests if the geometries are disjoint
	Intersect	Tests if the geometries intersect
	Touch	Tests if the geometries touch each other
	Cross	Tests if the geometries cross each other
	Within	Tests if a geometry is within another geometry
	Contain	Tests if a given geometry contains another geometry
	Overlap	Tests if a given geometry overlaps another given geometry
	Relate	Returns TRUE if the spatial relationship specified by the 9-Intersection matrix holds
Spatial Analysis Operators	Distance	Returns the shortest distance between any two points of two given geometries
	Buffer	Returns a geometry that represents all points whose distance from the given geometry is less than or equal to a specified distance
	ConvexHull	Returns the convex hull of a given geometry
	Intersection	Returns the intersection of two geometries
	Union	Returns the union of two geometries
	Difference	Returns the difference of two geometries
	SymDifference	Returns the symmetric difference (i.e. the logical XOR) of two geometries

In this classification, spatial operators are placed into three categories, namely basic operators, topological operators, and spatial analysis operators. Basic operators allow a user to access the general properties of a geometry such as its location, shape and boundary. Topological operators are similar to those in the Clementini and Di Felice classification, forming predicates expressing the spatial relationships between geometries. Spatial analysis operators allow a user to construct analytical spatial queries using a single geometry (for example, buffering) or a multitude of geometries (for example, convex hull, union, and intersection).

4.3.2   Spatial Operations and Filtering

Because of the large size of a typical spatial database and the complexity of spatial operations, spatial queries are not normally applied directly to a database. Instead one or more methods of filtering are used to expedite the process of accessing the database.

Figure 4.13 shows the two tier approach used by Oracle Spatial to satisfy queries of spatial data. In this query model, the primary filter uses the concept of approximation to reduce the number of candidate geometries by means of the spatial index of the database, as explained in Section 4.2.2. The data set obtained from the first part of the filtering process, which is many times smaller than the original data table, is then subject to the secondary filter that is made up of one or more of the spatial operators noted above.

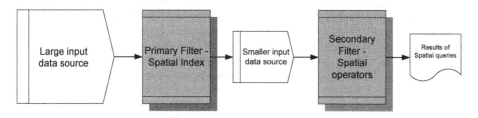

Figure 4-13. Spatial query using the method of two-tier filtering

Filtering enables spatial queries to work much faster because it reduces the need for intensive computation at the second pass by applying the spatial operations of a query to a relatively limited number of geometries only.

4.3.3   Topological Relations and Predicates

In the various classification schemes for spatial operators noted above, topological relations play a significant role in spatial queries. Topological operators analyse the relative positions of geometries in space. They have formed an integral part of the spatial analytical functions of GIS since the very early days of their development and they continue to figure prominently in applications of this technology.

In spatial database systems, topological relations are normally used with predicates, which by definition are Boolean functions that determine whether a specific relationship exists between a pair of geometries. A predicate returns the value of 1 (TRUE) if a comparison meets the function's criteria, or 0 (FALSE) if the comparison fails. A predicate can test for a spatial relationship of geometries that are of different types (for example, a point inside a polygon, or a line crossing a polygon) or dimensions (for example, a building footprint within a parcel).

A basic problem with the use of topological predicates in spatial queries is to define all possible relationships. Egenhofer and Herring (1990) noted that for two objects there are four intersection sets when comparing their boundaries and interiors. As each of the two objects might be empty or non-empty, there are $2^4 = 16$ combinations. They noted that since eight of these combinations are not valid, and two of them are symmetric, only six relationships remain significant. These topological relationships are *disjoin*, *in*, *touch*, *equal*, *cover*, and *overlap*.

This approach of defining topological relations has been extended by different researchers in different ways. One of the extensions was the *Dimensionally Extended 9 Intersection Model* (DE-9IM), proposed by Clementini et al. (1993). This extended model took into consideration the dimensions (0-dimension for points, 1-dimension for lines and 2-dimension for areas) and resulted, in principle, in $4^4 = 256$ combinations. As many of the relationships were not valid, a total of 52 topological relationships among point, line and area features remained. However, 52 topological relationships are practically too many to be implemented in a spatial database system. In practice, most spatial systems include only a relatively small subset of the possible topological relations, such as those shown in Figure 4.14.

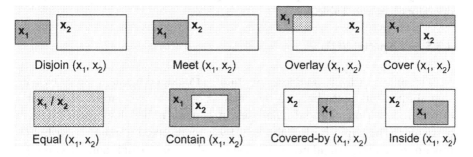

Figure 4-14. The most common topological relations in spatial systems

4.3.4   Spatial Joins

A spatial join is a spatial query that compares two or more geometries with predicates according to their locations. Functionally, a spatial join serves the same purpose as a regular join in relational database query (see Chapter 2) except that the predicate involves a spatial operator such as overlay, intersect, contain, union, difference and symmetric difference, all of which are listed in Table 4.2. In practice, a spatial join is probably the most important of all spatial queries because it provides the mechanism for comparing two or more database layers in spatial analytical functions such as overlay analysis using spatial data and their associated attributes.

A spatial join can be used in various applications involving spatial data of different geometric types. One example of using a spatial join is to answer a spatial query such as "Find all 4-lane King's Highways that are patrolled by different detachments of the Ontario Provincial Police (OPP)". This query can be resolved by joining a layer that stores the geometries of OPP detachments with another layer that stores provincial highway geometries. The system will first use a primary filter to identify the geometry identifiers of OPP detachments and the geometry identifiers of highway segments that cross in the spatial index. The resulting data sets are then subject to a second filter that will determine the intersection of the boundaries of different detachments and the highways that cross them. Finally, the system will identify from the attributes stored in the database all of those highway segments that have four lanes. The final results of this particular spatial join would be a table and a map showing the OPP detachments and the 4-lane highways under their jurisdiction.

Spatial joins are computationally intensive by nature. However, they can be used to create very powerful spatial analysis and modelling tools. Database applications would be very limited without the flexibility provided by spatial joins and, as a result, they have conventionally been mandatory functions of GIS. All spatial database systems are expected to support the spatial join function with spatial indexing and join algorithms for the spatial operators listed in Table 4.2.

4.3.5   Spatial SQL

The interest of the database community in extending standard SQL with spatial operators dates back to the early 1980s. However, most of the early efforts focused on pictorial databases in general rather than spatial data in particular. The paper by Egenhofer (1994) represented a major contribution toward extending standard SQL specifically for spatial data. Egenhofer examined the characteristics and requirements of spatial operations. He proposed that a spatially extended SQL should consist of two parts, namely a *query language* to define what data to retrieve and a *presentation language* to specify how the results of a query are displayed.

The query language aspect of spatial SQL should ideally be developed as a minimum extension to the interrogative part of standard SQL and characterised by the following:

- Preservation of SQL concepts, that is, to retaining the SELECT-FROM-WHERE construct as the framework for database query (Figure 4.15a).
- High-level treatment of spatial objects, that is, to be able to define the complex abstract data type spatial and its sub-types of different dimensions (Figure 4.15b).
- Incorporation of spatial operations and relationships using standard SQL and spatial SQL to query non-spatial and spatial data respectively and give instructions to a presentation component to manipulate and display the results (Figure 4.15c).

The presentation language of spatial SQL, on the other hand, is required to have several important features, including for example:

- Graphical combination or overlay or multiple queries by adding or removing the layers resulting from different queries in a display.
- Display of context by interpreting the results of a query and then selecting and displaying relevant background information of this query (Figure 4.15d).
- Generating graphical presentations by assigning different colours, patterns and symbols to distinguish one object class from another.
- Use of a legend to provide an explanation of the graphical presentation used in a display (Figure 4.15e).
- Automated label placement by selecting attributes from a database and display them as labels within a graphical presentation.
- Scale selection by determining the scale of graphical presentation with respect to application and cartographic generalisation.
- Sub-area queries by restricting attention of a particular area for query (Figure 4.15f).

Spatial SQL has been extensively researched (see for example, Rigaux et al., 2002). Proposals made to extend the SQL92 standard with spatial operators aim at integrating capabilities to handle spatial data directly in the SQL3 standard. The most notable proposal is that undertaken by OGC (OGC, 1999). Spatial SQL is now available in several spatial extensions to conventional database systems such as Oracle Spatial and IBM DB2 Spatial

Extender. It will probably not be very long before spatial operations are formally an integral part of the SQL standard.

```
SELECT parcel.name
FROM parcel, subdivision
WHERE within (parcel.loc, subdivision.loc)
AND subdivision.name="cranebrook"
```

(a) Preservation of the basic SQL SELECT-FROM-WHERE construct

```
CREATE TABLE parcel
      (parcel.ID                char(20)
       geometry                 ST_polygon)
```

(b) Defining a spatial object type "parcel" as a spatial data type "ST_polygon"

```
SELECT city
FROM ontario.city
WHERE geometry = PICK;
SELECT city
FROM ontario.city
WHERE city.name = "Waterloo"
```

(c) Query by location "PICK" (by means of a mouse) and by attribute value "Waterloo"

```
SET CONTEXT
FOR parcel.geometry
SELECT parcel.geomemtry, building.geometry, road.geometry, easement geometry
FROM parcel, building, road, easement
WHERE parcel.ID = "LONDON00221122145678"
```

(d) Setting the background information (building, road, easement) for a spatial object "parcel"

```
SET LEGEND
      COLOUR            green
      LINE.TYPE         dashed
FOR SELECT boundary.geometry
FROM parcel
```

(e) Setting the property of a legend

```
SELECT parcel
FROM parcel.layer
WHERE geometry = ZOOM.WINDOW;
SELECT parcel
FROM parcel.layer
WHERE geometry = PICK
```

(f) Restricting a query to a specific area

Figure 4-15. Examples of using spatial SQL

5.   SUMMARY

This chapter examined the concepts and methods of spatial database systems as a special class of database systems. Data structure and data management requirements of spatial database systems were emphasised. Different types of spatial data were defined and classified from a functional database perspective, rather than from the conventional cartographic or mapping perspective. The new approach to representing spatial features by means of "geometry" was introduced. The concept of "topology", as it applies to the representation of spatial relationships, was revisited and the growing acceptance of non-topologically structured spatial data was discussed. The concept of the geo-relational model, which dominated the spatial database world in the 1990s, was explained in conjunction with the geodatabase, which is one of the most commonly used contemporary spatial database models for organising and managing spatial data using a relational or object-relational database system.

Güting's definition of spatial database systems was adopted as a particular class of database systems that is capable of performing all the functions of a typical database system, such as indexing and joins, but has additional functionality to handle data with spatial attributes. In this context, the characteristics and requirements of spatial databases were discussed with reference to the concepts of spatial data integrity constraints, long transactions, spatial access methods by *R-tree* indexing, and spatial data processing using the concepts and methods of *spatial joins*. The emphasis on data rather than their applications is maintained in the following chapters. In particular, the next chapter discusses spatial data standards and metadata.

6. REFERENCES

Adler, D.W. (2001) "IBM DB2 Spatial Extender – Spatial Data within the RDBMS", *Proceedings*, VLDB Conference, Rome, Italy.

Brinkhoff, T., Kriegel, H.-P., and Seeger, B. (1994) "Efficient Multi-step Processing of Spatial Joins Using R-trees", *Proceedings of the ACM SIGMOD Conference*, Minneapolis, MN, pp. 237-246.

Clementini, E. and Di Felice, P. (1997) "A global framework for qualitative shape description", *GeoInformatica*, Vol. 1, No. 1, pp. 11-27.

Clementini, E., Di Felice, P. and Van Oosterom, P. (1993) "A small set of formal topological relationships suitable for end-user interaction", Proceedings, 3rd International Symposium on Large Spatial Databases, Singapore.

Cockroft, S. (1997) "A Taxonomy of Spatial Constraints", *GeoInformatica*, vol. 1, no. 4, pp. 327-349.

Cockroft, S. (1998) "User Defined Spatial Business Rules: Storage, Management and Implementation – A Pipe Network Case Study", paper presented at the 10th Colloquium of the Spatial Information Research Centre, University of Otago, New Zealand.

Corbett, J.F. (1979) *Topological Principles in Cartography*, Technical Paper 48, Washington, DC: U.S. Department of Commerce, Bureau of the Census.

Egenhofer, M.J. (1994) "Spatial SQL: A query and presentation language", *IEEE Transactions on Knowledge and Data Engineering*, Vol. 6, No. 1, pp. 86-95.

Egenhofer, M.J. and Herring, J. (1990) "A mathematical framework for the definition of topological relationships", *Proceedings*, 4th International Symposium on Spatial Data Handling, Zurich, Switzerland.

ESRI (1998) *Shapefile Technical Description*, Redlands, CA: Environmental Systems Research Institute, Inc.

ESRI (1999) "Getting Started with SDE", ESRI White Paper, Redlands, CA: Environmental Systems Research Institute, Inc.

ESRI (2002) "ArcGIS 8.3 Brings Topology to the Geodatabase", *ArcNews*, Vol. 24, No. 2.

Güting, R.H. (1994) "An Introduction to Spatial Databases", *VLDB Journal*, No. 3, pp. 357-399.

Lo, C.P. and Yeung, A.K.W. (2006) *Concepts and Techniques of Geographic Information Systems*, 2nd ed., Upper Saddle River, NJ: Prentice-Hall, Inc.

Lopez, X.R. and Gopalan, A. (2001) "Managing Long Transactions Using Standard DBMS Technology", in *Proceedings of Annual Conference*, Geospatial Information and Technology Association (GITA), Sydney, Australia.

Morehouse, S. (1985) "ARC/INFO: A Georelational Model Spatial Information", *Proceedings*, AutoCarto 7, Washington, DC.

Morehouse, S. (1989) 'The Architecture of ARC/INFO', *Proceedings*, Auto-Carto 9, Falls Church, VA: American Society for Photogrammetry and Remote Sensing.

OGC (Open GIS Consortium) (1999) *OpenGIS Simple Feature Specification for SQL* (OpenGIS Project Document 99-049), Revision 1.1, Wayland, MA: Open GIS Consortium, Inc.

Oracle (2002) *Oracle Spatial User's Guide and References*, Release 9.2, Redwood Shores, CA: Oracle Corporation.

Orfali, R., Harkey, D. and Edwards, J. (1999) *The Essential Cleint/Server Survival Guide*, 3rd ed., New York: John Wiley & Sons.

Peuquet, D.J. (1984) 'A Conceptual Framework and Comparison of Spatial Data Models', *Cartographica*, No. 21, pp. 66-113.

Rigaux, P., Scholl, M and Voisard, A. (2002) *Spatial Databases with Application to GIS*, San Diego, CA: Academic Press.

Roberts, S., Hall, G. B. and Calamai, P. (2006) 'A Pre-Categorical Spatial Data Meta-Model', *Environment and Planning B*, forthcoming.

Strand, D. (1998) 'Shapefiles Shape GIS Data Transfer Standards', *GIS World*, Vol. 11, No. 5, pp. 28-29.

Theobald, D.M. (2001) 'Topology Revisited: Representing Spatial Relations', *International Journal of Geographical Information Science*, Vol. 15, No. 8, pp. 689-705.

Ullman, J.D. and Widom, J. (1997) *A First Course in Database Systems*, Upper Saddle River, NJ: Prentice Hall.

West, B.H., Griessback, E.H., Taylor, J.D. and Taylor, L.T. (1982) *The Prentice Hall Encyclopedia of Mathematics*, Englewood Cliff, NJ: Prentice Hall.

Zeiler, M. (1999) *Modelling Our World: The ESRI Guide to Geodatabase Design*, Redlands, CA: ESRI Press.

Chapter 5

SPATIAL DATA STANDARDS AND METADATA

1.   INTRODUCTION

This chapter considers the concepts and methods of data standards and metadata as they apply to spatial database design and implementation. The discussion focuses on spatial data standards and metadata initiatives in general, rather than the details of specific spatial data and metadata standards. Particular attention is given to the interrelationships among spatial data standards at the international and national levels, as well as the deployment of these standards and metadata in spatial database design and implementation. The discussion starts with a general examination of the concepts and methods of developing and implementing standards. Spatial data standards as part of an overall spatial database strategy are then explained in Section 3, which also considers the work of various spatial data standards organisations and their activities in key areas of relevance to spatial database systems. Spatial metadata are discussed in Section 4, and their relationship with data standards is examined in Section 5 with special reference to spatial database systems. Section 6 summarises the importance and use of data standards and metadata in spatial database design and implementation.

2.   STANDARDS AND STANDARDISATION

The concepts and techniques used for spatial data standards are drawn largely from standards developed originally for general applications in the computer and IT industry. Hence, it is logical to start this discussion with an overview of standards by defining what they are, explaining how they are typically developed, and discussing how they are used to achieve the goals of standardisation in practice. Where appropriate, examples that relate to spatial data standards and standardisation are used.

2.1   Definition of Standards

The term *standards* is defined differently by different standards organisations. This chapter combines the concepts and notions of standards as adopted by the International Organisation for Standardisation (ISO), the American National Standards Institute (ANSI), the British Standards Institute (BSI), and the Standards Council of Canada (SCC). Using these organisations and their interpretations, a standard can be best defined as a document or collection of documents, usually but not always published, that establishes a common language, terminology, accepted practices and levels of performance, as well as technical requirements and specifications, that are used consistently for the development and use of products, services and systems.

A standard can be documented differently, implemented in different ways, and used to describe different things at different levels of detail. Hence, a better way to understand the multi-faceted nature of standards is to define them using a three-dimensional matrix, the cells of which are defined by the above factors and explained below (Figure 5.1).

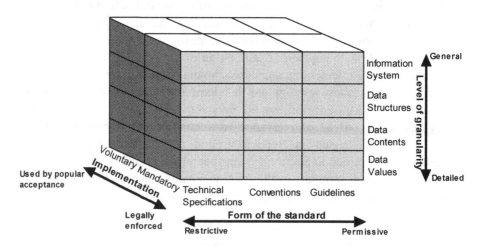

Figure 5-1. A Three-dimensional matrix definition of standards

2.1.1 Forms of Standards

Standards generally take three forms that vary from being very restrictive and specific to relatively permissive and general in application:

- *Technical specifications* are the most rigid and exacting of all standards and, if followed correctly, will yield consistent and identical results.
- *Conventions*, also called rules or protocols, are more flexible and more accommodating of variations in implementation, and therefore will yield similar but not necessarily identical results.
- *Guidelines* are the most flexible type of standard and they aim to provide a broad set of criteria against which the quality and performance of products, services or systems can be measured and evaluated.

2.1.2 Approaches to Implementation

Standards can be implemented using two approaches:

- *Mandatory*, where standards are implemented through government legislation and regulations, and
- *Voluntary*, where standards are implemented through consensus and popular acceptance by the user community.

Further explanation of these approaches is given in Sections 2.3 and 2.5.

2.1.3 Level of Granularity

This dimension of the standards matrix, which describes how detailed the subject of a standard is described, is product- or service-specific. In the case of spatial databases there are four levels of interrelated standards that progress from information system to data values, namely:

- *Information system standards*, which apply generally and broadly to depict the overall architecture of a system, including the roles of and interrelationships among its hardware, software and data components.
- *Data structure standards*, which define the way by which data are physically and logically represented in the system.
- *Data content standards*, which provide the rules for representing each element defined in the data structure covering, for example, the formats for dates, times, quantities and addresses, punctuation and capitalisation, and required and optional inclusion of specific items in the standard.

- *Data value standards*, also called object or feature catalogues, which provide lists or tables of terms, names, classification codes and other types of permissible values that are allowed for a particular element defined in a data content standard.

The above multi-dimensional definition of standards is adopted in this chapter as a general conceptual framework for explaining the classification, development, and implementation of standards.

2.2   Classification of Standards

Standards can be classified in different ways from different perspectives (Figure 5.2).

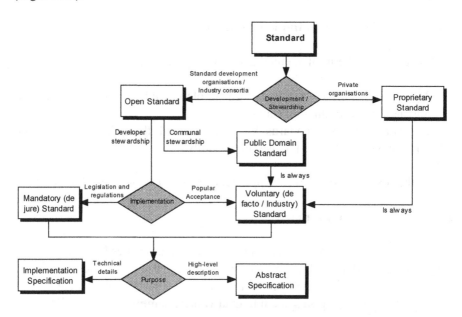

Figure 5-2. Classification of standards

Based on their development process and stewardship, standards can be classified as open or proprietary. The concept of open standards often means different things to different people. Some people perceive an open standard as one that is documented, generally available for use, and free of charge. Others put less restriction on the definition of "open" by including those standards that are generally available but require a licence fee to use. *Open standards* are defined liberally in this book to include all standards that were developed using some or all of the principles of open standards as described by Krechmer (1998). These include the ability of stakeholders to participate voluntarily in standards development, the use of consensus in the review and standards approval process, providing public access to all development documents and ultimately providing access to the completed standards. Legally, the developer of an open standard retains all related patents and intellectual property rights but third party users are free to support and create products that conform to it.

There are open standards that are not owned by a particular organisation. Such standards are developed by members of a user community collectively rather than by an identified organisation. They are usually called *public domain standards* and are commonly stewarded by volunteers or a standards organisation on behalf of the user community. In a sense, standards developed by government agencies and released for public use are also public domain standards. By nature, a public domain standard is always an open standard, but the opposite is not necessarily true. Standards developed or approved by the ISO, ANSI, BSI and SCC, including those spatial data standards to be discussed in Section 3.3, are for the most part considered to be open standards.

Proprietary standards are developed by organisations for internal use in the manufacturing of products or delivery of services. Developers of proprietary standards have sole ownership and retain all patents associated with the standards. Third party users are sometimes allowed to use a proprietary standard, either by paying a prescribed licence fee or free of charge, but they do not play any part in its development and maintenance. There are many proprietary standards in use within the computing industry, most notably the Microsoft Windows operating system. In the spatial data domain, examples of commonly used proprietary standards include a number of files formats such as the DXF (Drawing Exchange) format of AutoDesk, and the E00 (Arc Export) file format and the ArcView shapefile format, both of the Environmental Systems Research Institute, Inc. (ESRI).

Another way of classifying standards is to use the two methods of implementation noted in Section 2.1. As a rule, open standards affecting public health and safety are always implemented as *mandatory standards* through government legislation and regulations. Such standards are commonly called *de jure standards* because they are in effect laws and are strictly enforced as such. Open standards in other areas are implemented by popular acceptance as voluntary standards. These open standards, together with proprietary standards that are open to third party users, are commonly referred to as industry or *de facto standards* to distinguish them from de jure standards imposed by government legislation and regulations.

Conventionally, data standards were originally implemented voluntarily. However, as governments realised the importance of spatial data in effective and efficient governance, and the possible consequences in public health and safety caused by the abuse or misuse of spatial data, some countries began to make spatial data standards mandatory in undertaking government business. The United States Government, for example, has implemented the Spatial Data Transfer Standard (SDTS) as a Federal Information Processing Standard (FIPS) and, as such, SDTS is mandatory for all federal agencies and in business with the United States Government (see Section 3.3.2).

It is also possible to classify standards according to their purposes into *abstract specifications* and *implementation specifications*. An abstract specification is a conceptual description of the components of a standard and their interrelationships. Such a specification does not make any assumption on technologies used for or associated with the implementation of the standard. It is mainly intended for people who need to understand what a standard is, and the resource implications its implementation may have for an organisation. Spatial data standards developed by the ISO, for the most part, are written at the abstract specification level.

Implementation specifications, on the other hand, are technical details of how to use standards. An implementation specification specifies how the conceptual framework described by an abstract specification can be put into practice by using specific interface requirements, data encoding methods and structures, programming language constructs, and so on. In this regard, an implementation specification is in essence a technical manual used by professional or technical personnel responsible for the implementation of standards.

Spatial data standards of the Canadian Geographic Data Infrastructure (CGDI) and the Federal Geographic Data Committee (FGDC) of the United States are implementation specifications because they are primarily concerned with applying standards to data products. The OGC publishes standards both as abstract and implementation specifications. The role of the ISO, CGDI, FGDC and OGC standards as they relate to the concept of abstract and implementation specifications are discussed in more detail in Section 3.3.

2.3   Standards Organisations

Conventionally, standards are developed by international or national standards organisations. It is important to understand that there are two types of standards organisations that play relatively distinct roles in the standards development process, namely *standards organisation accreditation* and *standards development*. In Canada, for example, the Standards Council of Canada (SCC) is a standards accreditation organisation. It is a Federal Crown Corporation with the mandate to promote efficient and effective standardisation. The SCC itself does not develop standards but has the authority to accredit standards development organisations and approve standards proposed by these organisations under the *Standards Council of Canada Act*. Of the standards development organisations that the SCC has accredited, the Canadian Standards Association (CSA) is responsible for developing standards in the field of information technology and the Canadian General Standards Board, Committee on Geomatics (CGSB-COG) is responsible for standards relating to spatial data and geomatics (see Section 3.3.1).

In the United States, the ANSI plays a similar role as a standards accreditation organisation. Operating as a non-profit organisation that administers and co-ordinates standards development in the United States, the ANSI does not develop standards itself but approves standards proposed by standards development organisations such as the National Institute of Standards and Technology (NIST), the American Society of Testing and Materials (ASTM), the International Committee on Information Technology Standards (INCITS), and the FGDC (see Section 3.3.2).

In Europe, each country has its own national standards accreditation body, such as the BSI in the United Kingdom. A pan-European standards body called the European Committee on Standardisation (Comité Européen de Normalization, or CEN) operates through the European Commission with the mandate to co-ordinate standards development activities in its member countries. A technical committee, TC287, exists within the CEN for standards development. Standards development initiatives and projects have also been undertaken by other organisations within the European Commission (see Section 3.3.3).

With representation from about 150 countries, the ISO is the most important standards organisation internationally. The ISO has different technical committees (TCs) that are responsible for the development standards of international interest and relevance. Technical Committee 211 (ISO/TC211) was formed to develop standards in areas relating to geographic information and geomatics. Over the years ISO/TC211 has produced a suite of close to 30 spatial data standards (see Section 3.3.4). It should be understood that standards approved by the ISO will not automatically become national standards of any of its member countries. Different countries have different rules governing the adoption of ISO standards. In Canada and the United States, for example, it is necessary for ISO standards to be ratified, usually as profiles that take into consideration the specific local needs, by the SCC and ANSI respectively before they become national standards in these two countries.

Industry consortia are playing an increasingly important role in standards development, both nationally and internationally (see Section 3.3.5). Typically, an industry consortium in standards development is a membership organisation formed to ensure performance and quality in a particular industry or economic sector. Examples of consortia associated with standards for the IT industry include, among others, the World Wide Web Consortium (W3C), OMG and the Open Group. Memberships in industry consortia are open generally to hardware and software vendors, government agencies including national standards accreditation organisations, as well as academic and research institutions around the world. Although industry consortia are seldom accredited by national or international standards organisations, the standards that they propose always have strong support within a relatively short span of time because they are developed by members of the industries concerned, by consensus of stakeholders, and by using an open review and approval process as described in the next section.

2.4   Standards Development

Developing a standard is a complex process, no matter whether it is undertaken by an accredited standards development organisation or by an industry consortium. An open standards development project is typically initiated with a need for a standard that, in turn, results in the submission of a proposal to a standards development organisation. The proposal is then formalised into a standards document which is subsequently developed by a technical committee of subject experts through iterative consultation, evaluation and refinement, both internally and externally with stakeholders at large, until the standard is ultimately accepted for implementation. The process of standardisation, therefore, is characterised by open participation and consensus building among representatives of stakeholders. Consensus in standards development is defined as substantial agreement by those involved in the process. This is much more than a simple majority, but is not necessarily complete unanimity.

Standards development is an on-going process with a definite review and revision cycle. As user needs and technology change, existing standards will be refined and expanded, and may eventually be superseded by new standards. Maintenance and on-going revision are an integral part, rather than an option, of the standards development process. Therefore, the approval of a proposed standard does not mean an end of the development process. Instead, it signifies the beginning of a review and revision cycle. The ISO, for example, requires that all its standards be reviewed, and updated if necessary, at least every five years. Other standards organisations also have similar review and revision policies, although the durations between the reviews may vary from one another.

Ideally, a standard is capable of defining all elements of a product or service in a single document. In practice, however, this is not always possible because of the variability that may exist among different products and services. A standard that attempts to embrace every characteristic of variable products or services will inevitably become too complex to manage and implement effectively. On the other hand, a standard that attempts to specify only the commonality of all products or services will defeat the purpose of setting it up in the first place because it will leave many loose ends uncovered. As a compromise to satisfy the varying needs of different products and services, while at the same time containing the complexity of their contents, standards are often developed through a modular approach using *profiles*.

By definition, a profile is a subset of a standard that addresses a specific and distinct aspect of a product or service. A profile-based standard is typically made up of one or more modules of *basic standards*, also called *core profiles*, covering common elements, and multiple modules of *application profiles* each covering a set of requirements or characteristics specific to a particular part of the product. As an example, a natural resource data standard may contain one or more basic standards that define the characteristics common to all types of natural resource data such as scale, accuracy, map projection and co-ordinate system, updating frequency, encoding methods, and so on. In order to address the specific data needs of particular applications in natural resource management, independent profiles can be created separately for data pertaining to forestry, wildlife habitat, fisheries, hydrology, access roads, land use and land ownership. These application profiles contain distinct elements that are specific to themselves but irrelevant to other profiles. For example, classification of tree species, methods of wood measurement, and status of timber management plans are important for forestry data sets, but are of no relevance to fishery and hydrology data sets.

From the perspective of standards development, the concept of implementing standards through profiles is a divide-and-conquer strategy. Using such an approach to standards development, it is not only possible to keep a standard from becoming excessively complicated and taking too long to complete, but it also allows a standard to be developed incrementally (that is, one profile at a time) and collaboratively by having different groups of subject experts working on different profiles concurrently. In this way, a standard can be developed more expediently, be more accommodating to specific or local user needs, be more flexible to maintain and revise, and be more responsive to advances in technologies, as noted above. Many spatial data standards in use today, including the SDTS to be discussed in Section 3.3.2, were developed using a profile-based strategy.

Standards development is a very time-consuming and resource-intensive undertaking. Despite the advantages of using the profile-based approach to standards development described above, standards organisations tend to look for and harmonise existing standards through a process of inter-organisational co-operation and collaboration, rather than by the development of new standards in isolation. Section 3.3.4 examines how international and national standards organisations have collaborated in the development of spatial data standards.

2.5   Standards Implementation

As noted in Section 2.2, there are essentially two approaches to implementing standards, namely voluntary and mandatory. Generally speaking, voluntary standards are implemented by popular acceptance and agreement on the part of both the producers and users of a product or a service. Failure to comply with voluntary standards may lead to inefficiencies or incompatibility in the use of products and services or, more importantly, the potential loss of business opportunities, since consumers may be unwilling to buy a product or use a service that does not conform to an accepted standard.

An important aspect of voluntary standards implementation is conformance certification. Conformance testing services are usually provided by standards development organisations or by independent laboratories accredited by standards organisations. Product vendors and service providers apply for conformance certification by submitting their products or services for the prescribed tests of compliance. Products and services that have passed the tests can carry the stamp of conformance. Electrical appliances and industrial products have a long tradition of bearing a physical stamp of compliance of the testing laboratories or standards organisations. In the spatial database arena, OGC conformance certification has gained considerable credibility as more and more software vendors submit their products for certification, and advertise their products as certified in technical and promotional materials after the products pass the required tests.

Mandatory standards are implemented and enforced by government or government-appointed inspection, investigation and enforcement agencies authorised to ensure compliance in their respective applicable areas. Failure to adhere to mandatory standards will similarly result in inefficiencies or incompatibility in the use of products and services. In some cases, it may lead to extremely serious consequences, including damage of property and loss of life. Non-compliance with mandatory standards is a criminal offence that carries legal liability and penalty. In Section 2.2 it was noted that the

SDTS is designated as a mandatory spatial data standard for agencies of the United States Government. This form of mandatory standard implementation is becoming commonplace as governments begin to realise the inconvenience and, more importantly, the cost of using spatial data that conform to different standards.

3.   SPATIAL DATA STANDARDS

Standards are a crucial and indispensable part of spatial database design and implementation. The following discussion first explains the importance of standards in the spatial database environment. It then provides a brief but comprehensive overview of the various standards applicable to spatial database systems and describes the characteristics of important spatial data standards adopted in Canada, the United States and internationally.

3.1   The Importance of Spatial Data Standards

In spatial database systems, standards provide for consistency in the interfaces between data, applications and users. The use of standards in database design ensures that the resulting database system will be implemented according to accepted architecture, functionality and best practices in the IT industry. Specifically, the advantages of standards and standardisation in spatial database systems include:

- *Quality assurance and control.* Standards and standardisation ensure consistency in data modelling, collection, storage, management, application and presentation, thus helping to maintain the intrinsic quality of data throughout the SDLC of a database (see Chapter 8).
- *Accountability in spatial design and implementation.* Standards provide the yardsticks against which data quality and data processing practices can be measured and evaluated. This enables designers, developers and administrators of a spatial database system to understand clearly and precisely their respective responsibilities and the quality of the deliverables. It also allows project managers to identify who should be held answerable if and when something goes wrong during the development and operation of a spatial database system.
- *Accessibility and interoperability.* Standards provide usability information for potential users and allow them to exchange spatial data among database systems at different locations, thus eliminating the duplication of effort that might otherwise occur for data collection.

- *Best practice in spatial data management.* Standards document the principal properties of spatial data sets in a comprehensive and structured manner to help avoid information loss and assist in the transfer of knowledge between data supplier and end users.

- *Equal opportunity for all spatial data suppliers and users.* Standards nowadays are developed and approved by user participation and consensus, which implies that no data suppliers or users will possibly be able to dominate the market at the expense of others. This provides a level playing field for large and small organisations to compete with one another in a fair and equitable manner in the market.

- *Technological innovations.* Standards were traditionally developed to recognise and codify existing technology and maintain the status quo in applications and practice. Increasingly, standards are used to define the direction, requirements and use of new technologies for spatial database systems. In doing so, standards help to ensure that methods for the collection, management and application of spatial data are kept abreast of technological advances in other branches of IT.

- *Synergy and scale of economies in the use of spatial data.* Standardisation helps to minimise the variations among different data sets in the same application domain, to lower the cost of interoperation between different systems, to increase the number of people and applications that can interact with one another, to reduce the need for data conversion and translation (and the inevitable loss of meaning) and to exchange information and create greater economies of scale in providing support services and training.

In spatial database design and implementation data standards can play a crucial role in the selection and adoption of other categories of standards. For example, when an organisation evaluates existing DBMS standards for use in a project, a candidate standard must support the organisation's established data standards. Data standards are directly related to DBMS query standards such as SQL, and are referenced in the development of DBMS interface standards and Web API protocols. It is also essential for any professional practice standard to include a thorough understanding of data standards in general, and a working knowledge of one or more data standards for specific professional applications. The various standards that affect spatial database systems are discussed in the following section.

3.2   Standards for Spatial Database Systems

As spatial database systems are typically configured using a relatively complex combination of hardware, software, data sets, applications and management procedures, the design and implementation of these systems require a variety of standards applicable to different components. These standards can generally be grouped into the five categories shown at the top level in Figure 5.3. The characteristics and practical use of these standards are explained in the following sections.

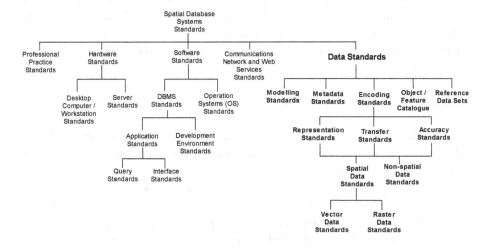

Figure 5-3. A typology of standards applicable to spatial database systems

3.2.1 Professional Practice Standards

Professional data standards are, generally speaking, guidelines that aim to set out the required knowledge and skills of people working on spatial database systems. It also includes best practices that spatial database practitioners are expected to follow at every stage of the development of a spatial database system. In recent years, several professional certification programs have been established by organisations such as the Urban and Regional Information Systems Association (URISA) and the American Society of Photogrammetry and Remote Sensing (ASPRS), which aim to establish standards for professional practices in spatial information systems. Another form of professional certification is membership in licensing bodies for land surveyors. For example, the Royal Institution of Chartered Surveyors (RICS) in the United Kingdom and the Association of Ontario Land Surveyors (AOLS) in Canada, now admit spatial database practitioners who satisfy their academic and professional experience requirements as members. In their professional services, these spatial database practitioners are required to maintain the practice standards established by their respective organisations.

3.2.2   Hardware Standards

Standards for hardware cover the selection of computers for desktop and client-side applications, and for data and application servers in a client/server architecture. For spatial database systems, the predominant standard for desktop applications is the Windows-based personal computer. In the case of server standards, there is now more freedom of choice than ever before as database managers can choose between personal computers and workstations using Microsoft Windows, UNIX or Linux-based server and application software.

3.2.3   Software Standards

Software standards include Operating Systems (OS) and DBMS. Conventionally, Microsoft Windows and UNIX were the dominant OS standards for spatial database systems. However, Linux has become increasingly popular in recent years as the computer industry as a whole has moved toward the adoption and use of open systems and open standards. On the DBMS side, proprietary standards still prevail. Commonly used software including Oracle, IBM DB2, and Microsoft Access and SQL Server can all be classed as proprietary 'standards' in the sense that organisations may choose a specific DBMS product in order to be product- or standard-compliant in pursuing a particular IT strategy.

Most, if not all, industry-standard DBMS are compliant with the use of open standards such as SQL and open development environments by allowing for custom applications development using software tools that are, themselves, compliant with industry standards. These tools include Visual Basic (VB), Visual Basic for Applications (VBA), C++, .NET and Java. Additionally, all DBMS include software extensions that support spatial data types specified by OGC standards, and can make use of open standards middleware such as CORBA for the APIs that are used to connect the database to GIS software (see Chapter 3).

3.2.4   Telecommunications Network and Web Services Standards

These standards include the transfer of data over global and local computer networks, such as the Transmission Control Protocol/Internet Protocol (TCP/IP) and Hypertext Transfer Protocol (HTTP), as well as API protocols for using spatial data on the Web, such as the Web Feature Services (WFS), Web Map Services (WMS) and Web Coverage Services (WCS) of the OGC, and the Arc/Info Internet Map Service (ArcIMS) of ESRI, among many others.

3.2.5 Data Standards

Data standards are by far the most complex of all spatial database standards from the perspectives of functionality, development and implementation. A spatial data standard typically consists of five components covering separate aspects of spatial data that require standardisation. These components are:

- *A spatial database model* that provides a specific abstract view of the real world represented by a database, including the location and characteristics of and the relationships among individual objects or features that are identified in the constituent data sets.
- *An object or feature catalogue* that contains lists of tables and lists of terms, names, classification codes and other types of permissible values that are used to describe the characteristics of an object/feature or a class of objects/features in the data sets of a particular application domain.
- *A set of encoding specifications* that govern the digitising of spatial data for storage in the computer and also for the transfer of data between different applications, databases and computer systems. In essence, these specifications form the core of a data standard. Individual encoding specifications define the digitising of data into bits and bytes, the accuracy at which the data are collected and used, the method of classifying graphical and attribute data, as well as the representation of the data in specific raster and vector formats.
- *One or more reference or framework data sets* that form the foundation on which other layers or themes of spatial data are geo-referenced and integrated. Geo-referenced data sets usually include layers of topographic, hydrological, transportation, geodetic control and digital elevation model (DEM) data, among others, and a geographical name gazette.
- *A collection of information about the data*, which collectively forms the database metadata, including the content and production information of data sets (see Section 4).

Data standards are explicitly declared as a cornerstone of the Global Spatial Data Infrastructure (GSDI), the Canadian Geospatial Data Infrastructure (CGDI) and the American National Spatial Data Infrastructure (NSDI). This reflects the paramount importance of data standards in spatial database systems. The next section concludes the discussion on spatial data standards by examining selected examples of national data standards currently used in Canada, the United States, and the European Union, and ISO and OGC stewarded international spatial data standards.

3.3   Examples of Spatial Data Standards

It is not within the scope of this chapter to provide a comprehensive study of all spatial data standards in use today. Hence, the discussion here is limited to demonstrating how spatial data standards are developed and implemented in practice using examples from Canada to illustrate the use of standards in a national spatial data infrastructure program, the United States to illustrate the development and use of a spatial data transfer standards by government agencies, and the European Union (EU) to shown a typical regional effort toward spatial data standardisation. The discussion also considers ISO and OGC work because of their impacts on the global spatial data user communities.

3.3.1   Spatial Data Standards in Canada

Spatial data standards in Canada are developed by the CGSB-COG. The CSGB is funded by the Department of Industry of the Government of Canada and receives its accreditation to develop standards from the SCC, as noted in Section 2.2. The CGSB-COG committee maintains three associated committees, namely National Standards, Map Library Cataloguing, and the Advisory Committee on ISO/TC211, which is the ISO technical committee for geographic information and geomatics. Spatial data standards developed and approved by the CGSB-COG are maintained in the National Standards Database of the SCC, and are used by Government of Canada agencies and the CGDI. Figure 5.4 shows the working relationship between the CGSB-COG, its associated committees, the SCC and other standards-related organisations in Canada and internationally.

The CGSB-COG has developed or approved several national spatial data standards, including the following:

- CGI-SAIF (Canadian Geomatics Interchange Standard – Geomatics Spatial Archive and Interchange Format), which corresponds to the British Columbia Crown Lands Specification for the transfer of spatial data.
- CGI-Cataloguing Rules (Canadian Geomatics Interchange Standard – Geomatic Data Sets Cataloguing Rules), which provide rules for the cataloguing of geographic information products in map libraries.
- CGI-Directory Information (Canadian Geomatics Interchange Standard – Directory information Describing Geo-Referenced Data Sets), which provides a printed form that can be used to describe digital geo-referenced data sets.

- The CGI-DIGEST, which corresponds to the DIGEST standard of the Digital Geographic Information Working Group (DGIWG, a standards development organisation affiliated with the North Atlantic Treaty Organisation (NATO)).

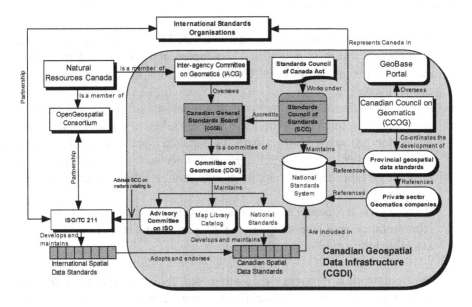

Figure 5-4. Standards development in Canada

The CGSB-COG has also endorsed the use of several international spatial data standards in Canada, including:

- ISO/TC211 – Geographic Information/Geomatics, which is a suite of separate spatial data standards covering the modelling, encoding, management and conformance testing of spatial data.
- ISO/TC204 – Transportation Information and Control Systems, which is a suite of standards and technical reports relating to transportation information and control systems, including Technical Report ISO/TR 14825: 1996 Geographic Data Files (GDF).
- ISO/JTC SC24 – Computer Graphics and Image Processing, which defines the Basic Image Interchange Format (BIIF).
- ISO/JTC SC32 – Data Management and Interchange–SQL/MM–Spatial, which defines the use of SQL for accessing spatial data in multimedia formats.
- IHO S-57, which is a spatial data standard developed by the International Hydrographic Organisation (IHO) for data used in the hydrographic application domain.

The CGDI is a joint federal-provincial program that was set up to promote and facilitate the use of spatial data and related technologies in Canada. In order to help achieve its objectives, the CGDI has made standards and standardisation one of its highest priorities and has established a standards structure that is made up of the following components:

- *A national registry of profiles*, that is created and maintained under the authority of the SCC and which is adopted and customised from ISO standards for application in Canada with the maximum degree of compatibility.
- *A set of national base profiles* that consists of a general feature catalogue, a suite of predefined spatial schema, quality rules and evaluation procedures, a metadata file for data discovery, a selection of encoding schema for interchange and interface, a set of portrayal symbology for specific applications, and a Web access service specification.
- *A set of product specifications fo*r CGDI *framework data* that define the representation and encoding of the reference data sets that are made available to the public free of charge.

The framework or reference data sets noted above are provided through an Internet portal called GeoBase (http://www.geobase.ca). This portal is overseen by the Canadian Council on Geomatics (CCOG), a sector membership organisation that co-ordinates spatial data standards development nationally and provincially. When the GeoBase portal was launched in November, 2003, there were six reference data sets, including:

- National Road Network (NRN) data.
- Canadian Digital Elevation Model (CDEM) data.
- Geographic names of Canada.
- Administration boundaries.
- Canadian geodetic network data.
- Landsat-7 orthoimage data.

The launch of GeoBase represented a major milestone in Canada's effort to establish a national data standards infrastructure within the framework of the CGDI. It also set an example that signified the importance of standardised data sets as a key component of a spatial data standard.

3.3.2 Spatial Data Standards in the United States

Development of spatial data standards in the United States is co-ordinated by the ANSI. Actual development and stewardship of spatial data standards are undertaken by government agencies, standards development organisations accredited by the ANSI, and industrial consortia such as the OGC, as noted earlier in Section 2.3. The International Committee for Information Technology Standards (INCITS) is an ANSI-accredited standards development organisation in the area of information and communications technologies. The INCITS Technical Committee L1-GIS develops national standards for GIS and spatial metadata in the United States. This particular committee is also the Technical Advisory Group (TAG) from the United States to ISO TC211. On the government front, the FGDC is the Federal agency mandated to administer and co-ordinate policies relating to all aspects of spatial information in the United States, including the development of spatial data standards. Figure 5.5 shows the standards organisations in the United States, the working relationship between them, and with the ISO.

Since it is practically impossible to cover all of the numerous spatial data standards in use in the United States, the following discussion describes the two most important standards developed and stewarded under the umbrella of the FGDC, namely the SDTS and the Content Standard for Digital Geospatial Metadata (CSDGM). This choice is based on two grounds, namely the importance of these two standards to spatial data users in the United States, and possibly to users outside of the country as well, and second, the different natures and purposes of these standards (that is, the SDTS for spatial data transfer, and the CSDGM for spatial metadata). The SDTS is discussed here and the CSDGM is explained later in Section 4.3 when spatial metadata standards are discussed.

The SDTS has its root in the FIPS program that was established by the United States Government in the 1960s to standardise the use of computers in its agencies. This program started a series of standards development initiatives covering every aspect of using computer-based information processing within the United States Government. As the use of computers in producing map data grew, the need for standards became apparent. In 1980, the United States Geological Survey (USGS) was designated the lead agency to develop earth science and related data. In the following years, the USGS worked with the academic community, industry, and government agencies at federal, state and local levels to come up with a STDS that was approved as FIPS Publication 173. This standard was updated and ratified by the ANSI in 1998 as ANSI standard NCITS 320-1998.

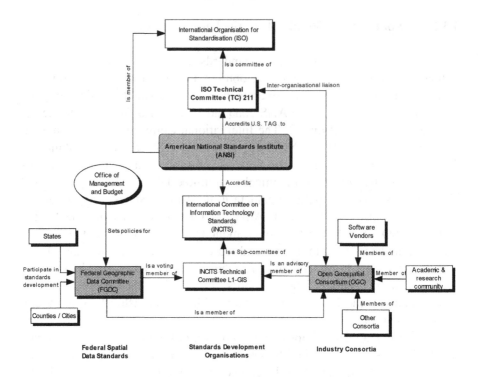

Figure 5-5. Standards development in the United States

The SDTS is available for use as a voluntary standard by state and local governments, the private sector and the academic community. However, when non-Federal Government organisations do business with federal agencies involving the use of spatial data, all data sets must be compliant with the SDTS. Translators are available for converting spatial data using the SDTS to and from proprietary formats commonly used in the spatial data industry.

The SDTS is implemented through the profile-based approach that was introduced in Section 2.3. Such an approach to standards implementation makes the SDTS a neutral, modular, growth-oriented, scaleable and extensible standard suitable for application domains where data sets for specific applications tend to exhibit considerable variations among themselves. As a profile-based standard, the SDTS is organised into three base specifications and multiple data profiles, as follows:

- *Base Standards* (containing elements common to all types of data sets).
- Part 1 – Logical Specifications.
- Part 2 – Spatial Features.
- Part 3 – ISO 8211 Encoding.

- *Data profiles* (containing elements applicable to specific types of data sets).
- Part 4 – Topological Vector Profile.
- Part 4 – Raster Profile.
- Part 6 – Point Profile.

The relationship between the base specifications of data profiles for the SDTS and the types of spatial data sets that each of the data profile describes is shown in Figure 5.6. This figure also illustrates the idea of extending the SDTS by developing new profiles in response to changing user needs in the future, for example a computer-aided design and drafting (CADD) profile for spatial data sets in CADD formats such as AutoDesk's exchange file format (.dxf).

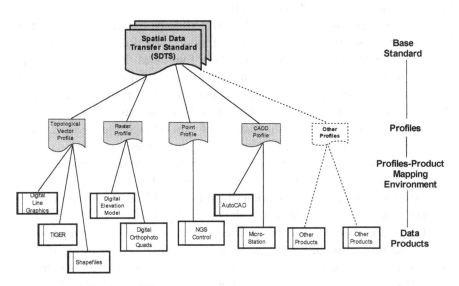

Figure 5-6. Implementing the STDS

3.3.3 Spatial Data Standards in the European Union

The status of spatial data standards within the EU is less clear than those in nation states such as Canada, the United States, Australia and New Zealand (the latter under the stewardship of the Australia-New Zealand Land Information Council [ANZLIC]). The primary reason for this is that conditions vary greatly between national standards agencies within Europe. Further, in terms of spatial data quality and quantity there is considerable variation not only among the current member states, but also other countries in the process of joining the EU. This international fragmentation has made it difficult to formulate, adopt and adhere to spatial data standards consistently in Europe.

In addition to participation in ISO 19100 standards series (including TC211), a number of individual initiatives launched within Europe seek to coordinate a strategy for managing geographic information (for example, the European Umbrella Organisation for Geographic Information [EUROGI], the European Territorial Management Information Infrastructure [ETeMII], the Geographic Information Network in Europe [GINIE], and the Infrastructure for Spatial Information in Europe [INSPIRE]), all of which are operating concurrently. In addition, the European Commission has established a Committee on Geographic Information (COGI) within the commission whose mandate is to co-ordinate the use of geographic information and assure the application of common technical standards within the commission services (Annoni, 2001).

Of the various initiatives noted above, perhaps the most important is INSPIRE because it is a legal initiative of the EU. INSPIRE's mandate is to address technical standards and protocols in Europe as they relate to spatial data, as well as address organisational, data policy and coordination issues including data access and spatial data maintenance and creation. In principle, INSPIRE is very similar in organisation and function to the FGDC in the United States in that it has a number of working groups charged with producing standards position papers on various themes that follow typical ISO versioning and revision. Moreover, there is explicit harmonisation between the nascent INSPIRE specifications and ISO 19100 series of standards for geographic information (including ISO TC211) as well as other relevant standards, such as those developed by the OGC.

The context model proposed by INSPIRE for spatial data management, standardisation, harmonisation, integration and use is shown in Figure 5.7. This model is easily translated into a generic open architecture reference model (Figure 5.8) that distinguishes between four major groups of closely related functions or components, namely a user applications, geo-processing and catalogue services, catalogues, and content repositories The ultimate success of this reference model is dependent on establishing and adhering to respective constituent standards for all twenty member states of the European community.

The INSPIRE spatial data reference model is open in the sense that it is able to accommodate new geospatial data products and services as they are developed and it allows the inclusion of ISO or OGC spatial data standards, depending on their applicability within the model's structure. The model is sufficiently flexible to be able to accommodate base standard specifications that are implementation-neutral, such as ISO 19103 Conceptual Schema Language, or those that are platform-specific but implementation-neutral, such as CORBA, SQL and Extensible Markup Language (XML) for Web services (see Chapter 10).

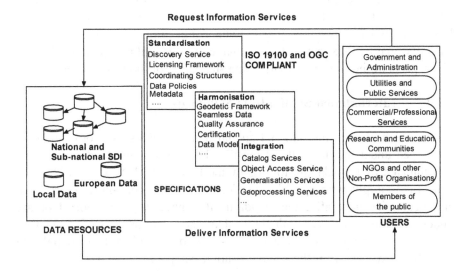

Figure 5-7. The INSPIRE spatial infrastructure high level model

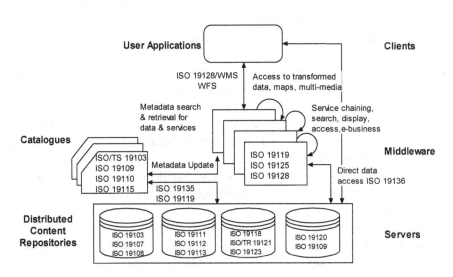

Figure 5-8. The INSPIRE architecture reference model including example ISO19100 standards

Efforts to develop coherent spatial data standards in Europe continue to coalesce around collaboration with ISO 19100 series working standards and the complementary work of the OGC. The difficulties of working with multiple countries, languages, datums, projections, and significantly different levels of spatial data completion pose technical challenges. However, there is optimism that a common spatial data infrastructure will emerge within Europe during the next few years. Even if this proves impossible, it is at least reasonable to expect that individual countries will accept and adhere to key aspects of the emergent spatial data standards for their own databases.

3.3.4   International Spatial Data Standards

As noted above, international spatial data standards are conventionally developed under the umbrella of the ISO 19100 series of standards. TC211 was initiated by Canada and formed by the ISO in 1994. Since TC211 is not concerned simply with spatial data alone but with geographic information and geomatics generally, the standards that have been developed or are under development cover both the systems and data aspects of geomatics. This means that only part of the TC211 standards is applicable to spatial data directly. However, from the standardisation perspective, the TC211 suite of standards probably represents the most comprehensive spatial data standards produced to date. It defines virtually every aspect of spatial data, including cataloguing, modelling, encoding, portrayal, metadata and quality, as well as aspects of developing, structuring and implementing spatial data standards.

TC211 standards can be classified into four categories, as follows:

- *Guidance*. This category is designed to control the overall standardisation process (for example, Reference Model, Conceptual Schema Language, Conformance and Testing).
- *Components*. This category defines subsets of data sets applicable to a specific area of use (for example, Spatial Schema, Spatial Referencing by Co-ordinates, Metadata).
- *Rules*. This category defines rules or protocols to be applied to specific areas of use (for example, Cataloguing, Quality Evaluation Procedures, Portrayal, Encoding).
- *Technical Reports*. These are not spatial data standards *per se*, but are reports of technical studies resulting from standards development projects (for example, Functional Standards, Image and Gridded Data, and Qualifications and Certification of Personnel).

Besides TC211, the ISO has other standards development projects that are important to spatial data users. These include, for example:

- *ISO/JTC SC32 – SQL/MM.* This is an information technology standard that defines basic elements to query and process multimedia data using SQL. Part 3 of this standard addresses spatial data in multimedia formats.
- *ISO/JTC SC 24 – BIIF.* This is Part 5 of an ISO standard called "Information Technology – Computer Graphics and Image Processing – Image Processing and Interchange (IPI) – Functional Specification" and describes a mechanism for exchanging raster-structured spatial data.
- *ISO/TC204 – Transport Information and Control Systems.* This is a standard for information, communication, and control systems in the field of urban and rural surface transportation, including intermodal and multimodal aspects, public transport, commercial transport, emergency services and commercial services in the transport information and control systems (TICS) field.
- *ISO/FDIS 14825 – Intelligent Transport System – Geographic Data Files (GDF).* This is an overall data standard for the development of a computer-based system for transport planning and management.

ISO TC211 standards are widely referenced and adopted for use as national spatial data standards. As noted earlier, TC211 also works in partnership with OGC in the development and harmonisation of spatial data standards previously stewarded by the two organisations, as illustrated by the examples used in the following discussion of OCG standards.

3.3.5   OGC Spatial Data Standards

In recent years, the OGC has played an increasingly important role in the development of standards pertaining to the field of geomatics. The objective of the OGC is to deliver publicly available specifications for the development of spatial data interfaces and services. The OGC Reference Model (ORM) provides the framework for the development of the consortium's existing specifications and candidate specifications under development.

As noted in Section 2.2, OGC standards are published as abstract and implementation specifications. The OGC abstract specifications are designed to provide a conceptual framework for the development of interfaces and services. The following abstract specifications are of particular interest to users of spatial data:

- Topic 1 – Feature Geometry (Also published as ISO 19107).
- Topic 6 – Coverage Types.
- Topic 8 – Relationship between Features.
- Topic 10 – Feature Collection.

- Topic 11 – Metadata (Replaced by ISO 19115).
- Topic 13 – Cataloging Service.
- Topic 14 – Semantics and Information Community.

OGC implementation specifications, which are intended for professional and technical personnel responsible for implementing spatial data standards, can be generally classified into four categories according to their purposes, as follows:

- Spatial data services and application – Catalog Interface, Co-ordinate Transformation, Filter Encoding.
- Spatial data encoding tool – specifically, Geography Markup Language (GML).
- API – Simple Feature/CORBA, Simple Feature/SQL, and Simple Feature/OLE/COM.
- Internet-based mapping – Styled Layer Description, Web Feature Service, Web Map Context, Web Coverage Service and Web Map Service.

From a data standards perspective, the roles of the OGC and ISO are different as well as complementary. The focus of TC211 standards is mainly on abstract component definitions whereas that of the OGC is more on practical implementation. However, these two organisations are working in partnership on several standards development projects, as exemplified by the concurrent publication of the abstract specification for feature geometry, and the replacement of the OGC metadata specification by the ISO standard on the same area.

The increase in cooperation and collaboration between international and national standards has created a global spatial data standards infrastructure as shown in Figure 5.9. This has resulted in the internationalisation of many spatial data standards in use today, which in principle will lead to a higher degree of compatibility among different international and national standards. In essence, this is also the foundation of the concept of interoperability through standardisation that is introduced in Chapter 6. However, the move toward the creation of global spatial data standards is fraught with difficulties, given the widely differing readiness of countries to adopt and implement these standards. While the model of co-operation shown in Figure 5.9 is a good starting point from which to begin to harmonise activities, with the roles of the ISO and OGC prominent in conjunction with national bodies, the practicalities of achieving true harmonisation in spatial data standards infrastructure mean that this outcome is still some way ahead of the current level of international harmonisation.

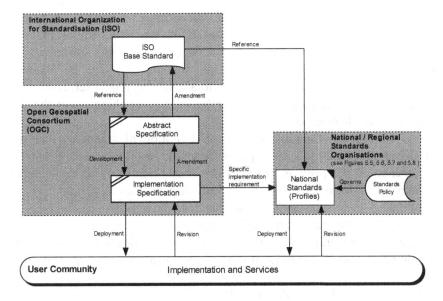

Figure 5-9. A model of a global spatial data standards infrastructure

4. THE CONCEPTS AND METHODS OF METADATA

Creating standards only helps data producers and users to manage spatial data more effectively and efficiently. Two interrelated conditions must be fulfilled to facilitate data sharing. First, the data must be made widely known to potential users and, second, users looking for data must be able to find as many candidate data sets as possible relative to their needs. Users can then evaluate the goodness of fit between the data and their intended use and access the data that are deemed to be most suitable. Metadata serve the dual purpose of allowing data producers to advertise their data products and data users to find relevant data. The following discussion defines metadata and explains how they are created, implemented and used.

4.1 A Definition of Metadata

The term *metadata* is usually taken to mean "data or information about data". However, this view is overly simplistic, and potentially misleading, since the concepts and methods of metadata also apply to services (applications and data processing procedures) and systems (hardware and software specifications) that interact with databases. Hence, metadata is best defined as a formally structured and documented collection of information about data that reveal minimally what is in the data, where the data originated from, who produced them, when they were produced and modified, why they were produced, and how the data can be obtained. "Data" in this definition refer generally to a database, a data set, and even a data element. Hence, metadata include basically everything about the data except the data themselves.

The concept of a multi-dimensional definition of standards that was introduced in Section 2.1 applies equally to metadata. From an implementation perspective, metadata documentation can also be voluntary or mandatory. Metadata can serve different purposes including guidelines for use of data or technical specifications that describe how data can be used. As well, they describe data with varying degrees of granularity, ranging at the most general level from an entire database or data series level, specific data sets, down to specific individual data elements, as noted in the definition above.

A good general example of metadata is found in a library catalogue which is used to look for information about a book or other materials in its holding using "author", "title", "keywords", and "subject" as criteria or key words for a search. These keywords are generally called *metadata elements* or *tags*. However, searching for information using individual metadata elements is not very efficient and searching for information is normally conducted by grouping metadata elements into *metadata schemas* or *standards*.

For spatial metadata schema, the most prominent examples include the three standards noted earlier, namely Content Standard for Digital Geospatial Metadata (CSDGM, developed by the FGDC), ISO 19115 – Metadata, and the GEO Profile of Z39.50 (the application profile of Z39.50 for spatial metadata). These three standards are discussed further in Section 4.3. Metadata that follow such accepted metadata standards, are commonly called *formal metadata* to signify their status in the information infrastructure of an organisation, and to distinguish them from casually collected "information about data" or *discrete metadata*.

Spatial metadata are a special type of metadata that are associated with a spatial database, a spatial data set, or a particular class or instance of spatial features. The idea of spatial metadata is not new. The annotations on a map (that is, the title, source, scale, accuracy, producer, legend, warning on use, and all the information found on the map border) are elements of map-based metadata. Map catalogues in libraries are another traditional form of metadata that relate to spatial information products. Typically, catalogue metadata are limited to information such as geographical coverage, series identifier, publication dates, and distribution information. The importance of this information is explained in the next section.

4.2   The Importance of Metadata

In order to understand the importance of metadata it is important to think beyond the simplistic notion of 'information about data' and adopt a conceptualisation of metadata that embraces all aspects of using data in a database environment. With particular reference to spatial data, metadata are important for the reasons identified in the following sections.

4.2.1   Uniformity of Data Collection

As metadata contain a comprehensive and structured list of definitions pertaining to the encoding, attributing, classifying and structuring of data, they represent a detailed technical specification for data collection. Data collection activities following the descriptions of a particular metadata standard will generate data sets of consistent and predictable qualities no matter where, when or by whom a particular data set in a database system is collected. Such an application of metadata in data collection is particularly useful for database projects involving multiple parties at different locations and spanning several years.

4.2.2   Data Management

Data management is a much more complex and critical task today than it was in the past. Organisations not only collect data at a significantly faster rate from more sources and in larger volumes than ever before, but they are also required to analyse the data using more sophisticated tools, for a greater variety of operational and executive decisions, for internal as well as external users. Metadata, with the wealth of information that they contain, are an indispensable management tool in this regard that allows data managers and database administrators to manage data as a corporate asset, in the same way that financial, human and facilities assets are managed. Metadata help an organisation safeguard its investment in data as an asset by maintaining their integrity, and facilitating exchange and selling of data to external users. In this regard, data management can be severely hindered by a lack of well documented and up-to-date metadata, while current and well documented metadata will ease the management of large and complex data resources.

4.2.3   Data Use

In the past, spatial data sets were collected for specific applications. Data collectors were also data users, or at the very least they belonged to the same organisation as the users and shared the same objectives. Thus, users usually had intimate knowledge of the purposes and quality of the data in use. The increasing use of spatial data has led to the separation of data collectors from data users. With the commercialisation of spatial data, it is also possible that there exist value-added data re-sellers between the data collectors and users. Proper documentation by means of metadata provides those not involved in the collection or value-added production of data with the information that is necessary to evaluate the goodness of fit of the data for their intended applications.

4.2.4   Data Understanding

Spatial data are an abstraction of the real world and, as a rule, there are approximations, generalisations and omissions during data collection and processing. Metadata contain information, by direct description or by inference, about the assumptions and limitations affecting the collection and processing of spatial data. Such information is particularly useful for spatial data users who do not have the necessary training in geomatics and information technology to ascertain technical aspects of a dataset for themselves. As the number, complexity and diversity of spatial datasets continue to grow, so does the importance of providing an understanding of all aspects of data by means of metadata.

4.2.5   Data Sharing

The ability to share spatial data depends very much on the awareness of existing data resources by potential users. To assist users in their search for spatial data, metadata generally use a more controlled vocabulary and provide the context of the words. As a result, they provide more scope for locating useful information with precision and high recall rate. For example, metadata can indicate whether an article containing the name "John Doe" is by him or about him, which can often be valuable to searchers. Also, since metadata are usually documented using human-readable labels or names rather than binary coding, they allow people to find information the way they want it, rather than the way applications "see" and manage it. The ease and efficiency of searching for spatial information around the world has in fact been one of the major driving forces behind the thrust of spatial data sharing that is evident in current database systems.

4.2.6   Data Archiving and Warehousing

The growing use of data warehouses has greatly heightened interest in the archiving and value of legacy data which, by definition, are data retired from active or operational use in organisations. Metadata provide the necessary information for users to combine spatial data from different sources and collected at different points in time to be used for spatial data mining for high-level decision making and business planning. Legacy data play a significant role in data warehousing and data mining, and metadata provide the means to validate the quality of legacy data. New developments in spatial warehouses and the use of metadata for the mining of legacy data are discussed further in Chapter 6.

4.3   Spatial Metadata Standards

Metadata facilitate information search by using a predefined structure and a controlled vocabulary. However, search efficiency cannot be maximised unless the metadata themselves are standardised. There are several principles for the development of metadata standards, for example Duval (2002) notes the following:

- *Modularity*, which is a key organising principle for the user environment of metadata characterised by a high degree of diversity in terms of sources, contents, and approaches to resource description.
- *Namespaces*, which are defined as formal collections of terms managed according to a policy or algorithm that provides the mechanism to ensure global uniformity in the vocabulary used by a metadata standard.
- *Extensibility*, which allows for profiles to be developed so that particular needs of a given application can be accommodated by a standard without unduly compromising the functionality of the base metadata schema.
- *Granularity*, which allows a metadata designer to choose a level of detail appropriate to a given application.
- *Multilingualism*, which is the ability of a metadata standard to accommodate documentation that accommodates the linguistic and cultural diversities of its users.

A number of spatial metadata standards are in use today, although there is reasonable uniformity between most of them. As noted earlier, this chapter focuses on three widely accepted standards including the Z39.50 GEO Profile, the CSDGM, and ISO 19115 – Metadata. The characteristics of these standards are explained in the discussion below.

4.3.1   Z39.50 Application Profile for Geospatial Metadata

The Z39.50 profile is an ANSI standard for information retrieval in libraries. Its Application Profile for Geospatial Metadata, commonly referred to as the "GEO Profile", was designed to provide the specifications for the overall standardisation of aspects of the FGDC's CSDGM that lie outside the scope of Z39.50. The GEO Profile is made up of two parts, namely specifications for metadata and requirements for the server where the specifications are installed, which is called a GEO server, in the Internet environment. The GEO Profile supports search and retrieval of both metadata entries and related spatial data sets accessible to GEO servers through TCP/IP networks, local area networks based on the Ethernet protocol, and wide-area networks using a consistent host addressing scheme. Some of the data resources pointed to by GEO entries may also be available electronically through other communications protocols, including the common Internet protocols that facilitate electronic information transfer such as remote login (Telnet), File Transfer Protocol (FTP), and electronic mail (SMTP/MIME). Metadata elements in the GEO Profile are classified as mandatory or optional, but optionality is dependent in several cases on the type of geospatial data being described. For example, as the numbers of rows and columns in raster data are not an applicable property for vector data, rows and columns are mandatory for a raster data set but are suppressed in descriptions for a vector data set.

4.3.2   Content Standard for Digital Geospatial Metadata (CSDGM)

This standard, as noted earlier, was developed by the United States FGDC in the mid-1990s for defining and identifying metadata elements used to document digital spatial data for different purposes. Its purposes include:

- Availability of a set of spatial data.
- Fitness the set of spatial data for an intended use.
- Means of accessing the set of spatial data.
- Transfer of the set of spatial data from the producer to the user.

The CSDGM establishes the names of metadata elements and compound elements to be used for these purposes, the definitions of these data elements and compound elements, and information about the values that are to be provided for the metadata elements. However, the standard does not specify the means by which this information is organized in a computer system or in a data transfer, nor the means by which information is transmitted, communicated, or presented to the user.

A key feature of the CSDGM is the ability of spatial data users to develop new profiles of the base standard. Many of these profiles have extended the base standard by adding metadata elements to meet their own specific metadata requirements. There are two FGDC-endorsed profiles, namely the Biological Data Profile and Metadata Profile for Shoreline Data. The CSDGM has been adopted by Canada and many other countries as their respective national spatial metadata standard. However, the standard is undergoing harmonisation with the ISO's metadata standard and in effect will become a profile of the latter as explained below.

4.3.3   ISO 19115 - Metadata

At the time of writing this book, this metadata standard is under development within Working Group 3 of ISO TC211. It aims to define and standardize a comprehensive set of metadata elements and their characteristics and the schema necessary to document spatial data fully and extensively. This standard applies to all collections of spatial data ranging from data series and data sets to individual spatial features, including their attributes. The initial ISO metadata standard was based on various existing standards, including, for example, the ANZLIC Working Group on Metadata - Core Metadata Elements; the Canadian Directory Information Describing Digital Geo-referenced Data Sets; the European Committee for Standardisation (CEN) Standard for Geographic Information - Metadata, the FGDC CSDGM, as well as transfer standards which provided metadata elements such as the Digital Geographic Information Exchange Standard (DIGEST), the International Hydrographic Organization (IHO) Special Publication 57, and the SDTS.

As an international standard that has to satisfy the diverse requirements of a large global user community, the ISO metadata standard defines over 400 metadata elements. However, most of these elements are listed as "optional". Individual user communities, nations, or organisations are allowed to develop "community profiles" of the base metadata standard by selecting elements that are mandatory for their own applications or special needs. They are also allowed to establish additional metadata elements which are not in the base metadata standard. However, these added elements will not be known outside the community unless they are published.

Besides providing flexibility with the metadata *content*, this standard is also flexible with *implementation* by allowing for two levels of conformance:

- *Conformance Level 1 metadata.* This contains a minimum number of metadata elements (50) that support the cataloging of datasets for discovery, especially in on-line catalogs and clearinghouses. It is actually a profile or subset of the full metadata set;

• *Conformance Level 2 metadata*. This contains a complete inventory of metadata required to describe spatial data fully. Many of these elements are optional and, when used, standardise metadata down to the lowest level of detail. For ease of understanding, metadata are divided into the following sections as shown in Figure 5.10, namely Identification Information, Data Quality Information, Lineage Information, Reference System Information, Spatial Representation Information, Feature Catalog Information, Distribution Information, and Metadata Reference Information.

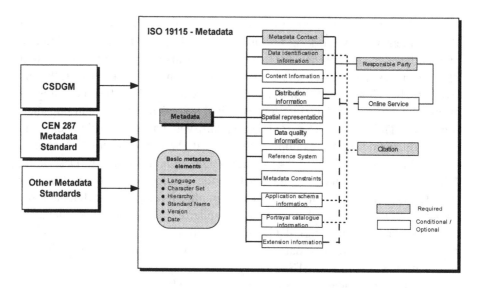

Figure 5-10. Organisation of ISO 19115 metadata standard

For several years INCITS L1-GIS studied the possibility of harmonising ISO and FGDC spatial metadata standards (see Section 3.3.2 and Figure 5.5 for the role of this particular working group in standards development in the United States). In mid-2003, the ANSI decided to accept the recommendations of INCITS L1-GIS, and adopted the ISO standard without change. It used the ISO standard as a basis of follow-on technical amendments designed to bring the ISO standard into alignment with US national requirements. These technical amendments aim to bring the ANSI-adopted ISO standard into harmony with the FGDC version 2.0 metadata standard and allow data sets to be both ISO- and FGDC-compliant. The technical amendments are designed to protect current investments in data sets while allowing for both ISO and FGDC compliance. This work involves mapping ISO elements to FGDC elements (and vice versa). The results of this activity will be evaluated to determine if more elements must be added or subtracted or obligations changed.

ISO and FGDC mapping, commonly called *crosswalk* in standardisation parlance, will follow the FGDC public review process. This procedure and the actual crosswalk between the standards are available for download from the FGDC website (http://www.fgde.gov/). The technical amendments will require making several existing ISO elements/entities more restrictive to harmonize the adopted standard to existing FGDC requirements. Additional technical amendments currently undertaken will result in additional changes to the FGDC version 2.0. These include, for example:

- Imagery and gridded data requirements.
- Existing FGDC profiles and extensions.
- United States Geospatial One-Stop Base Standard and extensions.
- North American Profile (Canada, Mexico and the United States).

4.4 Spatial Metadata Tools

Numerous tools exist to assist with the capture and formatting of metadata, and the delivery of metadata services. Figure 5.11 shows how different metadata tools are used in practice. Between the original data resources and the end user, at the top and bottom of the figure respectively, there are many possibilities for documenting source data such that searching and discovery can occur efficiently. The tools that sit in the body of Figure 5.11 can generally be classified into four categories according to their functions and operating characteristics. While it is not possible given the scope of this book to provide practical examples of these tools, the main classes are described in the following sections.

4.4.1 Metadata Capture and Documentation Tools

This category of tools is designed either to extract either automatically or manually, through user intervention, metadata elements from data sets and document them in the structure of an accepted metadata standard such as ISO 19115 or the CSDGM standards among others. There are four principal types of metadata capture and editing tools that vary according to the degree of user intervention required to extract the metadata elements. In general, the approaches vary between between purely manual documentation, some degree of automated documentation but primarily manual, and purely automated documentation. Each of the associated types of tools is somewhat different from the others, although a number of hybrid tools have been developed that share characteristics and functions across categories.

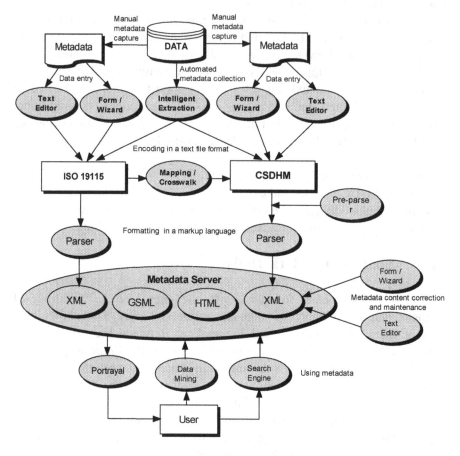

Figure 5-11. Implementing spatial metadata

The categories of tools are as follows:

- *Intelligent metadata extractors* which are capable of automatically scanning a spatial data set and extracting prescribed metadata elements. These tools are typically product-specific, and therefore must be implemented using the scripting or macro language of the particular GIS software products using the spatial data. Examples of this category of metadata tools include FGDCMETA, BLMDOC and METALITE, all of which are metadata extractors written in Arc Macro Language (AML) to extract metadata elements from Arc/Info software coverages.

- *Form-based metadata entry wizards*, which provide a graphical user interface (GUI) that helps guide the user through the documentation process. Unlike intelligent metadata extractors, form-based metadata entry wizards are stand-alone applications that are not necessarily tied to any specific GIS software products. Examples of form-based metadata entry wizards include the NOAA FGDC Metadata Toolkit, MetaMaker, Metadata Collector, Metagen, among many others.
- *Hybrids* that provide metadata pushing capabilities as well as custom programmed wizards to create, according to a variety of commonly accepted standards, metadata records from direct user input. Examples of hybrid tools include ESRI's ArcCatalog and an FGDC compliant spatial metadata profile developed at the University of Waterloo for publishing university-based spatial datasets and metadata on the Internet (http://www.fes.uwaterloo.ca/computing/spatialdata).
- *ASCII and word processing templates* which can be used to facilitate the documentation of metadata in a standard format.

4.4.2 Metadata Utilities

This category of metadata tools include the following types of software programs designed to process and display metadata:

- *Parsers*, which are utility programs that validate the compliance of metadata with an accepted standard such as the CSDGM and generate an output file in formatted text using Internet-ready formats. Parsers are standard-specific, for example, mp (metadata parser) for the CSDGM.
- *Pre-parsers*, which are software programs designed to convert a poorly-formatted metadata file that cannot be validated by a parser into one that can be used by a parser. Pre-parsers are, therefore, parser-specific. A good example is cnc, which is a pre-parser for the mp parser.
- *Metadata servers*, which enable the delivery of metadata on servers by providing access to published metadata files. Examples include Isite (by FGDC) and ArcIMS Metadata Server (by ESRI).

4.4.3 Metadata Solution Toolboxes

A metadata solution toolbox is a suite of software programs that is designed to provide a complete chain of spatial metadata services from creation to publication. The Spatial Metadata Management System of Intergraph is one such tool box that consists of software applications to extract metadata elements from spatial data sets, export the metadata elements in XML with stylesheets (see Section 4.4.4 below), and create an enterprise metadata database in a commercial database system such as Oracle.

4.4.4   Encoding Tools

Metadata files can be encoded in text files using ASCII. However, for delivery of metadata over the Internet, XML is the preferred mode of delivery. XML is an ISO-compliant subset of the Standard Generalized Markup Language (SGML). It is designed to ease the use of SGML for delivering structured information on the Web. Unlike the Hypertext Markup Language (HTML), which specifies how data are displayed in a Web browser, XML provides strictly for the content and structure of information entities using beginning and ending tags. In order to display XML data in a browser, an accompanying *stylesheet* developed using the XML Style Language (XSL) or Cascading Style Sheets (CSS) is required. Since different style sheets can be associated with a given XML file, data encoded in XML can be displayed in different forms. This separation of data from presentation enables the seamless integration of data from different and diverse sources, making it an excellent choice for the encoding of metadata in a Web-based user environment.

4.5   The Process of Implementing Spatial Metadata

Spatial metadata are typically implemented in four progressive stages that include planning, capture, integration and publication. The planning stage is concerned with the preparation for a metadata implementation project. The primary objective of the planning stage is to identify and define the metadata requirements of an organisation. These requirements are determined by, or related to the following tasks enumerated in Figure 5.12:

1. To estimate the financial, technical and human resources that are required for the collection of metadata, its publication and the on-going maintenance of a metadata repository or database.
2. To identity the metadata standard to be used, which must always be an accepted standard such as the Z39.50 GEO Profile, the CSDGM or ISO 19115 standard.
3. To prioritise metadata capture activities if metadata are required for more than one data series or data set.
4. To seek approval and support from senior management of the organisation.

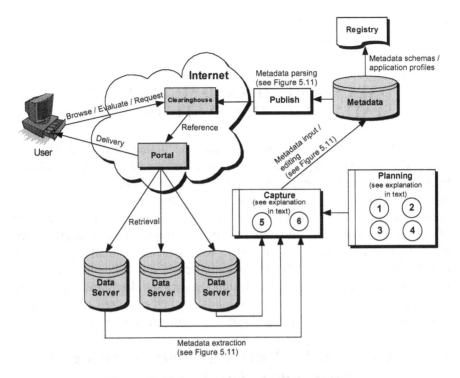

Figure 5-12. Implementing spatial metadata

Subsequently metadata must be captured in some way. Typically, this second stage of work involves two principal tasks, as enumerated in Figure 5.12:

5. Selecting and acquiring metadata capture tools and provide training in the use of these tools. In the selection of tools, it is necessary to take into consideration the budget approved, the expected time frame for delivery, and the metadata standard that will be used.
6. Gathering information about the datasets to be described and document the metadata using the selected standard.

After metadata have been created they must be integrated into the organisation's information management infrastructure, thus subjecting the metadata to the same rules and protocols for the maintenance, change management, use and archiving for other forms of information resources of the organisation. Finally, activities must be undertaken that seek to advertise the availability of metadata to all stakeholders and potential users, the registration of the metadata with one or more Internet service providers, spatial data clearinghouse administrators, and creating or updating Web pages associated with metadata and related data sets of the organisation.

5.   DATA STANDARDS AND METADATA IN SPATIAL DATABASE SYSTEMS

As noted in Section 2.1, metadata are one of the five components of standards. However, experience has shown that metadata have not always been treated seriously or given a high priority in organisations concerned with the collection and use of spatial data. In many organisations metadata coding and maintenance are often afterthoughts rather than an integral part of the project planning process. Despite the tremendous increase in the amount of spatial data collected and in use, it is still a relatively daunting task for users to find readily usable spatial data. The following discussion focuses on the major issues in implementing data standards and metadata in spatial database systems. A practical model is then proposed for using data standards and metadata in spatial database design and implementation.

5.1   Issues with Implementing Standards and Metadata in Spatial Database Systems

Organisations that use spatial database systems are, for the most part, users rather than developers of standards. Thus, the issue of standardisation is primarily concerned with implementation rather than development of spatial data standards and metadata. In this context, one of the major challenges facing spatial database designers and administrators is the task of educating senior managers about the importance of standardisation in securing the funding required for implementing standards. Another challenge is the selection of a corporate data standard that is compatible with the organisation's information technology infrastructure and business requirements of users in different departments.

Proprietary standards prevail in the spatial database arena. Different software products associated with proprietary data standards tend to have different strengths and weaknesses with respect to different application domains, such as natural resource management, facilities and utilities management, and engineering. The problem of translation between data formats is largely solved at the technical level (see Chapter 6), however considerable difficulty remains in dealing with the semantic or linguistic differences between data sets from different sources. From a standardisation perspective, this means that unless it is absolutely necessary, there should always be one and only one spatial database within an organisation. It is all too common for people to complain that there are too many standards in use, but on the contrary too little standardisation will affect the ability to make data sets in different standards functionally interoperable.

There are relatively few issues with implementing metadata that need to be understood. One of the most contentious is concerned with the relationship between metadata and the data sets that they describe. Generally speaking, there are two common approaches to resolving this problem:

- *Embedding metadata within the data set.* The advantage of this approach is the tight coupling between metadata and data. The down side of it is that embedded metadata are collected at the time of collection by the data collector, who may not have the expertise to implement metadata correctly.
- *Storing metadata in a separate database or registry.* The separation of metadata from data sets has the advantage of managing the metadata without affecting the content of the data set. However, this approach requires more complex and stringent management procedures to ensure compatibility between the metadata files and the data sets. For example, when a data set is revised, the relevant metadata elements must be updated accordingly.

There are arguments both for and against the above two approaches. The FGDC recommends that metadata be stored in a separate database for data sets that are subject to frequent changes, or if part of the metadata is common to many data sets. This approach requires a special output report generator to produce CSDGM-compliant data for submission to a NSDI clearinghouse to make the metadata searchable.

Another issue that needs to be addressed is the limitation and inadequacy of search services that enable users to find spatial data resources quickly and effectively. At present, practically all search services are based on the direct association of metadata elements with the values that the users want on them. This approach, used by the NSDI clearinghouse, makes use of an interface with some of the most common metadata search elements and enables users to specify values. This kind of service requires users to have considerable knowledge of metadata elements and their associated values in order to use it. While this may not pose a problem for users with a background in GIS and related technologies, the difficulty of use may be a deterring factor for non-specialist users from taking full advantage of metadata to find relevant data sets for their applications.

The tremendous growth in the use of spatial data in recent years has brought many new users from disciplines beyond traditional areas in geomatics, resource planning and management, business and the health sciences. Also, international co-operation in scientific research, such as global warming studies, renewable resources and bio-diversity research, has led to the construction of large trans-national or regional spatial databases where data are acquired by different organisations, from different sources, using different methods, and at different times. The use of standards and metadata, as noted earlier, is the only way to ensure consistency in the data used in such multi-national or multi-partisan applications.

As noted earlier in this chapter, the ISO, together with several national standards organisations, is spearheading many initiatives to harmonise international and national spatial standards. However, standardisation is a long process that can only accomplish its goals through consensus and co-operation. It is common for many years to pass before a given country ratifies an ISO-approved standard. Many of the important standards harmonisation efforts, such as the example of the CSDGM and ISO 19115 noted earlier, are far from becoming a reality. Individual organisations or users in the spatial data community are unlikely to be involved in the standards harmonisation process. However, there is a real need for these organisations to always consider internationally sanctioned standards in the first place when they are selecting a data standard for a particular spatial database system.

5.2   A Model of Using Standards and Metadata in Spatial Database Design and Implementation

On the basis of the clear importance of standards in all aspects of geospatial data activity, a standards-based approach to spatial database design and implementation can be proposed, as illustrated in Figure 5.13. Within the conceptual framework of this model, decision making at all levels of an organisation plays an important part in the standardisation of spatial data. Although the roles of different decision makers vary, the common objective of using standards is to unify the workflow from enterprise resource planning (ERP) to the daily operation of a spatial database system.

ERP standards are located in the upper tier of the model where they can be adopted relative to the strategic vision of the organisation, its corporate information technology policy, as well as its overall business plan. The ERP standards govern not only the nature of systems development and management standards but also systems operating standards. These standards are selected and adopted with due consideration to the flow of information from one process or business area to another, as well as the principles of systems engineering that guide the development and management of a spatial database system in a structured and systematic manner. Clearly, these standards do not operate in a vacuum and they must address the potential constraints of the standards that relate to the lower tier of the model.

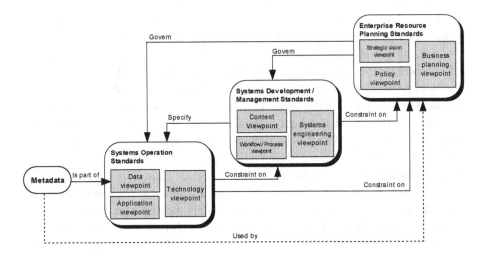

Figure 5-13. A model of a standards-based spatial database system

The lower tier standards in Figure 5.13 focus on standards implementation. In this context, data standards are one of the principal suites of standards to be implemented. The selection of specific data standards must be considered together with standards for applications and hardware/software components of a complete spatial database system. However, they also impose certain constraints or restrictions when the two higher tiers of standards are selected and adopted. Within the lowest tier, a metadata standard is implemented as a component of the overall data standard. Metadata are particularly important as a reference framework for the decision makers who will decide upon the exact ERP standards that will be implemented.

The proposed standards-based approach to spatial database systems provides a rigid conceptual framework for best practice in spatial database design and implementation and facilitates the interoperability of database systems through standardisation. Hence, spatial data standards in effect are the unifying mechanism that brings the spatial database industry together.

6.   SUMMARY

This chapter discussed the concepts and methods of data standards and metadata as they apply to spatial database design and implementation. First, standards were defined and classified in general in order to establish terms of reference for discussing their important role in spatial database creation and deployment. Selected national and international standards organisations were then introduced along with a description of the iterative process of creating standards and ultimately applying them in relevant domains. The conditions for voluntary and mandatory standards were differentiated, and the concept of conformance certification or accreditation was introduced.

In principle, the terms of reference for general data standards are equally applicable to spatial databases. However, given the recent profusion in availability of spatial databases, their diverse origins, their often unclear lineages, their use of different datums and projections, and their relative open access and download (from GIS data portals), there are aspects of spatial databases that make the need for and nature of their standards somewhat different from other cases. General standards for spatial database systems and several examples of contemporary spatial data standards were reviewed, including initiatives in Canada, the United States, the European Union, and international standards. In addition, the role of the OGC was noted in developing spatial data standards.

A comprehensive definition of metadata was provided at the outset of Section 4. The various roles of metadata were explained to reinforce their importance in all areas of database activity, including current convergence in international spatial metadata standards and the variety of tools that exist to build metadata for spatial databases and metadata presentation. Section 5 explained the importance of data standards and metadata in spatial database systems, and proposed a model for using spatial data standards and metadata in practice. Information discussed in this chapter provides the groundwork for the study of spatial data infrastructures, sharing and warehousing in Chapter 6.

7.   REFERENCES

Annoni, Alessandro (2001) "Geographic Standards of the European Commission", land Use Land Cover Unit, Institute for Environmental Sustainability, European Commission – DG Joint Research Centre, Ispra, Italy.

Bicking, B. and East, R. (1996) "Towards Dynamically Integrating Spatial Data and Its Metadata", *Proceedings*, 1st IEEE Metadata Conference, Silver Sprint, MD.

CGDI (2003) The Geospatial Standards Thrust of the Canadian Geospatial Data Infrastructure, Ottawa, ON: Canadian Geospatial Data Infrastructure.

CGSB (2002) *Procedures Manual for the Development and Review of Standards*, Ottawa, ON: Canadian General Standards Board.

Duval, E., Hodgins, W., Sutton, S. and Weibel, S. (2002) "Metadata Principles and Practicalities", *D-Lib magazine*, vol. 8, No. 4.

Fadaiem K., Habbane, M. and Tolley, C. (2002) "Canadian Geospatial Standards in Action", *Proceedings*, Symposium on Geospatial Theories, Processes and Applications, Ottawa, ON.

FGDC (1996) *FGDC Standards Reference Model* (Revised June, 1998, FGDC-STD-001-1998), Washington, DC: Federal Geographic Data Committee Secretariat.

FGDC (1998) *Content Standards for Digital Geospatial Metadata* (Revised June, 1998, FGDC-STD-001-1998), Washington, DC: Federal Geographic Data Committee Secretariat.

GeoConnections Secretariat (2001) *The Canadian Geospatial Data Infrastructure - Access Technical Services Manual*, Version 1.1, Ottawa, ON: GeoAccess Division, Natural Resources Canada.

Gill, T. (2000) "Metadata and the World Wide Web", in *Introduction to Metadata: Pathways to Digital Information* by Baca, M. (Ed.), Los Angeles, CA: Getty Research Institute.

Gilliland-Swetland, A. (2000) "Setting the Stage", in *Introduction to Metadata: Pathways to Digital Information* by Baca, M. (Ed.), Los Angeles, CA: Getty Research Institute.

IGGI (2001) *The Principles of Good Metadata Management*, Version 1.0, London, UK: Inter-governmental Group on Geographic Information (IGGI) Secretariat.

ISO (1999) *Geographic Information – Part 2: Overview (ISO/TC211/SC/N732)*, Oslo, Norway: ISO/TC211 Secretariat, Norwegian Technological Standards Institute.

ISO (2000) *Geographic Information – Reference Model (ISO/DIS 19101)*, Geneva, Switzerland: International Organization for Standardization.

Krechmer, K. (1998) "The Principles of Open Standards", *Standards Engineering*, Vol. 50, No. 6, pp. 1– 6.

Lutzet, C. (2003) "EuroSpec – Providing the Foundations to Maximize the Use GI", paper presented at 9th EC GI and GIS Workshop: ESDI Serving the User, A Coruña, Spain, 25-27 June, 2003.

Østensen, O. (2001) "The Expanding Agenda of Geographic Information Standards", *ISO Bulletin*, July, 2001, pp. 16 – 21.

Nebert, D.D. (2000) "Z39.50 Application Profile for Geospatial Metadata or GEO". (http://www.blueangeltech.com/standards/GeoProfile/geo22.htm)

Schweitzer, P.N. (2003) "Plain-language Resources for Metadata Creators and Reviewers", Reston, VA: U.S. Geological Survey. (http://www.geology.usgs.gov/tools/metadata/tools/doc/ctc/overview.shtml)

Teng, Y. and Nickerson, B.G. (2000) *XML (Extensible Markup Language) for Geospatial Metadata*, Technical Report TR-131, Faculty of Computer Science, University of New Brunswick, Fredericton, NB.

Chapter 6

SPATIAL DATA SHARING, DATA WAREHOUSING AND DATABASE FEDERATION

1. INTRODUCTION

Data sharing is one of the fundamental concepts of contemporary spatial database systems. Spatial data sharing in itself is not new. Survey and mapping agencies at different levels of government have, for many decades, shared the responsibilities for mapping at different scales and for different purposes. At the same time, spatial data users are able to purchase digital map layers from government mapping agencies and commercial data suppliers. It is also common for engineers, surveyors, scientists and academic researchers to exchange spatial data amongst themselves and to use spatial data collected by others. These examples of spatial data sharing have always characterised the collaborative nature of collecting and using spatial data.

The growing use of spatial information technology has given spatial data sharing new meaning, new dimensions, as well as new opportunities for the spatial information community. Spatial data sharing is no longer perceived simply as the selling of maps or the occasional exchange of data among individuals or organisations. Instead, it has become both a commercial business and a standard practice in modern data processing that transcends different application and technical domains. Many organisations in both the public and private sectors now include spatial data sharing as an integral component of their corporate information technology strategy. In many countries, spatial data sharing has its own national information policy and is a cornerstone of the national information infrastructure.

This chapter concludes Part 2 of the book. It begins with an overview of the nature of spatial data sharing. The importance of and barriers to implementing a data sharing strategy in an organisation is also discussed and emerging database connectivity standards and technologies are described that will enable data sharing among systems, applications and users. Section 3 examines the important concept of database heterogeneity and explains solutions to address issues arising from it. These solutions are then applied in the two general approaches to data sharing in the database environment, namely *spatial data warehousing* and *database federation*. Finally, the implications of spatial data sharing in database design and implementation are discussed.

As the concepts and methods of data sharing span the entire system development life cycle (SDLC) of a spatial database system, this chapter also lays the theoretical foundation for the discussion of implementation and project management in Part 3.

2.   THE CONCEPTS AND METHODS OF SPATIAL DATA SHARING

Contemporary database systems are often designed to decentralise the processes and sharing of data sources in a geographically distributed network of databases, applications and users. Data sharing in this environment is a requirement, rather than an option. The following discussion considers the nature, concepts and methods of spatial data sharing in light of the importance this approach has in database system development.

2.1   The Definition and Nature of Spatial Data Sharing

Conventionally, the concept of *spatial data sharing* is used in association with the exchange, interchange or transfer of data between two or more users, organisations or computer systems. The process of spatial data sharing often involves a change of the format and geographical referencing of the data, which are commonly referred to as *conversion* and *transformation* respectively. However, the objectives of spatial data sharing go far beyond the simple process of exchanging data between different computer systems and the conversion of data from one format or geographical referencing system to another. Data sharing spans a continuum of progressively sophisticated processes that aim to make different database systems *interoperable* at the data, application, and business process levels, as summarised in Table 6.1.

Table 6-1. Levels of spatial data sharing

Levels of Data Sharing	Data Sharing Characteristics			
	Computing Environment	*Systems Architecture*	*Procedure*	*Purposes and Applications*
Infrastructure	Open computing standards (Internet, Web services, Java), distributed processing and distributed objects	A distributed network of databases connected by the global telecommunicat-ions system	Global/Universal information access and application through database mediation and information brokering	Seamless spatial database interoperability and integration using operational and legacy data
Enterprise		Federated databases and data warehouses connected to an organisation's communications network	Inter-departmental information access and application through database mediation and information brokering	Simultaneous on-line transaction processing (OLTP), on-line analytical processing (OLAP)
Domain	Distributed databases connected using TCP/IP, HTTP and open database connectivity standards	Three-tiered client/server computing or data mart in a wide area network (WAN)	Shared databases with sophisticated collaboration among different users or organisations	Sector-based data management and applications, multi-sourced spatial data analysis and modelling
Functional		Two-tiered client/server computing in a local area network (LAN)	Heterogeneous data exchange	Spatial data visualisation and overlay analysis using data from multiple sources
Connected	Peer-to-peer proprietary network and communications protocols	Desktop computer with simple network connection	Homogeneous data exchange	Electronic exchange of text files and graphics files of the same format
Ad Hoc	Stand alone computers and independent data files	Independent desk top computers	Manual data exchange with hard copy maps, diskettes, CD-ROM	Occasional exchange or sale of data from ad hoc requests

At the lower end of the continuum are *ad hoc* data exchanges between individual users or organisations. In the past, data sharing at this level was done manually using hardcopy documents and maps, magnetic tapes and diskettes. However, the preferred data sharing media are now optical data storage devices such as DVDs and electronic networks. The smallest unit of data to be shared at this level, which is called *data granularity*, is the data file (that is, the whole data file is moved from the host or data supplier's computer to the target or data user's computer). The smallest unit of the process used to execute the sharing, which is called *process granularity*, is a stand alone application program or interface that is used to copy the shared data from the host computer to the target computer.

Data sharing by *peer-to-peer connection* is similar to *ad hoc* spatial data sharing except that the shared data are transmitted electronically across a simple data communications network. Exchange of text and graphics files on an organisation's intranet falls into this level of data sharing. The loading onto a computer of data collected by means of a Global Positioning System (GPS) or electronic distance measurement (EDM) unit also belongs to this level of data sharing.

Both functional and domain spatial data sharing processes are based on the distributed method of computing using a local or wide area network (LAN/WAN) across which data are shared by means of various database connectivity standards (see Section 2.3). In functional data sharing, which aims to provide data for a particular application or project, users access the data server through a LAN for data files that are required for applications performed on a client computer. The spatial data to be shared may be stored in raster and vector formats. A good example of functional data sharing in this context is a three-dimensional spatial visualisation created by draping a digital aerial photograph stored in one data server onto a digital elevation model (DEM) and vector topographic data stored in separate data servers. Functional spatial data sharing is essentially file-based and carried out by stand alone computer programs.

Domain data sharing differs from functional data sharing in purpose and systems architecture. Instead of serving the data needs of a particular application or project, domain data sharing seeks to provide data in support of the operational needs of a particular business area or application domain such as public health, land parcel mapping or urban transportation management. Domain spatial data sharing has a finer data and process granularity than the first three levels of data sharing noted above. By storing data in a database, it is possible to access and share a specific record in a data table using a query command embedded in an application program. Data for domain data sharing include both operational and legacy data pertaining to a particular business area of an organisation and can be drawn from multiple data servers across a WAN. A database containing legacy data for decision support and sharing in a particular domain is commonly called a *data mart* (see Section 4.1).

The major interest of contemporary spatial data sharing focuses on *enterprise* and *infrastructural* data sharing. Spatial data sharing at these two levels is based on open computing standards and a similar distributed architecture. These levels are distinguished by their procedures and purposes. In essence, enterprise spatial data sharing aims to facilitate inter-departmental information access and application within an organisation. The data to be shared are stored either centrally in a *data warehouse* (see Section 4) or in a distributed *federated database system* (see Section 5). Since data are often stored in different formats and according to different specifications in an enterprise computing environment, data sharing at this level requires the use of relatively sophisticated database mediation and information brokering methods (see Sections 3.2 and 3.3).

Infrastructural spatial data sharing aims to provide global and universal access to databases both within and outside of an organisation. Hence, the word "infrastructural" carries three important meanings, namely:

- The objective of data sharing is to make data available for the general public to use just like other physical infrastructures that are used on a daily basis (for example, roads, water supply, electricity and telecommunications services).
- It conveys the concept of providing spatial data not only as physical artefacts but also as a critical and essential public service, in the same way social, educational and health services are provided in modern society.
- It implies the use of standard practices (see Chapter 5) and protocols for data documentation, management, publication, retrieval and delivery to ensure consistent and reliable spatial data sharing for the common interest of society as a whole, and not just for the benefit of individual users or organisations.

Infrastructural spatial data sharing is based on the concept of a *spatial data infrastructure* (SDI), the objective of which is not to establish a single central database, but to set up a distributed network of databases, based on accepted standards, that is managed and stewarded by individual organisations in the public, academic and industrial sectors at the local, state/provincial, national, regional and global levels (Nebert, 2001). Within this infrastructure, users can seamlessly or transparently use data from different databases regardless of the *syntax* (content and type), *structure* (schema and format) and *semantics* (domain-specific language and definitions) of the data stored in the constituent spatial databases. This free exchange of all kinds of data drawn from different databases is called *interoperability*, as defined by the OGC (Buehler and McKee, 1996). The ability to fuse spatial data with other forms of data to achieve some desired level of usability by different applications across different hardware and software platforms is called *integration* (Abel et al., 1994) and also *data fusion*. The creation of an SDI is arguably the ultimate aim of spatial data sharing.

In this book spatial data sharing is defined broadly to include the interoperability of databases and integration among different systems *thematically* (that is, transcending different application domains) and *hierarchically* (that is, from local to national, regional and global databases). The nature of spatial data sharing is substantially different from what it was in the past. Specifically, spatial data sharing is now *service-focused* rather than data-focused because the objective of sharing is not limited to solving the *ad hoc* data problem of a particular application, but to provide a feasible business model of using spatial data to meet individual and organisational information needs.

To this end, spatial data sharing is implemented as a proactive information technology strategy for an organisation, which is in sharp contrast to the traditional reactive method of data sharing characterised by the typical request-and-response relationship between data suppliers and users. This also implies that spatial data sharing is dependent decreasingly on the physical transfer of data between computer systems, and increasingly on *virtual* and *logical* sharing whereby an application on a client computer is able to use data simultaneously on multiple servers over a network. As the increasing use of standards has gradually eliminated many of the difficulties resulting from incompatibility of data structure and syntax, spatial data sharing at present is predominantly concerned with issues of access (including policy, legal liability (see Chapter 7), discovery and dissemination) and collaborated applications (semantics and interfaces), rather than with the technical exchange of data between disparate data sets as in the past.

2.2   The Importance of Spatial Data Sharing

There are numerous benefits of sharing spatial data, many of which are the direct results of the ability to share the same data sets, for example:

- Supporting the fundamental principles of best practice in data management that encourage local autonomy while at the same time promoting regional, national and global collaboration by enabling the development of sophisticated spatial applications that require a vast amount of data from multiple sources, collected at different times and stored in different formats.

- Reducing the cost of using spatial information by minimising or eliminating the need to collect or convert data when setting up a new spatial database.
- Reducing the cost of data maintenance by storing and managing data locally where they are collected, used and distributed.
- Increasing the quality of spatial data, and hence the degree of certainty in decision making, by enabling different applications to share and adhere to a standardised framework or to reference data sets.
- Reducing software costs because software programs can access and use spatial data from different sources and in different formats, thus eliminating the need to develop or purchase new application software when data from new sources or formats are required.
- Expediting application development and deployment as users are able to use data from different sources without the need to customise the data in the first place.
- Reducing the risk of "vendor lock-in" and "stranded" technologies by enabling the users to adopt new hardware and software platforms if and when the hardware or software in use is unable to meet the changing information needs of an organisation, or to keep abreast with advances in technologies.
- Making it easier to integrate spatial data with other forms of business data to improve decision making within an organisation.

In the context of an organisation's return on investment (ROI) in spatial information technology, spatial data sharing makes good business sense because it allows the organisation to recoup part of the cost of data collection by creating a commodity from their data and on-selling them to other users. The trend towards commodification of spatial data has actually created an active value-added data industry in which companies purchase base data from a spatial data collector (for example, a government mapping agency), improve their content or add one or more themes of application data, and then on-sell the value added data set commercially to users (other government agencies, private companies, education institutes, and the general public), sometimes through licensing agreements in partnership with the data collector agency. This industry also includes a growing number of consulting firms whose business is to help clients identify, assemble and customise spatial data from existing sources to create their own spatial databases. As a result of this activity, spatial data sharing has become a crucial and integral part of the information or knowledge-based economy.

From a broader societal perspective, spatial data sharing has led to the development of *participatory approaches* to using geographic information (Craglia and Onsrud, 2003; see also Chapter 12). The aim of participatory approaches to using spatial information is to move its use beyond the traditional exclusive groups of technical and professional users towards the popular and inclusive use by people in all walks of life. The foundation of these approaches is the ability of citizens to access spatial data held by government and other public and private sector organisations in an equitable manner, which in turn motivates them to participate in public debate over government policies that affect the well-being of society and the environment. In this way, spatial data sharing is not an end in itself, but a means for citizens to participate in public affairs and to exercise their civil rights and privileges in a democratic society.

The benefits and socio-economic ramifications of spatial data sharing as explained above strongly suggest that the importance of data sharing cannot be stressed enough in the design and implementation of spatial databases. Figure 6.1 summarises the importance of spatial data sharing by substantiating the contention that it is much more than the simple exchange of data between different users or applications. It is in fact a unifying theme for spatial database development that transcends implementation methodology, institutional policies and protocols, as well as hardware and software technologies that support its use.

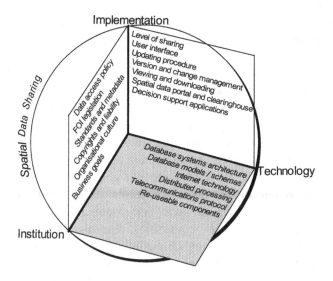

Figure 6-1. Spatial data sharing from the perspectives of implementation, technology and institution of database systems

The following discussion focuses on technical solutions to spatial data sharing in database systems. Important implementation and institutional issues of spatial data sharing are addressed in Part 3 of this book.

2.3 Barriers to Spatial Data Sharing

Although the need for spatial data sharing is well researched and understood, a wide range of circumstances have inhibited the implementation of a data sharing strategy in many organisations. It is beyond the scope of this chapter to present a detailed analysis of the causes and underlying nature of these problems, however several general barriers to spatial data sharing can be identified, namely:

- The inherently complex and diverse characteristics of spatial data, such as scale, geographical referencing, spatial and temporal resolution, generalisation and aggregation, feature coding or classification, data models and storage formats, which collectively make the technical and procedural sharing of spatial data much harder than the sharing of ordinary text-based data.
- The non-deterministic nature of human cognition and language often results in different people using different terminology to describe the same thing, and different people describing different things using the same terminology.
- Differences in data management policies, user access protocols, systems security measures, database partitioning and structuring methods, network bandwidths, and hardware/software standards among different organisations.
- Technical incompatibilities and systems dependencies that arise from different characteristics and technical specifications of hardware platforms, operating systems and software acquired at different times or for different purposes.
- Lack of data model and format standardardisation, resulting in incompatibilities in database structures and applications.
- Restricted availability of data, either due to inaccessibility to or inadequacy of metadata, which in turn results in the inability of potential users to be aware of existing data resources and thereby to evaluate the usability of existing data for intended applications.
- Unwillingness of organisations to share data either due to the lack of a formal data policy or concerns about infringement of copyright and intellectual property, and the legal liability and repercussions of damages that may result from potential errors in the data.
- Unwillingness of the individual departments of an organisation to share data due to the fear of losing autonomy, control and diminished systems performance of the local databases.

- Restrictions on releasing data to the public due to regulatory factors and considerations such as national security, protection of privacy, and archiving requirements for public documents.
- Diversity in the types of users and the disparities between the needs of high-end (for example, policy analysts, researchers, engineers) and low-end users (for example, occasional users of Web-based spatial information services).
- Lack of coordination between spatial data collectors, for example between local, state/provincial and federal mapping agencies, resulting in discrepancies about data needs, content, encoding standards, coverage, and revision cycles for data collected for the same or similar area.
- Lack of a supporting data discovery and delivery infrastructure, including, for example, sufficient network bandwidth for the high-speed transmission of large volume of spatial data in real time, organisational protocols to develop data sharing partnerships with potential users, expertise or mechanisms to enforce copyright and protect the organisation from legal liability, and the resources to develop and maintain a metadata system to broadcast the necessary information to potential users.
- The policies of many national governments require full recovery of the cost of data collection, which reduces the incentive for users to use data in public stewardship.

Advances in IT in general, and in spatial database concepts and methods in particular, have eliminated or greatly reduced the significance of some of the above barriers. At present, comprehensive theoretical and practical solutions are still evolving to address the inhibiting factors while at the same time promoting the benefits of spatial data sharing. However, since it is practically impossible to do away with all barriers completely, spatial data sharing will always remain an opportunity as well as a challenge for spatial data suppliers and users alike.

2.4   A Standards-based Framework for Spatial Data Sharing

Standards and standardisation play a significant role in spatial data sharing. Spatial data users have now generally endorsed the concept of interoperability through standardisation as advocated by the OGC (Buehler and McKee, 1996). Several important standards ease spatial data sharing among applications and database systems. These include database access standards such as Object Linking and Embedding (OLE), Open Database

Connectivity (ODBC), Java Database Connectivity (JDBC), and the Common Object Request Broker Architecture (CORBA), and Common Object Model (COM), as well as a myriad of Web-based services protocols that are used for spatial data sharing over the Internet. The working principles of these standards are explained briefly in the following discussion.

2.4.1 Object Linking and Embedding

OLE is the *de facto* standard for data access in the Microsoft Windows environment. It is a common data access method for Microsoft server software (that is, SQL Server), desktop application software packages (for example, PowerPoint, Word, Access, Excel) and application development tools (for example, Visual Basic, Visual C++, .NET).

Using an architecture that layers services over the Component Object Model (see Section 2.4.5), OLE allows users to access and exchange data freely among Microsoft software products, for example, creating a PowerPoint presentation with a linked or embedded Access database table. In 1995, Microsoft unveiled a component-based strategy for enterprise development that consisted of a variety of OLE-based technologies, including data integration software called OLE DB (also code-named Nile), Network OLE (for remote object access), OLE Transaction (for transaction processing), OLE Team Development (for repository management) and OLE Directory Services (for database directory management).

The addition of OLE DB was particularly important to inter-product data access in the Windows environment. It enables users to use an SQL-based data access approach to query a relational or object-relational database. It also allows users to define cross-component business rules and create objects that export an event model so that operations (for example, updating a spreadsheet) will be able to invoke events and make related updates. This capability, which is conceptually similar to a trigger in relational database processing, is useful for maintaining data integrity among databases and data files in an organisation and for reliable and consistent data sharing.

2.4.2 Open Database Connectivity

Open database connectivity (ODBC) is an open standard API for accessing one or more database from an application. Initially created by the SQL-Access Group (SAG, which is now part of The Open Group) in 1992, ODBC was extended and deployed by Microsoft as a standard Call-Level Interface (CLI) that allows client and server applications to exchange and share data without the need to know anything about each other. With the aid of ODBC, it is possible to access any data from any application, regardless of which DBMS is handling the data (Figure 6.2).

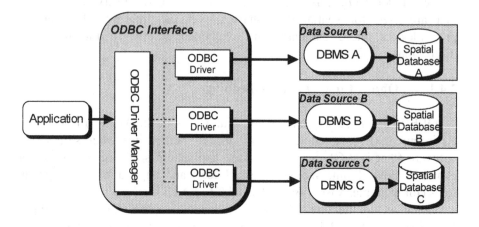

Figure 6-2. The working principles of open database connectivity (ODBC)

This multi-database access capability is made possible by inserting a middle layer, called a database *driver*, between an application and the *data source* which, in database terminology, refers collectively to the repository of data and its underlying DBMS. The function of the driver is to translate the queries from an application into commands that the data source understands. For this to work, both the application and the data source must be ODBC-compliant, that is, the application must be capable of issuing ODBC commands and the data source must be capable of understanding and responding to these commands.

Since different data sources work differently, it is necessary for each data source to have a specific ODBC driver for it to be accessed. The separation of applications from drivers has proved to be a flexible approach to inter-process data access. When an application is developed, there is no need for the developer to worry about the databases being accessed. It is up to the user to obtain and install the necessary ODBC drivers from database software vendors or third party software developers. The ODBC *driver manager* manipulates the installed drivers to access the appropriate data sources transparently during data processing.

There were two important developments of ODBC in recent years. One of these was the adoption of a SQL/CLI as part of the SQL-92 standard by the ANSI and ISO bodies. Unlike the original SQL standard that included only embedded SQL commands and used source-code pre-processing and compile-time data binding, SQL/CLI permits execution-time binding and, therefore, can be used for the development of interoperable applications more easily. ODBC Version 3 and higher are compatible with SQL/CLI and, consequently, can be integrated easily with SQL-based query and applications. The second development was caused by the proliferation of ODBC in the Windows-based database environment and the emergence of a CLI standard from the ANSI and ISO, which in turn contributed to the demand for ODBC on other platforms. As a result, there are now versions of ODBC for UNIX, Macintosh, PowerMac and Solaris operating systems.

2.4.3 Java Database Connectivity

Java database connectivity (JDBC) is an API specification for Java applets, servelets and applications to access data stored in databases, spreadsheet and flat or ASCII files. The development of this database connectivity tool is championed by SUN MicroSystems, and it can also be downloaded from the company's Web site. JDBC is commonly used to connect a user application to a data source, regardless of what DBMS is used to control the database.

In function and architecture, therefore, JDBC is very similar to ODBC, as described above (Figure 6.3). JDBC can be used with a one-tier driver in the same way as ODBC to access a data source directly. Alternatively, it can also be used with a two-tier driver, for example a JDBC-ODBC bridge to communicate with ODBC-enabled databases (such as Oracle and Microsoft SQL Server which have built-in ODBC functionality), or a JDBC network driver when the Java program is running in a network environment to access a remote database.

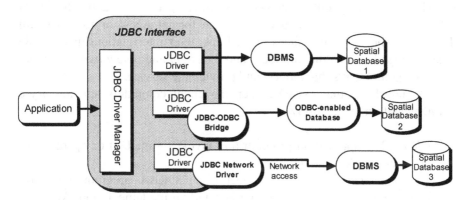

Figure 6-3. The working principles of Java database connectivity (JDBC)

2.4.4 The Common Object Request Broker Architecture (CORBA)

The *Object Request Broker* (ORB) is a middleware technology that manages communication and data exchange between objects in object-oriented programming and databases. Functionally, an ORB acts somewhat like a telephone exchange. It provides a directory of services (that is, the methods or operations associated with data objects), helps establish connections between these services and their clients, activates the required services on behalf of the client, and returns the results to the clients as shown in Figure 6.4. All of these processes are done transparently, which means that there is no need for the client application to know where the object is located, what its programming language is, what operating system is being used or any other aspects that are not part of its interface.

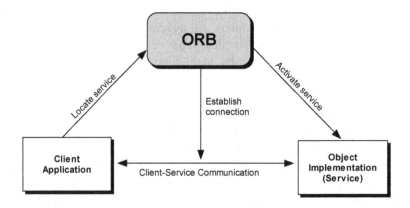

Figure 6-4. The working principles of object request broker

The CORBA specification was first developed in 1992 by the OMG. CORBA is an industry standard but not a formal ANSI or ISO standard. Figure 6.5 shows the interfaces of CORBA, their interrelationships, and the relationships between the interfaces and the *client* and *object implementation* software components that request and provide the service.

The interfaces that exist between the client computers and the object implementation are as follows:

- *Client.* The application program that invokes a method or operation on an object implementation.
- *Object implementation.* An object is a CORBA programming entity consisting of an identity, an interface and an implementation. An object implementation, which is also called a service or method in object-oriented terminology, defines the operations supporting an interface definition language (IDL) and can be written in a variety of programming languages including C, C++, Java, Smalltalk, and Ada among others.

- *Object Request Broker Core.* The CORBA run-time infrastructure for transparent communication between a client and object implementations as explained above.
- *Dynamic Invocation Interface* (DII). The mechanism for constructing requests at run time.
- *Client IDL Stubs.* Precompiled stubs or static interfaces that define how clients invoke corresponding services on the servers.
- *ORB Interface.* An interface containing help functions and APIs that can be used by a client or an object implementation to enhance or extend their functionality.
- *IDL Skeletons.* The server-side analogue of IDL stubs, which receive requests for services from the object adaptor, and call the appropriate operations in the object implementation.
- *Dynamic Skeleton Interface* (DSI). The server-side counterpart of DII, which is a run-time binding mechanism for an ORB to deliver requests to an object implementation that does not have compile-time knowledge of the type of the object it is implementing.
- *Object Adapter.* The mechanism that assists the ORB with delivering requests to the object and with activating the object.
- *Interface Repository.* A run-time distributed database that contains machine-readable versions of the IDL-defined interfaces and serves as a dynamic metadata repository for ORBs.
- *Implementation Repository.* A run-time repository of information about the classes a server supports, the objects that are instantiated, and their respective identities.

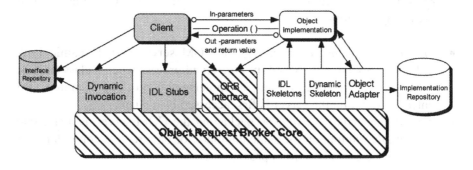

Figure 6-5. Interfaces and relationships of CORBA

2.4.5 The Component Object Model

The component object model (COM) approach is the ORB specification and implementation developed by Microsoft to provide a framework for integrating software components in the Windows environment (Figure 6.6). Interoperability of the COM approach is achieved by defining API components using a binary structure. As long as components adhere to this binary structure, they can interoperate with one another no matter which programming language they are written in.

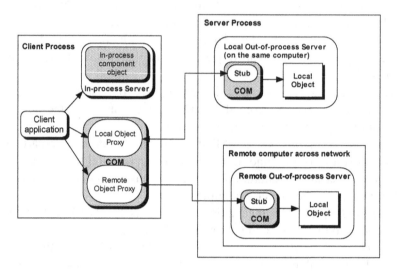

Figure 6-6. The working principles of the common object model and DCOM

Distributed COM (DCOM) is an extension of COM for network-based interaction between remote clients and services (see also Figure 6.6). The COM/DCOM architecture allows software components to interact in two different ways, as follows:

- *In-process serving*, by packaging a COM server as a Dynamic Link Library (DLL) that is loaded into the client process when a class (that is, a body of source code defining an implementation of one or more COM interfaces) within the server is first accessed by a client.
- *Out-of-processing serving*, by packaging the COM server as a separate executable running on the same computer or on a remote computer over a network using a DCOM approach.

Client applications interact with COM components through interface pointers. In the case of in-process COM servers, calls made by a client using an interface pointer go directly to a component object created in the client's process. On the other hand, calls to component objects in the case of out-of-process servers, which can reside on the same computer or a remote computer, go first to an object proxy that is responsible for invoking the requests using a *remote procedure call* (RPC). In the server, a "stub" receives each of the incoming requests and dispatches it to the appropriate component objects. The results of the request will return to the requesting client application through the same RPC communication channel.

As noted in Section 2.4.1, COM is the foundation technology for OLE. While COM represents a "low level" technology that allows components to interact, OLE represents high-level application services capable of "linking" and "embedding" component objects to generate compound documents (that is, documents created by different applications using data from different sources), and enable data transfer and inter-application scripting.

2.4.6 Web Services Protocols

As a result of the advances in telecommunications and related technologies, the Internet has now become the predominant mechanism for spatial data sharing. The Internet not only links data users and suppliers physically and logically together, but it also provides a crucial metadata service for data suppliers to advertise their data products as well as for the data users to search for and identify data sets that are suitable for their purposes. This tripartite relationship among data users, suppliers and metadata services fits well with the emerging concept of *Web services* that is destined to become one of the standards for software interoperability in the computing industry (Chung et al., 2003).

From a technical perspective, Web services are software components that can be accessed over the Internet through standards-based protocols. These software components include desktop application programs, Web browsers, Java applets, and software running on mobile devices such as cellular phones. From an application perspective, Web services provide a standardised way of integrating Web-based applications using open standards, as noted in the discussion below. A typical Web services-based architecture, as shown in Figure 6.7, consists of three types of *computing nodes*, namely the *client*, the *service*, and the *service broker*. Each of these nodes plays a different and distinct role as follows:

- The *Client*, is any computer application that makes a request to one or more other computing nodes for services and receives the results.
- The *Service*, is a computing process in a server node that receives and responds to a request from a client.
- The *Broker*, is essentially a service metadata portal that registers the services of server nodes and facilitates the discovery of these services by clients across a network.

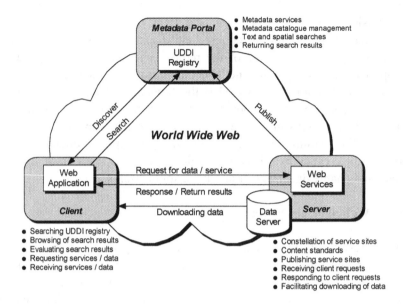

Figure 6-7. The web services-based framework for spatial data sharing

Conformance to standards is crucial for Web services in order to ensure that each computing node can interoperate and collaborate with one another, and to support the sophisticated communications among the computing nodes on a network. Important Web services standards include:

- *Extensible Markup Language* (XML*)*, which is the standard language for structuring the messages transmitted between computing nodes.
- *Geography Markup Language* (GML*)*, which defines a data encoding in XML that allows spatial data and their attributes to be transmitted between disparate systems.
- *Scalable Vector Graphics* (SVG*)*, which is an XML-based format for defining two-dimensional graphics, including vector graphic shapes and images, and their associated text attributes for Web-based and other applications.
- *Simple Object Access Protocol* (SOAP, also known as Service-oriented Access Protocol), which is the standard for transmitting messages between computing nodes.
- *Web Service Description Language* (WSDL), which is the standard language for describing services that are available on a network.
- *Universal Description, Discovery and Integration* (UDDI), which is the standard for registering available services and creating a directory service for users to locate and identify available services on a network.

The Web services-based architecture is a generic standards-based computing model that provides an open and interoperable environment for data sharing over the Internet. Web services are important building blocks for most of the operational and planned spatial data infrastructures. Many organisations, for example the CGDI, have developed this model for their spatial data services (GeoConnections Secretariat, 2001). Map services are also used by the Environmental Systems Research Institute (ESRI, 2002) to build the geography network or *g.net* framework that describes how this company's software products ArcGIS, ArcIMS, ArcSDE and other tools can be used in a spatial data infrastructure setting. In order to provide a standardised approach to Web services for the spatial information industry, the OGC (OGC, 2003(b)) has undertaken the development of a series of implementation specifications that define the architecture, creation, and operation of Web map services. The application of these specifications in the design and implementation of Web-based spatial databases is explained in more detail Chapter 10.

3. DATABASE HETEROGENEITY AND ITS SOLUTIONS

Heterogeneity is a term used to refer generally to the differences, disagreements and dissimilarities among members of an identified community of people or collection of ideas, phenomena and physical objects. In the context of spatial data, heterogeneity is associated with or caused by the inconsistencies between data sets resulting from different surveying techniques, errors in measurement, different definitions of precision, fuzziness of spatial objects, and variations in the use of terminology and nomenclature. As databases are set up by different users for different purposes, at different times and using different technologies, heterogeneity is an inherent property of database systems. The following discussion explains the nature of heterogeneity of spatial data and the methods used to overcome the problems of heterogeneity in spatial data sharing.

3.1 The Nature and Characteristics of Database Heterogeneity

Database heterogeneity as defined above occurs at the systems and data levels shown in Figure 6.8. The underlying reasons the existence of systems heterogeneity is easy to understand in terms of the diversities of the purposes of implementing and using databases, the wide varieties of software and hardware options available for each of these purposes, and the conventional approach to systems implementation that stresses local autonomy. The landscape of systems heterogeneity is further complicated by the rapid evolution of information technology that continuously produces new software and hardware products that differ drastically both functionally and architecturally from those in use.

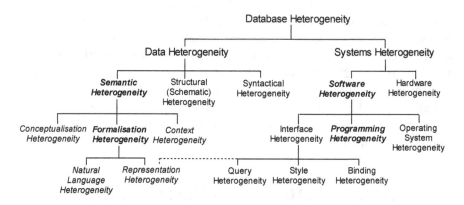

Figure 6-8. A typology of database heterogeneity

There are three aspects of data heterogeneity, namely:

- *Syntactic heterogeneity*, which is caused by the use of different data types (that is, characters, numerics and logicals), storage word lengths and precision, date format, permissible values, units of measurement, abbreviations and acronyms to represent data in different data sets.
- *Structural* or *schematic heterogeneity*, which is caused by the use of different data models to abstract the same real world features or phenomena which, in turn, results in the storage of data in different formats, for example, the various vector and raster formats and data transfer formats, as well as relational, object-oriented and object-relational classes of databases.
- *Semantic heterogeneity*, which is defined as the inconsistencies or disagreements between digital representations and corresponding real world features or phenomena within a certain context. It is caused by the interplay of a collection of factors that can be generally identified as:

    - *Conceptualisation heterogeneity*, also known as *cognition heterogeneity*, which occurs when different people abstract the same real world features using their own experience or from the perspective of their respective disciplines.

- *Formalisation heterogeneity*, which results from the use of different languages (*language heterogeneity*) and terminology (*representation heterogeneity*) to describe a conceptualisation.
- *Context heterogeneity*, which is related to the changing meanings of a conceptualisation or formalism under different conditions or circumstances.

Figure 6.9 illustrates the above three aspects of semantic heterogeneity using the examples of four data sets covering the same stretches of highways and local roads. Note how different data collectors conceptualise the real world for the purposes of building (a) a road network database, (b) a pavement management database, (c) a medium topographic mapping database, and (d) a small-scale topographic mapping database. These examples clearly show that different data collection methods and objectives can easily lead to different object classes, different attributes and different geometric representation by points, lines and polygons. Note also how different terms are used to represent the same conceptualisation, for example, "node" in network data in (a) and "intersection" in topographic mapping data in (c), both of which refer to the same point where one road meets another.

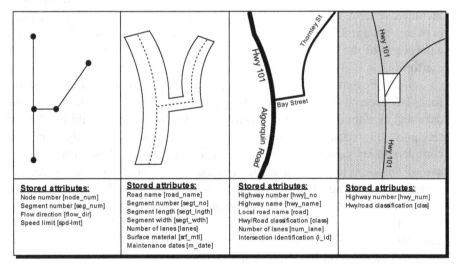

(a) Road Network Database   (b) Pavement Management   (c) Medium-scale map   (d) Small-scale map

Figure 6-9. Conceptualisation, formalisation and context heterogeneity

The term "segment" as it is used in (a) and (b) is a good example of how the meaning of a conceptualisation and formalisation may change under different circumstances. In a network data set, a "segment" is strictly defined as the line between two nodes and is used to model traffic flow directions and movement of people, goods and services. When this same term is used in a pavement management data set, it may not carry such a rigid definition. It is represented as a polygon and is used to calculate the amount of work and cost associated with the maintenance of the highway with no reference to traffic flow or movement.

The reconciliation of the differences among different systems has long been an important research area in database technology. Sheth (1999) identified three phases of heterogeneity research with relatively distinct focuses and objectives pertaining to the three aspects of heterogeneity noted above. The first phase commenced from the beginning of computer systems to the mid-1980s. Data used during this period were mostly structured, such as hierarchical and relational databases, and the communication was usually within a local area. The focus of research was the physical exchange of data among computer systems using different hardware architectures, operating systems and data management practices. The second phase was from the mid-1980s to the mid-1990s. During this period, data were not limited to structured data but also included semi-structured data, such as text files, and unstructured data, such as image or audio files. The focus during this time was more on the integration of data at the syntax and structure levels. This included schema transformation and data conversion to facilitate exchange between systems. The third phase began in the early-to-mid 1990s with the emergence of the Internet and the Web as the primary data communication platform.

As data become more varied and applications get more sophisticated, the focus of research moves toward the solution of semantic differences among data sets as a transitional step toward total database interoperability. Recent research on database heterogeneity has focused primarily on two complementary conceptual constructs of interoperability. One of these is the use of *ontology* to create a global vocabulary of terms in order to resolve problems resulting from semantic heterogeneity. The other is the reconciliation of the syntactic, schematic and semantic differences among data sets by means of *information mediation*. The following discussion considers these two conceptual constructs and how they can be used to implement a data sharing strategy in practice.

3.2   The Concept and Method of Ontology

The term ontology is used differently by different disciplines (Auxilio and Nieto, 2003). Combining the notions and concepts contained in the definitions of this term from computer science and related disciplines such as artificial intelligence and knowledge management, the term is used in the context of spatial database interoperability to represent the following interpretations:

- A *concept* of using formally and explicitly defined terminology and vocabulary to describe real world features or phenomena associated with a specific discipline, domain or application.
- A *systematic collection and specification* of spatial entities, their properties and relations, which are commonly stored in a hierarchical structure and used consistently by users in a particular discipline or domain.
- An emerging *approach* to designing spatial database systems that has several advantages over conventional methods of systems development, including:

  - Allowing the establishment of correspondence and interrelation among different domains of spatial entities and relations.
  - Contributing to create better information systems by improving communication between systems developers, managers and users.
  - Enabling a user-centred approach to systems development.
  - Providing the underlying concept and technology for interoperable database systems.
  - Designing spatial databases from a perspective beyond the map metaphor that views the real world as independent layers of information that can be combined and overlaid.

Ontologies are created by consensus among the users of data pertaining to a particular domain. These users are sometimes collectively referred to as an *information community*, using a series of ontology building activities (Auxilio and Nieto, 2003) (Figure 6.10a). These activities include extraction from existing database schemas and a formal data modelling process, called *semantic modelling*, that focuses on identifying and defining relevant terms. In the ontology building processes, it is often necessary to solicit the help of subject matter experts to ensure accuracy and precision of definitions.

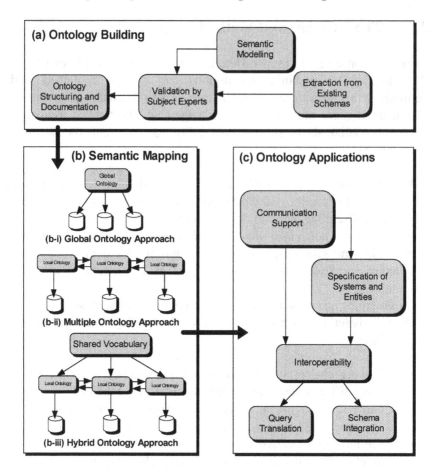

Figure 6-10. Creation and use of ontology in database systems

Ontologies can be documented using various markup languages, such as HTML, XML, and Ontology Interchange Language (OIL). Ontologies can also be recorded graphically using entity-relationship diagrams or UML. Since the definition of some terms may change over time, and new terms have to be added in response to changing application requirements, maintenance of ontologies is a relatively complex task. However, there are ontology tools available to support the editing, updating, and inspection of their content (Auxilio and Nieto, 2003).

From a database design perspective, the process of ontology building and documentation is comparable to conceptual data modelling because both processes aim to identify and define real world features and determine their relationships. However, although the processes are similar, the end products are not the same. While the purpose of a conceptual schema is to describe the intended database structure at a high level of abstraction, an ontology represents a consensual agreement on the meanings of and relations between the vocabulary of terms used to represent data. There is not necessarily direct correspondence between the structure of an ontology and the structure of the database as it is represented by a conceptual database model.

The ability of ontologies to provide unambiguous meanings of and structured relationships among the terminology used to describe the real world makes them a useful tool to address the problem of semantic heterogeneity in database design and application. There are three approaches to applying ontologies, namely:

- A *global ontology approach*, where all data sources are associated with one single common domain ontology (Figure 6.10(b-i)).
- A *multiple ontology approach*, where each data source is related to its own ontology that can be logically connected with one another (Figure 6.10(b-ii)).
- A *hybrid ontology approach*, where each data source has its own ontology that is logically connected with one another and to a common domain ontology (Figure 6.10(b-iii)).

Ontology as an approach to database design and implementation serves several useful purposes (Figure 6.10c). As explained above, its function is not limited to the role of enabling database interoperability, but it is also a crucial medium of communication by providing precise notions that can be used to describe an application domain. It also provides the means to help define the semantics of database fields in a clear and unambiguous manner. Of course, the greatest value of ontology is its role in supporting database interoperation strategies by means of *query translation* and *schema integration*. Query translation, as the name implies, is the process of translating or mapping heterogeneous field names used in different data sets to an ontology in order to query them simultaneously using a single operation, for example by one SLQ statement. Schema integration, on the other hand, makes use of the concept of ontology to combine the schemas of individual data sources into one global schema.

3.3   Information Mediation

Information mediation is a database interoperability strategy where queries against multiple heterogeneous data sources are communicated through a middleware medium known as a *mediator* (Wiederhold, 1992). In essence, a mediator is a collection of software components, database access optimisation rules and a catalogue of information about the data sources to be shared. It facilitates database interoperation by user applications that, together, address the problem of database heterogeneity. The following processes are used for this purpose:

- Re-writing queries, by optimising the queries using an execution plan created according to the database access rules.
- Fragmenting queries to individual data sources and, if necessary, mapping the heterogeneous data field names to their corresponding ontologies using the catalogue of information about the data sources.
- Dispatching query fragments to their respective target data sources.
- Assembling individual results into a composite response and returning it to the query user interface.

The number of functions that the mediator performs implies that it is relatively difficult to create and maintain, particularly when the number of data sources to be accessed is large. Given the relative difficulty of creating and maintaining a mediator, a more flexible approach to information mediation that has become quite widely used in current database architectures that involve some form of interoperability is shown in Figure 6.11

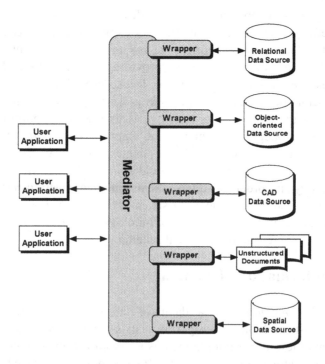

Figure 6-11. The concept of information mediation

In this figure another tier of middleware is placed between the mediator and the data sources so that five components exist between the mediator, the data stores and the client applications. The function of this tier of middleware, known as a *wrapper*, is to manage data source heterogeneity. By using wrappers, all heterogeneous data sources become homogeneous to the mediator that now mainly plays the role of a user interface. This means that when access to new data sources is required, it is necessary to add a new wrapper only. The mediator user interface is left completely intact. This two-tiered mediator-wrapper architecture represents a higher level of database interoperation because it is capable of addressing not only semantic heterogeneity, as in the case of ontology discussed in the previous section, but also syntactic and structural heterogeneity as well. The operational use of ontology and information mediation in the design and implementation of database systems with different data sharing objectives and requirements is discussed in Sections 4 and 5.

4.   DATA WAREHOUSING

There are two major approaches to implementing a data sharing strategy in the database environment, namely, data warehousing and database federation. The former seeks to merge data physically from several sources, whereas the latter emphasises simultaneous on-line access to multiple data sources. This section explains the concept and methods of data warehousing, and introduces advances in spatial database technology that aim to improve spatial data sharing using this approach. The nature and purposes of database federation are explained in Section 5. It is important to understand at the outset that although these two approaches are treated separately in this book, they are not mutually exclusive technologies. Both of them are dependent on the use of the techniques described in the previous section to address the problems of heterogeneity in databases, albeit in different ways. A data warehouse can be set up as an integral part of a federated database system, and serves both as a supplier to and consumer of data from other members of the same database federation. A comparative analysis of these approaches to spatial data sharing is provided in Section 5.3.

4.1   The Definition and Characteristics of Data Warehouses

A data warehouse is a special type of database system that is set up and maintained separately from an organisation's operational databases that support its daily business. There are three different approaches to implementing a data warehouse, namely a central repository, an enterprise data warehouse, and data marts. Although these approaches are all designed to facilitate data sharing, they have different purposes and systems architecture that are explained in the following sections.

4.1.1   Central Data Repository

A central data repository is a data warehouse that stores common reference data sets required by different applications running on different computers. Such data sets are typically relatively stable over time, too voluminous to be stored repeatedly on different servers, and contain data elements required by all of the intended applications of an organisation. Digital topographic databases maintained by government mapping agencies are good examples of this class of data warehouses. Once set up, a digital topographic database provides a common geo-referencing framework for spatial databases in the public, private and academic sectors. The use of a common topographic base not only saves users valuable resources required for the collection, storage and management of the data, it also helps to eliminate the problem of *spatial heterogeneity* (that is, positional discrepancies among corresponding spatial objects in different data sets) in spatial data sharing.

4.1.2   An Enterprise Data Warehouse

An enterprise data warehouse is a repository of data derived from operational data sources within an organisation. This class of data warehouse was developed from the realisation that useful information on issues such as changing land use patterns and trends and consumer purchasing behaviour can only be obtained adequately by cross-functional and integrated analysis of data from all relevant data sources. From a data sharing perspective, therefore, the building of an enterprise data warehouse is seen as an intermediate step to enable the use of enterprise-wide data for senior management or executive levels of decision making. In other words, an enterprise data warehouse is more a means to an end, rather than an end in itself.

This form of data warehouse typically contains hundreds of gigabytes or even terabytes of data. Before the data sets are entered, it is necessary to "clean" them in order to remove any syntactic, structural and semantic heterogeneity that may exist. It is also necessary to transform the data sets into a *multi-dimensional database model* that is designed to facilitate the analytical use of constituent data by *on-line analytical processing* (OLAP) and *data mining* applications (see Section 4.2). This means that the building of an enterprise data warehouse is a very complex task that requires months or even years of project planning, business modelling, systems design and implementation. As a result, the idea of developing smaller, scaled-down and more manageable warehouses called *data marts* was proposed.

4.1.3   A Data Mart

A data mart is a subject-specific data warehouse that is usually set up to meet the information needs of users of a particular department or functional unit within an organisation. The size of a data mart, therefore, is generally many times smaller than an enterprise data warehouse. The implementation cycle is likely much shorter as well.

A data mart can be implemented using a top-down or bottom-up approach. In the former, which is called a *dependent data mart*, data is drawn directly from an enterprise data warehouse. In the latter, which is called an *independent data mart*, individual data marts are built by capturing and transforming data from existing local operational databases in a department or business area. Occasionally, data can be purchased from external sources. Many organisations use independent data marts as building blocks for the construction of an enterprise data warehouse because, as noted above, the implementation cycle of a data mart is much shorter. By using such an incremental approach to data warehousing, it is possible to see the benefits more quickly and to spread the capital costs of implementation over several years. However, it may require complex integration in the long run if not enough consideration is given to corporation-wide data sharing during the development of the data marts.

4.1.4   Key Features of a Data Warehouse

Of the three classes of data warehouse, enterprise data warehouses are the most important and the most complex to implement in the database world. Coincidentally, the concepts and techniques pertaining to enterprise data warehouses also apply largely to central data repositories and data marts. Hence, discussion of enterprise data warehouses also covers the issues relevant to the other forms noted above. Unless stated otherwise, references to data warehouse in the discussion below imply a typical enterprise data warehouse.

Inmon (2002) identified four key features of the data warehouse environment, namely:

- *Subject-oriented.* A data warehouse is organised around major applications of an organisation. For a transportation company, for example, the major subject areas for a data warehouse may include customer orders, fleet management, human resources management, and financial management.
- *Integrated.* A data warehouse is usually built by integrating data from multiple heterogeneous data sources. The integration aspect is practically the most important of all of the four features both conceptually (integrating operational database models into a single corporate data warehouse database model) and technically (transforming and loading a large number of heterogeneous data sets on to a central data repository from physically distributed database servers).
- *Time variant.* The content of a data warehouse is drawn from different sources and at different times, which implies that all data sets pertain to, either implicitly or explicitly, a time element or dimension. This time element is called a time stamp, which applies to both individual records and individual transactions within the database itself.
- *Non-volatile.* A data warehouse is a separate physical construct from the data sources which its content is drawn from and created. Data in a warehouse are usually loaded en mass. They are accessed mostly by read-only applications that do not result in any change to the data warehouse content. When updating of the data is required, it must be done by reloading from the original data sources. Each update of the data warehouse content is a snapshot of the data sources in a static form and it is recorded in the history table of the warehouse database.

4.1.5   Differences between Data Warehouses and Operational Databases

Data warehouses and operational databases serve relatively distinct but complementary purposes in supporting the information needs of an organisation. In practice, these two types of systems must be set up and operated separately because of the following conflicts between them:

- *Conflicting processing objectives.* Databases are implemented to support the daily business operation of an organisation. They are designed for well-defined tasks and optimised for high availability (that is, they can be accessed by hundreds users simultaneously) using indexing methods and pre-established OLTP techniques. On the other hand, a data warehouse is designed to perform complex analysis of a large amount of legacy data using OLAP, as noted above. As the structure of data in a warehouse is optimised for high performance and is transaction-oriented, processing is not suitable for complex ad hoc queries and it is necessary to transform operational data into a structure suitable for fast scanning to detect hidden patterns and trends in OLAP and data mining.

- *Conflicting processing requirements.* Read/write processes in operational database systems require concurrent control and recovery mechanisms (see Chapter 2) to ensure consistency and integrity in transaction processing. Such mechanisms are designed for short and simultaneous read/write access to the database. If they are applied to long read-only OLAP processes, they will seriously jeopardise the high availability requirement of transaction processing by locking the data in use by the OLAP processes for an extended period of time.
- *Conflicting data requirements.* Data warehouses are designed for decision making use which requires data that are clean and that pertain to different points in time. Such data must be captured periodically over time, verified, and consolidated before they are loaded into the warehouse database. Data in ordinary operational databases are raw business data that may contain errors, may possibly be incomplete and may represent only a snapshot of the current state of the value of data. Using operational data, therefore, is likely to increase the degree of uncertainty in decision processes. Also it is not possible to obtain time-dependent information from such data.

4.2   Architecture of a Data Warehouse

Different approaches can be used to implement a data warehouse. The three-tier architecture in Figure 6.12 comprises the following components:

- *Data warehouse server.* This server, which is almost always configured using a relational database system, stores cleaned and transformed data that are extracted from the operational databases and external sources using a multi-dimensional data model. The communication between the data warehouse and the data sources uses database connectivity standards such as those introduced in Section 2.4. The data warehouse server also contains a metadata database that keeps information about the content of the data servers, as well as the database administration protocols and procedures for the management of the data warehouse.
- *OLAP server.* The OLAP server maps user queries to the stored data in the data warehouse server. There are two basic methods of implementing an OLAP server: (i) *a relational OLAP* (ROLAP) approach is an extended relational DBMS that maps standard relational operations to multi-dimensional data, and (ii) *a multi-dimensional OLAP* (MOLAP)

approach which is a special-purpose server that is capable of directly performing multi-dimensional operations.

- *Client applications.* The front-end interface of the data warehouse includes client applications such as query and reporting tools, data analysis tools and data mining tools. Of these three classes of tools, data mining tools, are particularly important because they provide the underlying technology for the detection of hidden patterns and trends in the data.

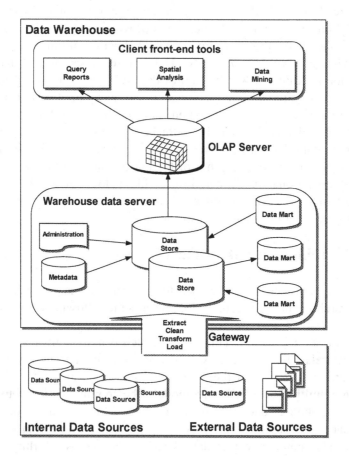

Figure 6-12. The architecture of a data warehouse

A key concept in the architecture of a data warehouse is the *multi-dimensional data model* that views data in the form of a *data cube* (see Figure 6.12 OLAP server). In data warehouse terminology, a data cube is a special data model that allows real world facts and features to be represented from several perspectives or dimensions. The concept of a multi-dimensional data model is illustrated more explicitly in Figure 6.13. This example shows the data model of a crime statistics data warehouse that is designed to keep track of the occurrence of different categories of crimes in the jurisdiction of different police divisions, during different time periods in a day, and in different months of a year.

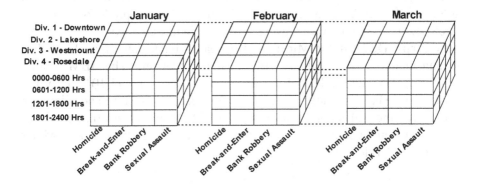

Figure 6-13. A Basic multi-dimensional data warehouse model

Hence, this data warehouse model is represented by a four dimensional data cube with each dimension having a specific *dimension table* that can be specified manually by subject experts, or automatically based on the data table structures of the contributing data sources. Such a data cube can be extended by adding further dimensions such as, for example, age and gender of offenders, convictions, and sentencing results. In practice, it is not uncommon for commercial data warehouses to be modelled by a data cube of ten or more dimensions.

4.3   Advances in Spatial Data Warehousing Technology

The ability to organise and store data in a data warehouse for subsequent multi-dimensional analysis makes them an excellent tool for spatial data applications that require the use of data from multiple disparate data sources. Spatial applications such as global climate change, environmental impact analysis, health care planning, crime analysis, and homeland security all rely on the use of a combination of data sources that are likely to be housed in geographically distributed data servers belonging to different organisations. It is practically impossible to perform senior level strategic decision making without the use of a data warehouse first to gather, clean and transform operational databases into a single well-structured and subject-oriented data repository.

The software industry has responded to the growing needs for data warehousing capacity in the spatial database environment by spatially enabling their conventional DBMS products. Oracle Corp., for example, has developed an extension called Spatial Data Option (SDO) that can be integrated with conventional GIS software products such as ESRI's Arc Spatial Database Engine (ArcSDE) to build spatial data warehouses. Similarly, IBM has enabled its customers to build spatial data warehouse solutions by combining conventional GIS software products, such as ArcSDE, with its data warehousing software modules, known as DB2 OLAP Server and DB2 OLAP Miner, and a spatial extension to its DB2 data management system, called DB2 Spatial Extender. The IBM spatial solutions also include IBM Informix Dynamic Server and IBM Informix Spatial DataBlade Module, which have an open architecture and allow the use of commercial mapping tools from ESRI, MapInfo and Intergraph, as well as custom applications provided by software developers such as Miner & Miner and Telcordia.

5.   FEDERATED DATABASE SYSTEMS

The concept of database federation was proposed around 1990 as a solution to the problem of database heterogeneity. The objective of database federation is to create a database architecture that provides uniform and simultaneous access to several heterogeneous data sources. Just like data warehousing, there are different approaches to database federation, leading to different systems architectures that have very little in common other than their name and the objective of solving the same problem. The following discussion briefly introduces the concept of database federation and explains the most commonly adopted database federation architecture currently in use. Since both data warehousing and database federation were developed to facilitate data sharing within and outside of an organisation, this chapter concludes with a comparison between them.

5.1   Approaches to Database Federation

There are three general approaches to database federation, namely:

- *Tight Database Federation.* A typical tight database federation architecture as proposed by Sheth and Larson (1990) makes use of a unified schema, also called an integrated or federated schema, as the access interface to member data sources of the federation. This unified schema can be built through an automated or semi-automated process whereby schemas of individual databases are progressively filtered and integrated semantically. Although this approach is conceptually simple, the process is by no means easy. Further, as the approach is based on the integration of database schemas, it is not useful for data that are not schematically organised (that is, semi-structured or non-structured data). From a data sharing perspective, tight database federation is not a useful approach and consequently it is seldom used.

- *Loose Database Federation.* Loose database federations do not offer a uniform schema for queries against the federation. However, they offer a uniform query language called multi-database query language (MDBQL) which abstracts from the query languages of the components and hides technical and language heterogeneity (Litwin, 1990). The limitation of the loose federation approach is that it can only be used if the data sources themselves offer query language access. This means that this approach is not applicable to semi-structured and non-structured data and its use as a solution for data sharing is rather restricted.

- *Mediated Database Federation.* This approach of database federation is used by IBM as a standard solution to the problems of database heterogeneity (Haas et al., 2002). Mediated database federation, as its name implies, is based on the principles and techniques of information mediation as explained above in Section 3.3. Within a mediated database federation, the data sources are federated by connecting together into a unified system using a special database management system called a *federated database system* (FDB). The FDB shields database users from the need to know what the sources are, where they are stored, how they are modelled and managed, what hardware and software they run on, and how they are accessed (that is, through what programming interface or language). In this way, a database federation appears to the users as a single database system. The users can search for information and manipulate data using the full power of SQL. It is possible to use a single query to access data residing in different data sources that include relational and object-relational databases, as well as other forms of semi-structured and non-structured data files, such as flat files, XML files, and graphic image files.

5.2   The Architecture of a Federated Database System

Figure 6.14 shows the architecture of a typical database federation following the IBM model. The architecture contains:

- The FDB is an ordinary DBMS that is configured to serve as the control centre of the database federation. Client applications communicate with the FDB using any supported database connectivity interfaces such as ODBC,

JDBC, CORBA and DCOM, as explained in Section 2.4. At the heart of the federated database system is the *federated database server*, which is in effect the mediator in the generic information mediation paradigm as explained in Section 3.3. It plays the same role to optimise the queries received from client applications and then dispatches the resulting query segments to appropriate wrappers using the information contained in the universal catalogue.

- *Data sources* of a database federation may include structured data in relational and object-oriented databases, geometric and attribute data in spatial databases, as well as a myriad of semi-structured and non-structured data sets such as XML programs, flat files and image files. A data warehouse can also be connected to a database federation.
- *Client applications* of a database federation include queries on data as well as requests for services (that is, database operations including transaction processing) on remote computers. This means that unlike client applications in data warehouses that are almost all read-only processes, client applications in a database federation can use both read-only and read/write processes in any data sources.

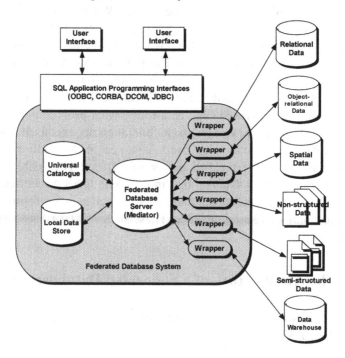

Figure 6-14. Typical database federation architecture based on the IBM model

Overall, database federation provides a conceptual framework and practical approach for data sharing among disparate heterogeneous data sources. This approach is transparent to the users and is scaleable with the aid of the federated database server and wrappers. The spatial data extensions of conventional database systems noted in Section 4.3 can also be usefully deployed in the building of database federations. Users can take advantage of these spatial data extensions to combine spatial and non-spatial data within a database federation to enhance their understanding of business and their ability to make more informed decisions by leveraging the value of existing data from different perspectives.

5.3 A Comparison between Data Warehousing and Database Federation

Data warehousing and database federation are two prevailing approaches to data sharing by database interoperation. Both approaches have strong support in the software industry and a number of standards and database management tools have been developed to assist in their implementation. Through partnerships between leading database software vendors and conventional spatial database software vendors, spatially enabled software extensions are now available for the construction of powerful spatial data warehouses and database federations that include one or more spatial data sources.

From the perspective of spatial database design and implementation, both approaches have their advantages and limitations. Which approach is better suited for a particular organisation depends on the interplay of a variety of factors, as shown in Table 6.2. In addition to the six categories of difference listed in the table, six further distinguishing characteristics of the two approaches can be summarised as follows:

- *Systems characteristics*. Data warehousing is based on a centralised data repository architecture which has a data supplier and data user relationship between the data sources and the data warehouse. Database federation, on the other hand, is a configuration of distributed and heterogeneous data sources that are connected by a federated database server.
- *Data characteristics*. Data in a typical data warehouse are legacy data loaded from different operational data sources. They are relatively free of error because data cleaning is an essential step before the transformation of the data to a specific model to facilitate their use in high-level decision making. A data warehouse normally contains a huge amount of data pertaining to different points in time, and the data tend to

remain unchanged for relatively long time periods. In contrast, data in a database federation include both legacy and operational data. Within individual data sources in a database federation, all data sets keep their original data models and structures. Data are mapped to a common syntax, structure and semantics on-demand as they are used. Generally speaking, the amount of data in a database federation is much smaller than that in a data warehouse.

- *Database characteristics.* A data warehouse is designed to support decision making by analysing large amounts of legacy data. Thus the database is optimised for high performance. Since practically all processes within a data warehouse are based on read-only access, there is no need for common database functions such as concurrent control, and the output is largely concerned with summarised views of the results of information analysis. A database federation is designed to support interoperation among distributed heterogeneous data sources. High availability is as important as high performance in a database federation. Inter-system or cross-platform transaction processing is an important function of this class of database systems. As a result, a database federation requires all the functions of an ordinary database system, and it is often configured as such.

- *Application characteristics.* A data warehouse is used primarily for subject-oriented information analysis and decision support. Its application focus is typically well-defined and is more concerned with high-level strategic planning than with the daily operation of an organisation's business. These aspects of the functions of a data warehouse are in sharp contrast to the applications of a database federation which is characterised by concurrent access, cross-functional and cross-platform collaboration processes in a wide range of application areas.

- *Data sharing characteristics.* A data warehouse facilitates "vertical" data sharing between operational data sources and decision support systems within an organisation. In contrast, a database federation seeks to improve data sharing between distributed heterogeneous data sources. This form of data sharing largely "horizontal" among members of an information community.

- *Spatial data sharing characteristics.* A spatial data warehouse serves a dual purpose, namely (i) as a repository of framework or reference spatial data that can be used by other spatial data sets, and (ii) as a data source for high-level spatial analysis and spatial decision support. Since a database federation is not necessarily subject-oriented, the need to implement this approach does not arise unless heterogeneous data sources are required by an organisation. A spatial data source can always be incorporated into a database federation, and interoperate with other data sources in a collaborative. This interoperability makes a federated approach very flexible in the sense that data stores can be added to the federation as required by an end user for a given task.

Table 6-2. Characteristics of spatial data warehouses and a federated database system

Characteristics	Data Warehouses	Federated Database Systems
General Description	A collection of subject-oriented data in a well-defined and tightly structured repository	A configuration of geographically distributed, autonomous and heterogeneous data sources and services, communicating using a standard protocol over a network either through schema integration or information mediation
Typical Systems Architecture	Central data server with distributed clients	Distributed data servers with distributed clients
Data Processing Characteristics		
- Local Autonomy	- High	- High
- Concurrent Access	- Easy to control	- Complex to control
- DB Size	- 100 Gigabytes to terabytes	- 100 Megabytes to Gigabytes
- Scalability	- Low	- High
- Modularity	- Low	- High
- Security	- Easy to enforce	- Difficult to enforce
- DP Overhead	- High	- Low
- DP Function	- Subject-oriented	- Application-oriented
- Access/Query	- Mostly read	- Read/write
- View	- Summarised	- Detailed
- Priority	- High performance	- High availability
Network Requirements	Generally high	Generally low
Interoperability Strategy	Pre-computed, data-oriented to merge data physically from several data sources	On-demand, application-oriented strategy to enable simultaneous queries on several data sources on-line
Applicability Scenarios	Relatively small number of structured core data sets, e.g. framework spatial data sets	Large number of distributed and heterogeneous data sources with structured, semi-structured and unstructured data
Application Focus	Subject-oriented data dissemination and OLAP in support of data mining and decision making	Application-oriented distributed services and OLTP in support of business operations
Spatial Database Application	Global, national, state and local reference data, spatial data mining and decision support, multi-dimensional (space, time and attributes) analysis and modelling	Collaborative spatial data analysis and modelling using multi-format and multi-media data, spatial and non-spatial data integration

6.   SUMMARY

This chapter has covered some of the most fundamental and important concepts underlying modern database systems, as most large scale databases have data sharing at the heart of their design and implementation. The chapter commenced with a comprehensive and practical definition of spatial data sharing within the concept of database interoperability at the data, application and business process levels of operation.

Following the discussion of data sharing and its attendant and compelling advantages for database development, the important problem of database heterogeneity was discussed. The concept of ontology was introduced and explained as a means of solving problems created by the incoherency that results from poorly specified and agreed to approaches to semantic modelling within the many database communities that exist even within the global spatial data domain.

The last two sections discussed data warehouses and a relatively new concept, federated database systems, introduced roughly a decade ago to address the problem of database heterogeneity. Data warehouses and data marts were first defined and then their functions and architectures were described. Database federations were then discussed and comparisons were drawn between them and data warehouses. Mastery of the material presented in this and the preceding chapters is a prerequisite for the study of the implementation and management of spatial database projects in Part 3.

7.   REFERENCES

Abel, D.J., Kilby, P.J. and Davis, J.R. (1994) "The Systems Integration Problem", *International Journal of Geographical Information Systems*, Vol. 8, No. 1.

Auxilio, M. and Nieto, M. (2003) *An Overview of Ontologies*, Technical Report, Center for Research in Information and Automation Technologies, Universidad De Las Américas Puebla, Puebla, Mexico.

Boehnlein, M. and Ende, A.U. (1999) "Deriving Initial Data Warehouse Structures from the Conceptual Data Models of the Underlying Operational Information Systems", *Proceedings*, ACM Second International Workshop on Data Warehousing and OLAP (DOLAP), Kansas City, MO.

Buehler, K. and McKee, L. (1996) *The OpenGIS Guide: Introduction to Interoperable Geoprocessing*, Wayland, MA: Open GIS Consortium.

Chung, J.-Y., Lin, K.-J. and Mathieu, R.G. (2003) "Web Services Computing: Advancing Software Interoperability", *Computer* (IEEE Computer Society), October-2003, pp. 35-37.

Craglia, M and Masser, I. (2003) "Access to Geographic Information: A European Perspective", *URISA Journal*, vol. 15, APA I: Special Issues on Access and Participatory Approach in Using Geographic Information, pp. 51 – 59.

Craglia, M and Onsrud, H. (2003) "Workshop on Access to Geographic Information and Participatory Approaches in Using Geographic Information: Report of Meeting and Research Agenda", *URISA Journal*, vol. 15, APA II: Special Issues on Access and Participatory Approach in Using Geographic Information, pp. 9 – 15.

Dodge, G and Gorman, T. (2000) *Essential Oracle8i Data Warehousing*, New York: John Wiley & Sons.

ESRI (1998) "Spatial Data Warehousing", An ESRI White Paper, Redlands, CA: Environmental Systems Research Institute, Inc.

ESRI (2002) "What is g.net?", An ESRI White Paper, Redlands, CA: Environmental Systems Research Institute, Inc.

ESRI (2003a) "Spatial Data Standards and GIS Interoperability", An ESRI White Paper, Redlands, CA: Environmental Systems Research Institute, Inc.

ESRI (2003b) "The Next Wave of Interoperability via Web Services", *ArcUser*, April-June, 2003.

Foote, Paul Sheldon and Malini Krishnamurthi (2001) Forecasting using data warehouse model: Walmart's experience, *The Journal of Business Forecasting Methods and Systems*, 20, 3, 13 - 17.

GeoConnections Secretariat (2001) *The Canadian Geospatial Data Infrastructure – Access Technical Services Manual*, Version 1.1, Ottawa, ON: GeoAccess Division, Natural Resources Canada.

Goodchild, M.F., Egenhofer, M.J. and Fegeas, R. (1997) *Report of a Specialist Meeting Held Under the Auspices of the Varenius Project*, National Center for Geographic Information and Analysis (NCGIA), Santa Barbara, CA.

Guarino, N. (Ed.) (1998) *Formal Ontology in Information Systems* (Proceedings of FOIS '98, Trento, Italy), Amsterdam, the Netherlands: IOS Press.

Han, J. (1999) "Data Mining" in Encyclopedia of Distributed Computing by Urban, J. and Dasgupta, P. (Eds.), Boston, MA: Kluwer Academic Publisher.

Inmon, W.H. (2002) *Building the Data Warehouse*, 3rd ed., New York: John Wiley & Sons.

Intergraph Corporation, Laser Scan, Autodesk and MapInfo (2003) "Open Interoperability with Oracle Spatial Technology", White Paper, Huntsville, AL: Intergraph Corporation.

ISO (2000) *Geographic Information Reference Model* (ISO/DIS 19101), Geneva, Switzerland: International Organization for Standardization.

Jhingran, A.D., Mattos, M. and Pirahesh, H. (2002) "Information Integration: A Research Agenda", *IBM Systems Journal*, Vol. 41, No. 4, pp. 555 – 562.

Kimball, R., Reeves, L. and Ross, M. (1998) *The Data Warehouse Life Cycle Toolkit: Expert Methods for Designing, Developing, and Deploying Data Warehouses*, New York: John Wiley & Sons.

Koperski, K. and Han, J. (1996) "Data Mining Methods for the Analysis of Large Geographic Databases, Proceedings, 10th Annual Conference on GIS, Vancouver, BC.

Nebert, D. D. (2001) *Developing Spatial Data Infrastructures: The SDI Cookbook*, Reston, VA: The Global Spatial Data Infrastructure (GSDI) Secretariat.

OGC (2003a) *Open GIS Reference Model* by Buehler, K. (Ed.), Wayland, MA: Open GIS Consortium Inc.

OGC (2003b) *Open GIS Web Mapping Server Cookbook* by Kolodziej, K. (Ed.), Wayland, MA: Open GIS Consortium Inc.

Roth, M.A., Wolfson, D.C., Kleewein, J.C. and Nelin, C.J. (2002) "Information Integration: A New Generation of Information Technology", *IBM Systems Journal*, Vol. 41, No. 4, pp. 563 – 577.

Roth, M.T. and Schwarz, P. (1997) "Don't' Scrap It, Wrap It! A Wrapper Architecture for Legacy Data Sources", *Proceedings of the 23rd VLDB Conference*, Athens, Greece.

SEI (Systems Engineering Institute) (1997) *C4 Software Technology Guide: A Prototype* by Foreman, J.T. (Mgt. Ed.), Pittsburgh, PA: Systems Engineering Institute, Carnegie Mellon University. (With updated information at http://www.sei.cmu.edu/technology/str/)

Sheth, A.P. (1999) "Changing Focus on Interoperability from System, Syntax, Structure to Semantics", in Interoperating Geographic Information System by Goodchild, M.F., Egenhofer, M.J., Fegeas, R. and Kottman, C.A. (Eds.), Boston, MA: Kluwer Academic Publisher.

Sheth, A.P. and Larson, J.A. (1990) "Federated Database Systems for Managing Distributed, Heterogeneous, and Autonomous Database Systems", *ACM Computing Surveys*, Vol. 22, pp. 183-236.

Visser, U. and Struckenschmidt, H. (2002) "Interoperability in GIS – Enabling Technologies", *Proceedings of 5th AGILE Conference on Geographic Information Science*, pp. 291-297, Palma de Mallorca, Spain.

Wiederhold, G. (1992) "Mediators in the Architecture of Future Information Systems", *IEEE Computer*, Vol. 25, No. 3.

PART 3

SPATIAL DATABASE IMPLEMENTATION AND PROJECT MANAGEMENT

Chapter 7

USER EDUCATION AND LEGAL ISSUES OF SPATIAL DATABASE SYSTEMS

1.   INTRODUCTION

The implementation of spatial database systems is concerned with much more than data, hardware and software. Human and non-technical factors sometimes play a more crucial role in making or breaking a spatial database project. This chapter serves two purposes relating to these factors. First, it provides an overview of the objectives and processes of setting up a user education and support plan as part of a spatial database implementation project. Today's spatial information users come from a broad range of disciplines and have different academic and professional backgrounds. Many of them do not have the same level of understanding of spatial data and concepts as traditional users of spatial data such as geographers, cartographers, surveyors, planners and resource managers. At the same time, spatial database systems are now generally much more complex to implement and use than conventional GIS, and applications are generally more sophisticated. Hence, user education is more important than ever before for organisations using spatial information technology.

The second purpose of this chapter is to explain the laws and regulations that apply, either directly or indirectly, to the implementation of spatial database systems. While user education is not directly related to this theme, there are important and subtle indirect linkages that involve educating data providers and data users as to the potential legal repercussions of improper data use. Laws and regulations are jurisdiction-specific by nature. It is practically impossible to cover the applicable laws of all countries comprehensively within the scope of this chapter. Thus, legal issues are approached primarily from a project management perspective, focusing on the nature and impacts of applicable laws and regulations as they apply to the implementation and use of spatial database systems, rather than on the content of individual statutes and regulations that may apply in one country but not in another. Where necessary, applicable Canadian and American legislation and regulations are used to illustrate and substantiate the discussion.

2.   USER EDUCATION IN SPATIAL DATABASE IMPLEMENTATION

This section explains the objectives and methods of a systematic approach to user education in support of the efficient operation of spatial database systems and the effective use of spatial information. First, the nature of user education is discussed from a project management perspective, with special reference to issues resulting from the diverse needs of different members of a user community. Following this, the concepts and techniques are explained that can be used to address these issues in practice.

2.1   The Nature of User Education from a Project Management Perspective

In order to understand the nature of user education from a project management perspective, it is necessary first to define who a "user" is and the scope of "user education" in the context of a typical spatial database implementation project.

The term *user* in this book includes not only the people who actually use the data in a spatial database, but also those who support the operation and use of the database. Four categories of users, who play relatively distinct roles with different requirements for education, can be identified for the purpose of developing a user education program. These categories include the following:

- *Project sponsors*. This category includes executives, senior managers and section heads who, by virtue of their position in an organisation, make strategic decisions regarding the direction of spatial information technology use and the allocation of resources required for spatial database projects. These individuals do not normally use a spatial database on a daily basis. Rather, they rely on regular summary business performance reports generated from the database for executive and managerial business decision making.

- *Systems staff.* These users include database administrators, business solutions consultants, system analysts and application programmers who provide technical expertise to operate and maintain the database system. They are there to support the use of the database by the following two categories of users, who are sometimes specifically referred to as end users.
- *Production and professional users.* These users of a database system require frequent access to the database. They include people who use the spatial database system for transaction processing, such as data entry clerks in a land registrar's office, and people who rely on the spatial database as the primary source of information to fulfill their job requirements, such as surveyors, engineers, planners, resource managers, business analysts, researchers and scientists.
- *Occasional users.* This category of users includes members of the general public who require access to the spatial data from time to time for a wide variety of purposes such as looking for driving directions, locating a friend's house or a business location, and querying an on-line land parcel database. Occasional users require spatial information primarily to inform and to enlighten themselves. This distinguishes them from both production and professional end users who use spatial information routinely as part of their jobs.

Table 7.1 shows how user education in spatial database systems can be generally divided into four levels or types of education and how each of these is related to the four categories of users noted above. Generally speaking, organisational user education is intended primarily for executives and managerial-level employees. The primary purpose is to ensure that these individuals fully understand the importance of a spatial database system as it pertains to the mission and goals of an organisation. This form of education clearly differs from the operational knowledge that is required to design, implement and maintain a spatial database system.

User education at the organisational level, which is delivered mainly by regular briefing notes and presentations at senior management meetings, is important to secure long term commitment and support of an organisation to the use of spatial information technology. Although the target audience is normally small, the impact of this form of education is often far-reaching in its implications for a spatial database system as it affects the direction of and the commitment to using spatial information as part of the organisation's business functions.

Table 7-1. Types of user education

Types of User Education	Description/ Purposes	Target Audience	Methods of Delivery
Organisational	To secure long-term commitment and support of corporate executives and senior managers by keeping them up-to-date on the relevance of emerging technology to the mission and goals of the organisation	Project Sponsors	Regular briefing notes and presentations at management meetings
Occupational	To provide or enhance short and long-term skill requirements to support the operation	Systems staff, and production and professional end users	Educational and technical training programs and courses at tertiary institutes, product-specific training, conferences, seminars and workshops, on-the-job training and mentoring
Individual	To provide or enhance immediate skill and knowledge requirements of individual members of systems staff and professional users	Systems staff, and production and professional end users	Educational and technical training programs and courses at tertiary institutes, product-specific training, conferences, seminars and workshops, on-the-job training and mentoring
Popular	To keep the general public aware of the availability and potential use of existing spatial databases	The general public	Mass communications media including broadcasting, brochures, spatial data clearinghouses, Internet portals

Both the occupational and individual types of user education in Table 7.1 are intended primarily for systems personnel and professional users. These two types of user education are commonly referred to as *user training* because of the focus on the acquisition of specific technical skills. They are also commonly delivered using similar formats that include formal courses at universities and colleges, product-specific training provided by hardware and software vendors, and conferences and seminars sponsored by professional and trade organisations. In the case of individual user education, it is possible to use an approach of on-the-job training and mentoring when the number of users requiring education is small. However, when the number of users is relatively large, more than a minimum of ten, then training is more likely to be in the form of a block course, using a vendor supplied generic curriculum and generic data.

There are, however, relatively clear distinctions between occupational and individual user education. Whereas occupational user education aims to raise the general skill and knowledge level of specific user groups within an organisation in order to meet the short- and long-term operational and application requirements of a database system, individual user education is more concerned with providing specific users with the skills and knowledge that are required to address some immediate or short term need of their database system. In practice, it is often preferable to separate occupational user education from individual user education because of the diversity of each user's academic background and work experiences. Hence, the variation in the needs for user education, even within a relatively small group of users, is likely to be diverse. Requiring every member of a user group to undertake the same educational program regardless of their individual needs is not necessarily a wise and prudent way of using valuable financial resources. Sometimes it can also be a hindrance to the effective delivery of user education programs.

Popular user education differs from the previous three types of user education in its intent and purpose. In essence, popular user education has no exclusive target audience but is open to anyone interested in spatial data and spatial databases. Popular user education programs are designed primarily to raise public awareness of the availability of and services provided by existing database systems. They are delivered through different mass communications media including broadcasting, brochures distributed through schools, libraries and community information centres, spatial data clearinghouses and other types of Internet portals.

User education from a project management perspective is a continuing capacity building process that seeks to ensure that all users are fully equipped, both technically and intellectually, to play their respective part in implementing and using spatial databases effectively. However, launching a user education program is not simply a matter of sending the users concerned to a particular course of study offered by a university, college or software vendor. To be effective, a user education program must be carefully researched and planned as part of the project planning process. Its objective must remain focused and be directed to address the specific needs of users, both individually and as a group, according to the immediate, short-term and long-term requirements of the systems development life cycle (SDLC). This is basically a *needs-based approach* to user education in project management and it is explained in the next section.

2.2   The Concept of Training Needs Assessment

The central concept of needs-based user education is *training needs assessment* (TNA), which is also commonly called *training needs analysis*. Witkin and Altschuld (1995, p. 4) define TNA as "a systematic set of procedures undertaken for the purpose of setting priorities and making decisions about program or organisational improvement and allocation of resources. The priorities are based on identified needs". The word *needs* in the context of a TNA generally refers to the discrepancy or gap between what is the users' current level of competency (which is defined as the knowledge, skills and ability of a user to perform a task or solve a problem, see for example Gaudet et al. (2003) for a detailed explanation) and what ought to be their level of competency in order perform their prescribed job functions proficiently and competently.

Figure 7.1 shows the process of TNA. In a TNA, the existing level of competency is determined by using one or more of the assessment methods to be described in Section 2.3. During the assessment, the assessor identifies the tasks that subject users perform on their job, how frequently individual tasks are performed, and how confident the users are in performing each task. The same information collected is used to establish the required level of competency by considering it together with additional information such as a user's job specifications, competency standards established by professional bodies, and industry and academic reports. A comparative analysis between the existing and expected levels of competency will generate the collective training needs of the users. If a comparison is made between the expected level of competency and the data generated for an individual user, the training needs of that individual will be identified.

As noted in the above definition, the concept of needs in TNA also includes the determination of the nature and causes of the needs for user education, as well as the prioritisation or timing of providing user education programs. These factors can be determined by further analysis of the assessment data collected, together with input from supervisors, the project manager and, if necessary, carrying out a supplementary survey.

The principles and techniques of training needs analysis are widely used in professional training settings in health care, engineering, manufacturing, human resource management and business. Typically, training needs arise when one or more of the following situations occur:

- New people are hired by the organisation.
- Existing members of staff are given new job assignments.
- New working methods, including new technologies, are introduced.
- New applications are implemented.

- There is a major upgrade of a software version.
- Lack of productivity is detected.
- Higher standards of job performance are required.

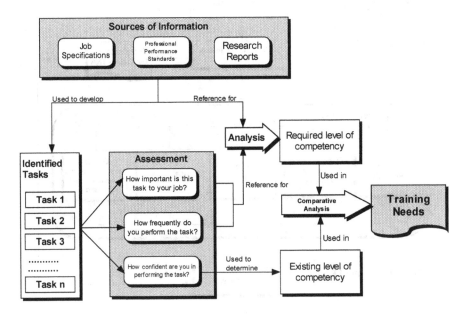

Figure 7-1. A three-phase approach to training needs assessment

Each of these situations will result in people doing different and new things, or doing the same things as before but in new and different ways. All people whose jobs are affected by change will require some form of user education to help them through the transition from the old to the new working environments. Hence, the general principles and techniques of TNA will apply invariably to the planning and development of user education programs in all spatial database projects independent of their scope and impact on the organisation.

2.3   The Method of Training Needs Assessment

There are different ways of carrying out a TNA. Figure 7.2 and Table 7.2 show a three-phase approach based on the conceptual framework originally proposed by Witkin and Altschuld (1995). This approach comprises a sequence of activities including pre-assessment (scoping and background information gathering), assessment (data collection), and post-assessment (data analysis and communication). The concepts and methods behind the

major activities and tasks that occur at each of these phases are explained in
the following discussion.

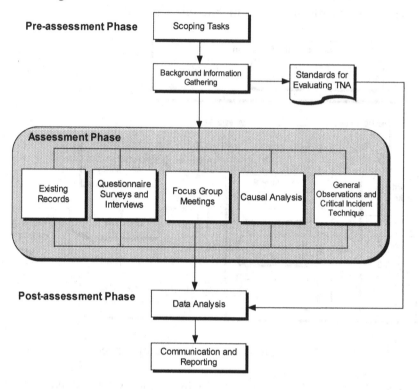

Figure 7-2. A three-phase approach to training needs assessment

2.3.1   The Pre-assessment Phase: scoping and background
             information gathering

The major tasks in this phase aim to define the scope of the TNA by
identifying who needs to be trained, at what level, when the training must be
completed by, and where training can be obtained. In order to answer these
scoping questions, it is prudent to establish supervisor's expectations of their
staff, and to gather background information such as each user's job
specifications, the opinions of subject matter experts within and outside of
the organisation, TNA reports of peer organisations, industry manpower
research reports, demographic data, and competency standards set out by
professional bodies. The information collected is used to produce an overall
TNA plan for the second and third phases, and a collection of proposed job
performance standards and required skill levels for evaluating the results of
the assessment.

Table 7-2. A three-phase approach to training needs analysis

	Training Needs Assessment Phases		
	Pre-assessment (Scoping and Planning)	*Assessment (Data Collection)*	*Post-assessment (Data Analysis and Communication)*
Major Activities/tasks	o Scoping general purpose of TNA o Identify major areas of needs and issues o Researching existing information regarding identified areas of need and issues o Identify data to be collected, sources, sampling methods and potential uses of data o Summarise findings of the above tasks in a TNA plan	o Collect data on needs using one of the methods shown in Figure 7.1 o Perform preliminary data analysis	o Refine expected level of competency o Determine training needs o Prioritise training needs o Explore causes of training needs o Develop action plans o Consider alternatives o Communicate results to project sponsors and other stakeholders
Outcomes	o Preliminary plan for Phases 2 and 3 o Proposal for the required level of competency	o Competency model o Preliminary action plan to implement user education according to identified priorities	o Action plans to implement a needs-based user education and support structure o Write TNA report

2.3.2   The Assessment Phase: data collection

Depending on factors such as the type of user education, available resources, project time frame and expertise of the training needs assessor, various methods can be used to collect information for TNA. In general, these methods can be grouped into the following categories:

- *Existing records.* These generally refer to an in-depth investigation of information resources identified during the background information research in the previous phase. The method of researching existing records is most suitable for identifying public user education needs noted in Section 2.1.
- *Questionnaire surveys and interviews.* These methods include a variety of mail-in and on-line questionnaires and face-to-face or telephone interviews that aim to solicit information directly from the users. Questionnaire surveys can cover a relatively large number of users within a short time frame and at a low cost. However, issues of questionnaire design and response rate can adversely influence the

quality and quantity of data obtained. On the other hand, individual interviews can be time-consuming but may yield more information and often unanticipated perspectives that questionnaire surveys do not pick up. Both methods are suitable for TNA associated with occupational and individual user education. Face-to-face interviews are particularly applicable for identifying the needs of individual user education.

- *Focus group meetings.* These are structured group meetings in which the training needs assessor, acting in the capacity of a facilitator, seeks to identify the job requirements of participants through interactive questions and answers. The format of focus group meetings is very similar to the *joint application development* (JAD) used in data modelling (see Chapter 8). Because of their interactive nature, focus group meetings promote and stimulate exchange of ideas between the training needs assessor and the participants. As a result, this method of data collection is probably the most commonly used method in TNA.

- *Causal analysis.* As its name implies, causal analysis is concerned with the identification of the underlying causes, factors or conditions that contribute to the existence or perpetuation of the needs of user education. This method of data collection is made up of qualitative questions such as "what factors cause the under-performance of users?" and "what conditions have prevented the current system from achieving its intended purposes?" Causal analysis is used to supplement quantitative methods such as questionnaire surveys in TNA. It can also be used in focus group meetings to help the training needs assessors uncover the cause and effect of particular aspects of user education needs.

- *General observations.* Listening and watching can also be an effective means of collecting information. Many training needs analysts prefer personally to observe the working environment of the users and to listen to their opinions about their jobs and related issues. General observations also include a more formal and structured approach referred to as a critical incident technique. This is a special survey method that uses direct observations or self-evaluation of a user's behaviours or reactions in a particular situation or when performing a particular task. General observations are usually carried out to verify and to obtain supplementary information collected by other methods.

2.3.3   The Post-assessment Phase

In this phase of a TNA, the training needs assessor interprets all the data collected in the assessment phase. The primary objective of the analysis is to identify existing strengths and weaknesses of the skills and knowledge of the users, and to determine the discrepancies or gaps by comparing the results against the required competencies for operating, maintaining and using the spatial database system, as explained in Section 2.2. At this stage, the assessor also estimates the budget and other resource requirements and develops a user education plan by prioritising the identified training needs with respect to the project needs, the criticality of the required training and the availability of internal and external training programs (see Section 2.4).

The final step of the TNA is to present formally the results of the analysis in a written report to the project sponsors and project manager. The final report serves several important purposes, including:

- It acts as a formal presentation of results that serve as the basis for reaching agreement and acceptance by users, their supervisors, as well as the project manager.
- It operates as a conceptual framework to match competencies of users to the requirements of implementing and using a spatial database system.
- It provides the basis for the development of a business case to seek the necessary resources for user education as an integral part of a spatial database project.
- It demonstrates the contribution of user education to the success of the spatial database project.

2.4   Implementing a User Education Strategy

The results of a TNA form the blueprint for the implementation of user education programs in support of a spatial database project. In developing a user education strategy, it is necessary to consider several important factors including:

- *Prioritisation of User Training Needs.* Determination of the priority of the different training needs identified in TNA is based on the judgment of several considerations such as the number of users requiring the training, the magnitude of the needs gap, the importance of individual training needs relative to the mission and objectives of an organisation, and the availability of training programs.
- *Time Frame of Training Programs.* Timing is critical to developing skills and maximising the effectiveness of training. Training is most effective when it is received immediately prior to the actual use of the new knowledge and skills that the training will build. If there is a substantial lag between the acquisition of new knowledge and skills and their use, recall will diminish and, as a result, productivity will suffer. Equally, users whose training is not complete when the expected

knowledge and skills are required will not be able to contribute fully within their workplace.

- *Location.* Training programs can be held locally within an organisation or at a university, college, a vendor's offices, or a neutral site. Training at sites away from the organisation is usually more costly because of expenses incurred for travel, accommodation and meals. However, this may be feasible when the number of user requiring training is small and options for alternatives are limited. Courses taught locally are most suitable when a large number of users require training. However, such courses can also be expensive because it is necessary to cover the travel and living expenses of the instructor(s).

- *Method of Delivery.* Training courses can be delivered differently, but essentially there are two broad types, namely instructor-led training and technology-based training. Instructor-led courses are most effective when hands-on training and personal interaction are required (for completely new users, for example), and the required knowledge and skills must be learned within a specified time period. Technology-based training, on the other hand, allows participants to learn at their own pace. Such courses are most effective in delivering consistent training using either on-line courses or CD-ROM-based instruction. The down side of this approach is that there is no opportunity for a trainee to ask and have answered questions that fall outside of the frequently asked questions (FAQ) anticipated or accumulated from prior experience by the company that is providing the training. Increasingly, however, end-user training is offered across the Internet. This method of delivery is highly accessible and is self-paced, but it is constrained as there is little, if any, opportunity for the trainee to interact directly with the instructor.

- *Train-the-Trainer.* This approach is most effective in large scale implementation projects where a large number of users at different locations must be trained at the same time. It involves the development of a training curriculum and methodology that is first provided to selected users. On completion of this training course, these users then return to their respective home offices and offer the same training locally.

- *Evaluation and Continuing Improvement.* Training is an ongoing process that needs to be evaluated regularly in order to ensure that the intended objectives are accomplished. There are different ways for evaluating the achievement of participants, including self-assessment, skill testing, and observation by peers or supervisors. No matter what method is used, it is important to refer the evaluation results back to the original user education action plan. This will enable decisions to be made as to whether the training program should continue, be modified, or cease.

As noted earlier, independent of the form, content and context of end-user education and training, an important and emergent issue that, in many ways, supersedes technical training concerns the legal issues that surround spatial database implementation and use. In the past, such legalities were less important as ease of information access was substantially more constrained than it is in the current era of data sharing and networked homes and workplaces. This new era has bought with it attendant concerns that any prudent spatial database manager must inculcate in his/her staff acitivities and business practices. The legal dimension of these concerns is discussed in the following section.

3. LEGAL ISSUES IN SPATIAL DATABASE IMPLEMENTATION

The legal complexities of modern society, in particular those involving government legislation and regulations, require users of spatial data to have a good understanding of the laws pertaining to IT in general, and spatial databases in particular. From a project management perspective, a good knowledge of applicable laws is essential for several reasons. On the one hand, it protects investment in a database system by preventing illegitimate or unauthorised access to and copying of the data. On the other hand, it helps to avoid inadvertent infringement of intellectual property rights of third-party data or software applications and, consequently, the legal repercussions that may possibly result. At the same time, it helps to ensure that the resulting database is in full compliance with prevailing legislation and regulations governing access to information and protection of privacy, thus avoiding potential legal actions by government authorities and disgruntled interest or advocacy groups.

The following sections discuss the key legal issues that need to be considered in spatial database implementation. It should be understood that laws by nature are jurisdiction-specific. They tend to vary not only between countries, but also within countries, such as between the provinces of Canada or the states of the United States of America. Laws are also constantly evolving in response to changes in social values, technology, government policies and international conventions and treaties. Hence, the remainder of the discussion in this chapter is intended more as a general introduction to the laws pertinent to spatial data and database systems, rather than to provide definitive legal opinions. Before any decision is made where legal issues are involved, it is necessary to consult legislation and regulations currently in force in an organisation's local jurisdiction and to seek the professional advice of qualified legal counsel.

3.1   The Legal Regime of Using Spatial Information

The term *legal regime* can be broadly defined as the structure of laws, regulations and professional practices that operate in society. The daily life of each and every individual in a society is governed by various aspects or components of a legal regime that include, to varying degrees, adherence to:

- *International conventions and treaties.* These are multilateral agreements that aim to promote the use of accepted protocols, procedures and practices of conducting business and other activities among signatory countries. International conventions and treaties cover a wide range of areas including social and economic development, health care, business and trade, scientific research, the environment, transportation and telecommunications. Some of the international conventions and treaties are global in their application (for example, the Berne Convention on Copyright, the United Nations Law of the Sea, and the United Nations Framework Convention on Climate Change, commonly called the Kyoto Protocol). Others are created and used by countries in a particular region only (for example, the North American Free Trade Agreement (NAFTA) between Canada, the United States and Mexico and other countries in the western hemisphere). International conventions and treaties are enforced by signatory countries individually after ratification by their respective national legislatures. Disputes between signatory countries are always resolved by arbitration or mediation through the appropriate administering organisations.
- *Statutes.* Also commonly known as *bills* and *acts*, statutes are formal, written laws enacted by legislative bodies at the Federal/central government level (for example, the Parliament of Canada or the United States Congress) and the Provincial/State level (for example, provincial/state parliaments and legislatures in Canada and the United States).
- *Administrative regulations.* These are rules and by-laws, established and enforced under the authority of statutory laws, in support of public administration by agencies of Federal, Provincial/State and Municipal governments.
- *Common law.* The term "common law" refers to past decisions of courts of law that are used as precedents or case law by judges in their rulings. The rules of common law have the authority of statutory laws unless they are specifically overridden by legislation. Common law is practiced in England, where the concept of common law originated in medieval times, and countries once colonised by England, including Australia, New Zealand and other countries of the British Commonwealth, all of

the Provinces of Canada except Quebec, and all of the States in the United States except Louisiana.

- *Civil law.* Civil law has its origin in ancient Roman law. It is based on a written legal code established through legislation and decree. Civil law underlies a unified legal system by working out, with maximum precision, the conclusions that can be drawn from its basic principles. Unlike judges in common law jurisdictions who base their decisions on precedents created by judicial decisions over time, civil law judges are bound by the provisions of the written code in their rulings. Civil law is practiced by most countries in continental Europe, and countries in Africa and Latin America that were former colonies of these countries. It is also the foundation for the law of Quebec and of Louisiana, as noted above. (Note: the term *civil law* also refers to laws governing relationships between persons or legal personalities such as corporations, as opposed to *criminal law* which regulates governmental sanctions, for example imprisonment and/or fines, as punishment for crimes against social order).

- *Standards for goods and services.* These are protocols and specifications for the production of goods and the provision of professional services that are established by international and national standardisation organisations and enforced through government legislation or regulations, as explained in Chapter 5.

All of the above aspects of the legal regime apply, to varying degrees and in different ways, to the use of spatial data. Statutes relating to land ownership and real property rights, for example, have resulted in the collection of significant volumes of spatial data by all levels of government. Spatial data are indispensable for the enforcement of administrative regulations governing land management, resource consent planning, mineral exploration and environmental impact assessment, among many others, by agencies at different levels of government. Increased data sharing opportunities among users in different application domains across international borders means that legal issues of spatial data are pertinent not only to national laws and regulations, but also to international conventions and treaties. As the role of spatial databases is rapidly expanding from conventional data management to become a crucial tool for public policy making and business decision support, the compliance with standards in order to avoid potential legal liabilities is no longer an option but a necessity for spatial data users.

The lack of understanding of the legal regime in spatial database implementation is an open invitation to legal repercussions that will jeopardise not only spatial databases themselves, but also the organisation sponsoring the database system. Cho (1998) highlighted the following legal risks that individuals and organisations run relating to the use of spatial data:

- Failure to secure intellectual property rights.
- Liability for infringement of intellectual property rights, whether intentional or not, including failure to control access to spatial data and application software tools, that would result in illegal use of the data or tools by others.
- Failure to secure accountability for defective spatial data or application software tools, including models, methods, and services based on the data and tools.
- Liability for breaching privacy or confidentiality obligations.
- Legal uncertainties involved in contracting out the tasks of spatial data collection, processing and dissemination, whether by a government agency or for such an agency or private organisation.

3.2   The Legal Issue Domains

The legal problems pertaining to spatial databases can be grouped into the following four major classes as they pertain to the four new approaches to spatial data that include data sharing, government information policy, policy making and decision support using spatial data, and spatial data as evidence in a court of law (Figure 7.3):

- *Intellectual property and copyright*, which is concerned with the ability to copy and reuse spatial data, spatial databases and software products as well as the terms of reference that govern whether or not a database is sufficiently original and creative to satisfy the legal definition of copyright under copyright law (see Section 3.3).
- *Access to information legislation*, which includes laws guaranteeing the public's right to know and policies for restrictive disposal of and access to personal information that is contained in or can be interpreted from spatial data (see Section 3.4).
- *Liabilities of supplying and using spatial data*, which are associated with a wide range of legal responsibilities for specific spatial data quality, disputes about contracted spatial data services, negligence and damages caused when using spatial data containing errors, or when used in inappropriate ways (see Section 3.5).
- *Evidentiary standards of spatial data in courts of law*, which are related to legal issues regarding the admissibility and integrity of spatial data extracted from database systems as evidence presented in a court of law (see Section 3.6).

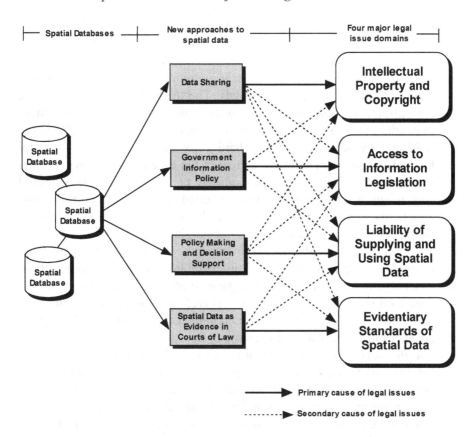

Figure 7-3. The legal issue domains of spatial database implementation

Although the types of legal issues in spatial database implementation are reasonably well-defined and well-understood, the applicability of laws to spatial data in practice can be very precarious and controversial due to a number of factors. These include:

- *The changing nature of spatial data.* Conventionally, spatial data are collected and used by government agencies in order for them to conduct government business. Spatial data are also collected and used by research and academic institutions. Within these contexts, the user environment for spatial data is rapidly changing as private sector organisations are playing an increasingly important role in the spatial data supply chain. Some organisations collect spatial data for sale commercially, others add value to existing data that may have originally been collected by government agencies by reformatting, refining and combining different data sets, in order to make them commercially attractive to end users. The commodification of spatial data, and the resulting separation between spatial data collection and their use, have led to a complex web of legal issues relating to ownership, access and liability that was seldom known to the spatial data users of the past.

- *The characteristics of spatial data in digital form.* Spatial data in digital form are not tangible material goods. The methods of their production, application, possession and exchange are different from ordinary industrial products. Spatial data are now largely collected digitally in the field and by scanning existing hard copy maps. Advances in digital technology have dramatically increased the ease and speed with which spatial data can be reproduced, modified, combined, distributed, abused and misused. This has in turn resulted in serious questions and arguments about the ability of existing laws, that were made to regulate physical industrial products, to embrace digital spatial data.

- *The inability of legal systems to meet the needs of rapidly evolving technology.* Spatial database systems are rapidly evolving both conceptually and technically. The passage of laws, in contrast, is a very time-consuming and slow process. Therefore, existing legal principles and laws always seem to be temporally offset behind technological advances in spatial data and database systems.

- *The inconsistency between legal systems of different jurisdictions.* The making and enforcement of laws are jurisdiction-specific by nature. It is not easy to reconcile legal differences when disputes occur between users residing in different jurisdictions, particularly when jurisdictions using the common law and civil law systems are involved. In Canada and the United States, for example, the use of spatial data may fall under the jurisdiction of the Federal, Provincial/State or even Municipal levels of government. Inconsistencies between the laws of different jurisdictions with respect to terminology, interpretation, and provisions and exceptions may often cause confusion to spatial data users.

- *The conflicting interests and requirements of different stakeholders.* Different people and organisations have different purposes in using spatial data that may lead to conflicting interests and requirements among different user groups. Data collectors and suppliers, for example, tend to favour more restrictive copyright laws in order to protect their ownership, whereas data users may like to see greater freedom in the use of spatial data products in order to minimise the costs of data acquisition. There is also a friction between commercial spatial data users, such as market researchers who prefer an open approach to using spatial data, and advocates of civil rights who see the protection of personal privacy as an uncompromising principle of a democratic society. These conflicting requirements, together with the sometimes contradictory characteristics of spatial data noted above, make it difficult for all stakeholders to agree that they were treated in an equal and fair manner.

The following discussion explains in detail the nature and characteristics of the four primary legal issue domains of spatial database implementation within the context of these factors. First, the fundamental issue of ownership is discussed as this aims to protect intellectual property rights when spatial data are collected, used and exchanged with other parties.

3.3 Intellectual Property and Copyright

In the context of spatial database implementation, *intellectual property* (IP) refers to the legal regime concerned with the ownership of data, databases and software. A good understanding of intellectual property is critical to spatial data users in two respects. On the one hand, it helps to prevent the infringement of the rights of the lawful owners of spatial data and software applications being used in spatial database projects. On the other hand, it enables users to protect their own intellectual property rights when their data or software products are shared and used by other users.

3.3.1 The Concept of Intellectual Property

IP is created through human intelligence or inventions and creations in art, industry, science and technology. In itself, IP is an abstract construct but it can be embedded in a physical entity which, as a result, enables it to be possessed, traded and used like real estate, vehicles, machinery, furniture and other types of physical goods. An *intellectual property right* is the term that embraces the various intangible rights, such as those listed in Table 7.3, that are granted by law to enable IP owners to restrict others from using their ideas or inventions without their prior approval.

The term *"public domain"* is often used to denote spatial data and software applications that are "free" or "available without financial transactions" involving a distributor and an end user. In the legal context, a data product or software application in the public domain implies that there are no IP rights or restrictions attached to their use or modification. Since spatial data products in the public domain are not copyrighted, any person can refine and reformat these data products, add new data values to them, and market the resulting data products commercially. When this occurs, the developers of the new spatial data products are given the IP rights to the new product, but the original IP status in the public domain remains unchanged.

Table 7-3. Intellectual property rights commonly used in association with spatial database implementation

Type of Right	Characteristics/Comment	Applicability to		
		Data	Software	Databases
Copyright	o Protects literary, artistic, musical and dramatic works o Does not protect ideas o Does not require registration o Respected across jurisdictions	No	Yes	Yes
Patent	o Protects novel inventions that must be capable of industrial application o Requires registration in individual countries	No	No	Yes
Database Protection	o Protects the content of a database o Does not require registration o Available in a relatively small number of countries only	Yes	Yes	Yes
Trademarks	o Names given to signs, symbols and logos used to distinguish individual goods and services from one another o Requires registration in individual countries o Do not apply to software itself but to software names, software companies and company logos	No	No	Yes
Trade Secrets	o Right relating to confidential information, including spatial information, ideas and other commercial know how that need not necessarily be inventive o May arise when one party (e.g. an employer) imposes a contractual obligation of confidentiality on to another (e.g. an employee)	Yes	Yes	Yes

Similarly, software applications in the public domain can also be refined and improved for reselling. The developers of the new software applications are often granted IP rights without affecting the original IP status in the public domain.

It is important to distinguish between "public domain" and "public sector" spatial data and software applications. Public sector spatial data are data collected and disseminated by a public sector organisation. Such data may be in the public domain or copyright protected, depending on governmental and institutional information policies. In the United States, for example, although the Federal government does not apply copyrights to its spatial data and software products, State and Municipal governments may do otherwise. Similarly, different levels of government in Canada may also adopt different policies toward the copyright of their spatial data and software products. Therefore, when a spatial data user has decided to use a particular spatial data set or software application, it is important to clarify their IP status and copyright, even though they are obtainable from public sector organisations.

In recent years, the concept of *open source* has emerged in the software industry as a means of fostering innovation and removing the barriers that tend to surround proprietary computer operating systems and software products. Open source software development subscribes to the concept of a community of developers and refers to the situation where software programmers collaborate freely (i.e. without financial transactions or often without legal agreement) and allow their IP in the form of readable versions of their programs to be used and modified by others. Many open source endeavours have resulted in the development of high quality software products and programming languages. The most visible examples of open source software in the current technology arena include the Linux operating system and the Apache web server software, as well as the server-side HTML embedded scripting *hypertext pre-processor* (PHP) language, all of which are successfully used to support spatial database processing.

3.3.2   The Concept and Doctrine of Copyright

Simply stated, *copyright* protects against the right to copy. Copyright laws are the legal instruments used to protect the IP of the creators of original literary, dramatic, musical and artistic works. The term "work" in the context of copyright laws suggests the use of a certain degree of skill and effort, originality and creativity to create the IP that is to be protected. Because of the enormous economic interest and political implications at stake, copyright legislation and enforcement are invariably among the most complex and hotly contested legal issues in modern judiciary systems.

There is no single universal copyright law that is applied globally. Different countries have their own copyright legislation. However, at a high level of generalisation, copyright laws in all jurisdictions are based on similar fundamental principles as exemplified by the copyright laws of Canada and the United States. These include, but are not limited to, the following:

- *The dichotomy between ideas and expressions.* Copyright does not protect ideas. Rather, it protects the expressions of those ideas. However, such expressions must normally be original and creative in nature. They must also be fixed in a tangible medium that can be perceived, reproduced or otherwise communicated. A mere recitation of facts, regardless how they are expressed, does not automatically qualify for copyright.

- *Transferability of right.* A copyright can be transferred by its original owner to third-parties either on an exclusive basis (that is, copyright ownership of one or more rights is transferred by the copyright owner), or on a nonexclusive basis (that is, the copyright owner retains ownership of the copyright and grants similar copyrights to others through licensing).

- *Authorship and copyright ownership.* Subject to certain exceptions, the author/creator of a work is the first owner of the copyright. One exception to this principle is where the work is created in the course of the author/creator's employment, the employer rather than the author/creator is the copyright owner. However, where a work is created under a contract of service, the author/creator retains copyright ownership unless transfer of ownership is specified in the form of a legal agreement by the requester of the service.

- *The first sale doctrine.* This doctrine prevents the owner of the copyright in a work from controlling subsequent transfers of that work. This means that the owner of a particular copy of a work is able to dispose of possession of that copy in any way, for example by selling it, leasing it, loaning it or giving it away. However, there is an exception to this doctrine with respect to two types of work, computer programs and sound recordings, which cannot be disposed of at the will of the licensees.

- *Non-perpetuity of copyright.* Copyright laws always state a prescribed finite period of time of protection. International conventions, such as the Berne Convention (see Section 3.3.3), normally provide for protection for the life of the creator plus 50 years as the minimum. The duration of copyright varies by individual countries but the duration of the life of the creator plus 50 to 70 years seems to be the norm. At the end of the period the work goes into the public domain and is available for use without permission or royalty.

- *The concept of fair use or fair dealing.* Copyrights are subject to certain limitations, such as exemptions for library use and the concept of fair use or fair dealing. The fair use concept provides exceptions in copyright laws for copying for personal use, research, education, archiving, reviewing and reporting. The scope and flexibility of these exceptions vary widely between countries, depending on such factors as the nature of the copyrighted work, the purposes (commercial or non-commercial), the amount of the work to be copied under the fair use provision to the copyrighted work as a whole, and the effect of the copy on the potential market for or value of the copyrighted work.

- *Civil and criminal liability for infringement.* People infringing copyright may be sued for either actual or statutory damages and, in addition, may be prosecuted under criminal laws generally without regard to whether the infringement is wilful or unintentional.
- *Registration and notification of copyright.* It is not necessary for any work to be registered to receive copyright protection, which normally distinguishes and separates copyright from *patents*. There is also no need for works to carry a conventional copyright notice, in the form of Copyright [date] by [author/owner] or © [author/owner], in order to be protected. Today, almost all major countries follow the Berne Convention to discard the practice of displaying copyright notification. In the United States, for example, almost everything created privately and originally after April 1, 1989, is copyrighted and protected automatically whether it carries a notice or not. This implies that whenever we want to use a data set or a software product, it is safe to assume that it is copyright-protected, unless we can verify that for some reason this is not the case.

The above concepts and doctrines form the basic framework of practically all international agreements and national laws to protect IP. The following discussion first considers the major international conventions and treaties governing IP. Then the key features of the copyright laws of Canada and the United States are discussed to illustrate how modern national copyright laws are affected by and relate to these international conventions and treaties.

3.3.3 International Conventions and Treaties on Intellectual Property

International conventions and treaties governing IP are not copyright laws *per se*. There is no such thing as an "international copyright law" that will automatically provide IP protection throughout the entire world. Protection against unauthorized use in a particular country depends, basically, on the national laws of that country. However, most countries do offer protection to foreign works under certain conditions, and these conditions are greatly simplified by international copyright conventions and treaties. These conventions and treaties also form the basis for arbitration and mediation in case of disputes between one country and another.

International conventions and treaties have resulted from multilateral negotiations, often extending over several years, under the auspices of international organisations such as the World Intellectual Property Organization (WIPO) and the World Trade Organisation (WTO). The

WIPO, which currently administers over 20 conventions and treaties relating to IP and copyrights, is by far the most important of all international organisations in this area. The rather long history of the WIPO can be traced to the Paris Convention for the Protection of Industrial Property in 1883 and the Berne Convention for the Protection of Literary and Artistic Works in 1886.

On the basis of these two conventions, the WIPO (2001) has, over the years, struck several international treaties and agreements that have significantly widened and strengthened the protection of intellectual property internationally and nationally. These include the WIPO Copyright Treaty (WCT), adopted in December, 1996, that aimed primarily to adapt legal principles that encompass emerging technologies and new areas of concern, such as the electronic reproduction of copyrighted material, distribution and use of copyrighted works over telecommunications networks and the Internet, as well as rights-management information (that is, codes which identify the work, author and terms of conditions of use of IP rights). Pursuant to the WCT, several signatory countries have enacted legislation against the circumvention of effective technological measures and devices to protect copyrighted works, for example, the Digital Millennium Copyright Act of the United States (see Section 3.3.4).

The WTO is the administrator of the Agreement of the Trade-related Aspects of Intellectual Property Rights (commonly known internationally as the *TRIPS Agreement*). This agreement covers both copyright and industrial property with particular reference to international trade. It offers a mechanism for the resolution of disputes between the WTO member countries, and non-compliance may lead to trade sanctions against countries that contravene its terms. In this regard, the TRIPS agreement is more powerful than the conventions and treaties administered by the WIPO. It has in effect implicitly forced all except a few countries in the world to follow a universal standard and approach to IP protection, which the Berne Convention of the WIPO failed to do. In 1995, the WTO and WIPO signed a cooperation agreement, which formalised their relationship by covering regular contact, information exchange and treatment of member countries of the two organisations.

At the regional level, NAFTA contains clauses pertaining to IP rights and access provisions to IP among the three signatory countries (Canada, the United States and Mexico). In Europe, the European Union (EU) Directive on Copyright came into effect in 2002. This directive offers a standardised approach to the protection of music, film and software companies, and extends protection to the Internet and other digital media within its member states.

3.3.4 Copyright Legislation of Canada and the United States

It is not within the scope of this chapter to provide a comprehensive treatise of copyright laws of different countries. As noted above, the discussion focuses instead on Canada and the United States as examples of modern copyright legislation with special reference to international conventions and treaties and application to spatial data and spatial database systems.

Canadian copyright law is governed by the Copyright Act (1921). The Copyright Act remained unchanged until 1988, when the Copyright Amendment Act was introduced. This amendment addressed copyright issues that are relevant to modern society such as computer programs, moral rights, copyright collectives and changes to the Copyright Appeal Board (now the Copyright Board), and remedies for infringement (CIPO, 2002).

There is no provision in the Copyright Act for the copyright protection of a single spatial data item such as a street name, the coordinates and elevation of a survey point, and the average household income within a census tract. Spatial data, therefore, can only be copyrighted collectively in "works" such as compilations or data tables (which are considered as literary works), and maps, survey plans, aerial photographs and remote sensing imagery (which are considered artistic works).

Spatial data collected by or on behalf of Federal, Provincial and Municipal agencies is protected under the Copyright Act. Permission must be obtained from the appropriate government agencies in order to use spatial data under their stewardship. Conventionally, royalties were required for the use of such spatial data. However, as the result of the establishment of the GeoBase spatial data portal (http://www.geobase.ca; Chapter 5 and Chapter 10) in late 2003, users are now able to access, over the Internet and free of charge, several spatial data sets under the provision of a Geobase Unlimited Use Licence Agreement. This portal is a component of the CGDI and is a joint Federal and Provincial/Territorial government initiative. Spatial data sources are all from these two levels of government. This means that the use of spatial data owned by Municipal governments still requires copyright clearance and the payment of a royalty if requested.

The Geobase Unlimited Use Licence Agreement grants the licensee a non-exclusive royalty-free right and licence to exercise all IP rights in the data, including the right to use, incorporate, sublicense (with further right of sublicensing), modify, improve, further develop, and distribute the data; and to manufacture and/or distribute derivative products from or for use with the data. The IP rights arising from any modification, improvement, development or translation of the data, or from the manufacture of derivative products, also vest in the licensee.

Spatial data users in universities and colleges can also access the Federal Government's digital topographic database through a licensing agreement signed in the summer of 1999 between Natural Resources Canada (NRCan), and the Canadian Association of Research Libraries (CARL) and the Association of Canadian Map Libraries and Archives (ACMLA). This allows professors, students and researchers to make greater use of the large inventory of socio-economic data from Statistics Canada (StatsCan), provided through the Data Liberation Initiative. Users can also use other sources of spatial and non-spatial data under the "fair use" provision of the Copyright Act as explained in Section 3.3.2. Similar agreements, most with some form of restrictive use covenants, are now commonplace within the spatial information technology industry between individual educational institutions and organisations that generate spatial data sets.

Canada is a signatory country of the Berne Convention, the TRIPS Agreement, the NAFTA, and a host of other international conventions and treaties relating to IP. This enables Canadian works to be protected in all countries that have signed these conventions and treaties. By the principle of reciprocation, works originating from these countries are also protected by Canadian copyright law in Canada. Canada also extends protection to certain non-signatory countries by way of notice in the Canada Gazette (CIPO, 2002).

The rights granted to owners of IP in the United States are derived from the United States Code, Title 17 – Copyright, Public Law No. 105-304, as last amended in 1998. The Copyright Act grants five exclusive rights to a copyright owner, namely the right to:

- Reproduce the copyrighted work.
- Prepare derivative works based upon the work.
- Distribute copies of the work to the public.
- Perform the copyrighted work publicly.
- Display the copyrighted work publicly.

Over the years, copyright legislation has been amended to reflect new technologies, changing perceptions of the rights/responsibilities of users and creators/investors and emergence of the United States as a major participant in the global economy. One of the most significant new pieces of copyright legislation in this regard is the Digital Millennium Copyright Act 1988. A provision in this particular Act makes it a criminal offence to break or de-scramble the encryption used to protect the digital or electronic content of a work, such as the Content Scrambling System (CSS) for DVDs. As a result of this provision, the copyright owner is granted the same rights against anyone trying to de-scramble a copyrighted work as that which is available against anyone who infringes the basic copyright protection in the work. In other words, the mechanism used to protect the work from copying is now protected to the same extent as the copyrighted work itself.

Internationally, the United States did not become a signatory of the Berne Convention until 1998, over 100 years after the initial convention ratified its agreement. Before that, the Copyright Act provided protection only for IP owners who were United States citizens or residents. At present, the United States is a signatory member of 15 of the conventions and treaties administered by the WIPO. It is also a signatory of the TRIPS Agreement of the WTO.

An important feature that distinguishes copyright legislation in the United States from those of most countries in the world is the prohibition of Federal Government agencies to impost copyrights on their data holdings. As a result, spatial data produced by the Federal Government are freely available in the public domain and through the Geospatial One-Stop portal (http://www.godata.gov; see also Chapter 10). State and local governments have the option to apply copyright to their public records if they choose to. Spatial data collected by the private sector, however, are copyright protected generally under the Copyright Act as artistic expressions on computer screens and in printouts, data models, integration with other sources, results of analysis, and the use of attribute and tabular data.

Since the United States is a signatory of various international conventions and treaties, spatial data products collected and distributed by American data suppliers are protected under the copyright laws of all signatory countries of appropriate conventions and treaties. Likewise, spatial data products originating in signatory countries of different international conventions and treaties are also protected in the United States by reciprocation.

3.3.5 The Concept of Database Rights

The concepts and practice of IP protection are constantly evolving in response to technological advances and changing geopolitical climates, as noted above. From the perspective of spatial database implementation, one of the most crucial developments in IP rights is the debate on copyright protection for databases.

The issue of database rights is raised because the use of conventional copyright laws to protect databases contravenes several of the fundamental principles of IP rights noted in Section 3.3.2. For example, since many databases are collections of pre-existing data, they do not necessarily meet the criteria of "originality" and "creativity" in conventional copyright legislation. Further, since many databases are constantly updated as a result of transaction processing, the concept of a finite duration for copyright protection is hard to apply. Although databases as a whole can be protected generally as compilations under copyright laws, the underlying data are not automatically granted protection because they may be regarded simply as a collection of basic facts that cannot be protected due to the lack of creativity in their compilation.

The primary objective of database protection is to provide a set of separate laws that remove the uncertainty and ambiguity of using existing copyright laws for the protection of databases. In 1996, the European Parliament, as the legislature of the EU, adopted the Database Directive that enabled a database to be protected in two distinct parts:

- A database, whether or not it is accessed electronically, is protected under copyright law where the selection or arrangement of the contents constitutes the database developer's own intellectual creation.
- The content of the database is protected by a unique *sui generis* (which in English roughly means "of its own type") right that is independent from any copyright existing on the database.

The *sui generis* right was designed to give shelter to database developers who have spent a substantial amount of time and great deal of effort in the compilation of data which could not otherwise be protected due to the creativity and originality requirements. This right prevents the extraction and/or re-utilisation of the whole or a part of the content of a database. In the Database Directive, the *sui generis* right is granted for a period of 15 years, running from the first day of January in the year following the date of the completion of a database, or the date of the release of a database for public use. However, substantial changes in the content of a database are considered a substantial investment that will qualify the database developer to start the 15-year protection duration from a new date. It should be noted that the Database Directive does not apply to the computer software used in the development and operation of a database, although supporting material for operation of a database such as a thesaurus and index system(s) are eligible for protection.

In a related development, the WIPO proposed a Treaty on Intellectual Property in Respect of Databases in its diplomatic conference held in Geneva in 1996. The proposal of this treaty focused primarily on the creation of a universal *sui generis* right similar to the European Database Directive. No agreement was reached at the conference in 1996 but participants resolved to continue the development of the treaty. In early 1997, the WIPO convened an extraordinary session to set up a new Committee of Experts to consider a proposal for a new treaty on database protection. Although no treaty is finalised as of the time of writing this book, the move of the WIPO has aroused considerable interest, and created heated debates in some cases, in its member countries that are interested in harmonisation of their copyright laws with the proposed WIPO treaty.

After a relatively comprehensive comparative study on copyright and database rights, Rose and Radcliffe (2003) concluded that copyright remains a vital source of protection for spatial data and spatial databases. The *sui generis* right as introduced in the European Database Directive has widened and strengthened the status of copyright as the primary legal instrument for the protection of IP in spatial data and spatial databases. As the methods available for representing spatial data electronically have diversified, so have the ways in which spatial data and databases can be infringed. Consequently, copyright as a legal instrument of protecting IP has become even more vital to spatial data users.

3.4   Access to Information and Protection of Privacy Legislation

Access to information is the ability to use information to know, to learn, to entertain and to make decisions. Spatial data, like other forms of data, are the key ingredient required to generate information through qualitative and quantitative interpretation, analysis and modelling. It is increasingly realised that access to information, especially that under the stewardship of government and public organisations, is a basic civil right of citizens. However, since spatial data may contain confidential personal information, it is necessary to have some mechanisms in place to prevent this information from being abused or misused. Personal privacy is, after all, a fundamental legal cornerstone of democratic societies and should be respected in legislation that differentiates public from private knowledge.

There are two apparently contradictory requirements at work, as far as data access and personal privacy are concerned, regarding access to a spatial database. On the one hand, the ability to use the database is required in order to improve access to information. On the other hand, the obligation not to compromise personal privacy as a result of improved access to databases must be enshrined. Thus, the legal regime governing access to databases must be built on the fundamental principle of balancing the right to know and the right to privacy. While it is easy to blur the boundary between this principle and right it is important to devise legislation that enshrines both perspectives. This section examines the legal issues associated with access to information from the perspectives of these two rights. The section also explains the process of handling a typical request for information in order to demonstrate how these two rights can be applied in practice. Where appropriate, examples from Canada and the United States are used to illustrate the discussion.

3.4.1   Access to Information Legislation

There are now approximately 50 countries that have access to information legislation (also commonly known as freedom of information legislation) in place (AIRTF, 2002). Although the structure of the legal regimes to access information in these countries tend to differ, the intent and spirit of the legislation are practically all based on similar first principles, namely:

- Government information should be available to the public.
- Necessary exceptions to the right of access should be limited and specific.
- Decisions on the disclosure of government information should be reviewed independently of government.

Under access to information legislation, citizens can make requests to access information collected by government agencies (see Section 3.4.3). In order to prevent the abuse of this right and to protect agencies from unreasonable demands, there are provisions to limit the right of access to existing information holdings, and to request payment to cover the expenses incurred in searching for and preparing the information for disclosure. There are also provisions to refuse access based on the following conditions:

- The information is easily available through other means (for example, published information).
- The requested information is considered to be of little value or importance, or have no reasonable ground or purpose.
- The request is deemed to be ill-intentioned and seeks to cause embarrassment to the agency concerned or third-party persons.
- The request will cause substantial and unreasonable interference with the work of the agency concerned.
- The requested information cannot be made available without substantial collaboration or research.
- The request will compromise the privacy of third-party persons and violate applicable data protection legislation (see Section 3.4.2).

In Canada, the Access to Information Act came into effect in 1983. This Act provides a right of access to general government information, identifies the institutions it applies to, the types of government information that may or must be protected in response to requests, and the types of information that are excluded entirely from its scope. The Act also delineates the process for making a request including the timelines and the procedures for notifying third parties. It establishes the Office of the Information Commissioner to receive and investigate complaints, and provides a further right of review by the Federal Court. The Canadian Access to Information Act does not cover private sector organisations, nor does it cover agencies of the Provincial/Territorial and Municipal levels of government. Access to information in the Provincial government agencies is regulated by the laws of individual Provinces and Territories and this gives rise to plurality across the smaller administrative areas within the country.

In the United States, access to information is governed by the Freedom of Information Act which was passed in 1966. This Act, like its Canadian counterpart, applies to agencies of the Federal Government only. The Freedom of Information Act requires Federal agencies to make available their records to the general public. Exceptions are created for data relating to national security and privacy rights which can be protected. Agencies of State governments are governed by the Open Records Act that defines the meaning and scope of the records that are open for public scrutiny, and specifies the conditions under which records can be made available.

3.4.2   Protection from Privacy Legislation

The term *privacy* means different things to different people. In the context of the discussion in this chapter, privacy is primarily concerned with *data protection* or *information privacy* that seeks to ensure through legislation and professional protocols the accuracy and confidentiality of personal data collected by organisations in both the public and private sectors. In the context of data protection legislation, personal data embraces a wide range of attributes about an identifiable individual, including but not limited to the following:

- The name, age, sex, race, national or ethic origin, sexual orientation and marital status of this individual (commonly called the *data subject*).
- The education, medical and employment history of the data subject or information relating to financial transactions in which the data subject is involved.
- Any identifying number, such those appearing on a birth certificate, driving licence, credit card, social insurance card, or other particulars assigned to the data subject.
- The address, telephone number, finger prints and blood type of the data subject.

- Personal opinions or views of the data subject, and the opinions or views of other persons about the data subject.

The notion of protecting privacy by legislation was in existence in Europe long before the advent of modern database systems. The right to privacy was included in the Universal Declaration of Human Rights proclaimed by the United Nations General Assembly in 1948. In 1981, the Council of Europe established a convention to guide the collection, storage and use of personal information. Around the same time, the Organization for Economic Co-operation and Development (OECD), a multinational trade organisation to promote trade and economic development based in Paris, developed a set of Guidelines on the Protection of Privacy and Transborder Flows of Personal Data. This document provided protocols for the collection and exchange of information in international trade.

Advances in digital data collection, storage and analysis technologies have greatly heightened awareness of the issue of data protection on the part of general public, civil rights advocacy groups, as well as government and non-government organisations that have a stake in the collection, dissemination and regulation of the use of data. Large amounts of data from different sources can now be assembled, manipulated, correlated and redistributed relatively easily. Address-based data sets can also be easily geocoded to enable them spatially for data integration and analysis. At the same time, spatial data pertaining to movement of people and goods can now be collected in real time using mobile data loggers and global positioning systems. Such advances in spatial technologies, together with the increasing integration of sensitive data such as policing, health care, real estate property registration and other financial transactions, have made the use of spatial information more intrusive of personal privacy than ever before.

The objective of data protection legislation is to create the necessary legal instruments to check the potential abuse and misuse in the collection, communication and disclosure of personal information. There are two aspects of data protection. The first involves giving rights to the individuals about whom data are collected and stored. The second aspect regulates the organisations that collect and process the data to ascertain their compliance with certain principles and rules. Data protection laws in different countries generally follow several common principles, as stipulated in the OECD guidelines noted above. These include, for example:

- Direct collection of information from the persons concerned (that is, the data subjects as used above).
- The use and disclosure of information only for the specific purpose for which it is collected.

- Informing the data subjects of the purpose of the information collected.
- Consent of the data subjects to the collection and use of their personal information.
- Access of the data subjects to the information about them, and the chance to correct this information if necessary.
- The independent oversight and monitoring of the application of the laws.

Unlike access to the information legislation noted above, which is concerned only with government agencies and public organisations, data protection legislation normally embraces these in both the public and private sectors. In Canada, for example, the Privacy Act of 1982 protects personal information collected by Federal government agencies and statutory corporations. This Act, which is overseen by the Privacy Commissioner of Canada, ensures that Canadians can access information collected about themselves and can challenge the accuracy of the information.

Private sector organisations are governed by the Personal Information Protection and Electronic Documents Act (often known as the PIPED Act). This Act initially applied to those private sector activities that are regulated by the Federal Government and personal information that is traded inter-provincially and internationally. Since January 2004, the Act covers personal health information and personal data collected in commercial activities including telecommunications, travelling and banking. The PIPED Act is also overseen by the Privacy Commissioner, who has the authority to investigate complaints regarding privacy issues. The Act also provides for individuals to take their complaints to the Federal Court. Organisations found to have violated the provisions of the PIPED Act are required to correct their practices, and complainants may be awarded damages if warranted.

A number of Canadian industry associations have developed privacy codes. One that has been particularly influential is the Model Code for the Protection of Personal Information developed by the Canadian Standards Association (CSA, 1996). The ten principles of fair information practice proposed in this document formed the foundation of PIPED Act, as well as the code of practice of organisations associated with medicine, banking, insurance, Internet services and marketing.

In the United States, the Privacy Act of 1974 focused on data collected and used by Federal government agencies. Unlike Canada, there is no independent agency to oversee issues relating to privacy legislation. Since the enactment of the Privacy Act, a range of Federal and State laws has been adopted that extended the coverage of data protection legislation to private organisations under different circumstances. These include, for example, the Cable Communications Policy Act of 1984, and, Electronic Communications

Privacy Act of 1986, the Telephone Consumer Protection Act of 1991, and the Children's Online Privacy Protection Act. In addition, the Federal Geographic Data Committee endorsed a policy on the protection of personal privacy that applied to all Federal government agencies that create and operate spatial databases (FGDC, 1998).

3.4.3   Processing a Freedom of Information (FOI) Request

Many organisations in the public sector have a special unit that sets freedom of information (FOI) policies and handles requests from the public. From time to time, managers of spatial database systems may be required to assist with the handling of FOI requests as subject experts. Although the FOI request procedures tend to differ from one organisation to another, the underlying principle is always the same, namely to ensure that the responses comply with both access to information and data protection laws. Therefore, a good understanding of application laws and internal FOI guidelines and procedures is important for spatial database managers.

Figure 7.4 shows the typical steps of handling and responding to a request for information. The process starts with the receipt of a written request specifying the information being requested. The request is then logged and date-stamped so that its status can be tracked during the process. It is also checked to ensure that it contains the necessary details and includes the prescribed fees. Once the FOI staff are satisfied that the request meets the basic submission requirements, it is defined either as a request for general or personal information. This will enable the FOI staff to determine whether or not the agency has the authority to disclose the requested information, and has reasonable grounds, as noted in Section 3.4.1, to deny access to the requested information.

The FOI staff will proceed to search for the requested information if it is decided that the disclosure of the requested information is within the jurisdiction of the agency and the request is justifiable and reasonable. All located records are then reviewed carefully with respect to the limitations and exceptions under the access to information legislation. Regulations and rules under the data protection legislation are also consulted if they are concerned about disclosure of information pertaining to third parties. Depending on the results of the review, a decision is made that grants access to the requested information in whole or in part, or denies access to the requested information altogether. This decision is then communicated to the requester in writing. In the decision letter, it is necessary to indicate the sections of the laws that are invoked, and to provide the opportunity to clarify the decision, including the mechanism of appealing the decision if only partial access is granted or the access is denied.

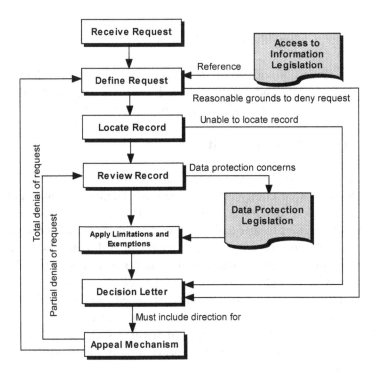

Figure 7-4. The steps of processing a FOI request

3.5 Legal Liability of Spatial Data Services

Spatial database creation is an inherently error-prone process. Errors tend to occur during data collection due to factors such as the difficulty of representing features in the real world exactly, imperfections in data collection technologies, omissions in data observation and measurement, and misidentification or misrepresentation of spatial features. Errors can also occur when existing maps are digitised and the resulting digital data sets are loaded into the database. When the spatial database is in operation, further errors will possibly be introduced when its content is changed during transaction processing and database updating, and when the data are extracted from the database for analysis. Since data in spatial databases are often generalised and aggregated according to the needs of their intended purposes, use of spatial data for unintended purposes or in unintended ways will inevitably lead to problems that are difficult to predict and to resolve.

The inherent errors in spatial databases and the potential abuse or misuse of spatial data mean that spatial data users can easily get involved in *liability* litigation when something goes wrong as a result of using spatial data. The potential liabilities become exacerbated as spatial data become more easily shared by different users and are increasingly used in policy making and decision support where the financial stakes are high and social impacts are significant.

The next section provides a brief overview of the concept of liability, and explains the principle and practice of exclusion and limitations of liability with special reference to spatial database implementation.

3.5.1   The Concept of Liability

The term *liability* in law means "the condition of being liable or answerable by law or equity" (Oxford English Dictionary, 908, Vol. VIII, 1989). Liability is a fundamental of *tort* law. By definition, a tort is a *legal wrong*. Tort law refers to that body of law where the plaintiff is the victim of an alleged wrong and the unsuccessful defendant is directed by the court to pay damages to the plaintiff (the usual remedy) or else to desist from the wrongful activity (which is called *injunctive relief*) (Coleman, 2003). The central idea in tort law is that liability is based not so much on acting wrongfully, but on committing a wrong that causes harm or damage. Therefore, torts require both wrong and harm, except in cases in which injunctive relief is awarded in order to prevent harm that is virtually certain but yet to happen.

When spatial data are used, various wrongs may occur that cause harm to third-parties (persons or organisations) and, consequently, will potentially lead to liability litigation. These include, for example (Figure 7.5):

- *Breach of contract.* A contract exists, either in writing or verbal, when a person or an organisation agrees to provide a product (for example, collecting spatial data or developing a software application) or a service (for example, to carry out a feasibility study for a spatial database project) to another person or organisation in return for a fee. A breach of contract occurs if the person or organisation contracted to provide the product or the service fails to deliver the service in time or according to the specifications in the contractual agreement. This is a breach of express terms, or the terms explicitly stated in a contract. Contract laws in some jurisdictions also contain implied terms, or terms that apply generally to all signed contracts unless they are deliberately excluded by mutual agreement of the parties concerned. Failure to honour the implied terms is also considered a breach of contract. If there is a breach of contract, then the provider of the product or service is potentially liable in damages to the customer. Depending on the degree of the failure to deliver, the customer is entitled to terminate the contract and seek full or partial damages to cover the loss resulting from the breach.

- *Nuisance.* A nuisance occurs when a person is found to be interfering with the rights of a private member of the public to enjoy a property or a service, for example, by repeatedly submitting an unreasonably large number of FOI requests with the intention of disrupting the regular operation of an organisation (see Section 3.4).
- *Infringement of third-party intellectual property rights.* This includes copyright, database rights and other rights as explained in Section 3.3.
- *Negligence.* Negligence is found against a person or organisation considered to have committed a wrong, regardless of whether it is intentional or not, that has caused harm to others. In order for liability from negligence to be established, a court of law has to find that three conditions comprising negligence apply to the plaintiff. These include (i) a duty of care was owed to the plaintiff by the defendant; (ii) there is a breach of that duty; and (iii) that damage was caused by that breach. In addition, two tests of reasonableness must also exist for the court to find negligence. These are (i) the event leading to harm must have been reasonably foreseeable; and (ii) the defendant must be proven to have not acted in the manner of a reasonable person to avoid the harm from occurring.
- *Non-compliance with legislation,* such as access to information and data protection laws discussed in Section 3.4, and regulatory standards of products or professional services.

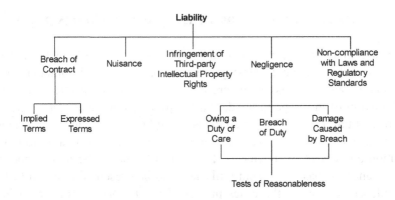

Figure 7-5. The issues of liability associated with the use of spatial data

Liability imposed on persons or organisations as a result of committing the above wrongs is called *fault liability.* It is also possible to impose liability where there is no wrong. In tort laws, such liability is called *strict liability.* The rule of strict liability applies if a user is able to show that a product is of an inherently dangerous nature. There is no need for a user to present evidence that the producer acted improperly in order to seek damage. In recent years, the concept of strict liability has evolved to *product liability*, under which heavy strict liability costs are imposed on makers of such varied items as foods, drugs and automobiles. Phillips (1999) argues that the concept of product liability applies to creators as well as distributors of spatial data products.

3.5.2   Exclusion and Limitations of Liability

Most commercial contracts include one or more clauses that attempt to limit and/or exclude liability. A common practice is to post a disclaimer to products excusing the producers from the consequences of using the product. It is also conventional for mapping agencies to place a disclaimer on printed maps by asking users to verify the accuracy of the information presented before it is used. However, independent of such disclaimers, liability laws generally do not allow the exclusion or limitation of liability in the following situations:

- Death or personal injuries resulting from negligence.
- Fraudulent misinterpretation.
- Statutory implied terms such as satisfactory quality, fitness-of-use requirements and conformance of products to regulatory standards.

3.6   Spatial Information as Evidence in Courts of Law

Printed maps are commonly used as exhibits in courts of law to depict the results of traffic accident investigations, scenarios in environmental impact assessments and other issues where they can provide evidential verification or refutation of the issue at hand. Hard copy maps are also commonly used as a referential resource in spatial decision and policy making. Occasionally, legal conflicts may occur when different stakeholders are unable to reach a common agreement over what is depicted on a map. The maps concerned and the analysis based on them will then also be presented in a court of law as evidence relative to the cases presented by the parties concerned in the dispute.

As digital spatial data become more readily available and digital maps gradually replace printed paper maps, there is growing concern about the ability of digital data to meet the legal standards of acceptable evidence. Onsrud (1993) suggested that information held or created in digital form has, for several reasons, more problems when used as evidence in the courts than paper map products. These include, for example, input errors, inherent inaccuracies caused by hardware and software, and flawed modelling concepts. Spatial data in digital form can be changed, deleted or otherwise modified relatively easily and alterations are hard to detect and trace. As a result, digital spatial data and maps generated from spatial databases tend to be less acceptable as evidence in courts of law than conventional hard copy maps, even although most conventional maps these days are themselves the products of digital spatial data.

For digital spatial data to be admissible legally in courts of law, three separate but interrelated tests must be completed. These include:

- *Authenticity.* This process seeks to verify claims associated with a spatial data set or spatial data objects in order to ascertain what they actually are and what is claimed by external metadata.
- *Integrity.* This seeks to establish that the content of a data set has not been altered or corrupted over time or in transit. It also addresses the accuracy and timeliness of the data.
- *Provenance.* This test is concerned with data origin, characteristics, history and the chain of custody and their relationships with other data sets. This last point distinguishes provenance from general metadata. Since provenance also includes claims about a data set, it is part of authenticity. Further, since provenance contains the history and chain of custody, it is an important reference source in establishing the integrity and authenticity of the data.

Various methods have been developed to secure spatial data sets and information products generated from them that are subject to the above tests. These include, for example, document locking, time-stamping, and the use of watermarks. The legal issues of using spatial data, therefore, are not solely concerned with laws and regulations. There is a strong element of technology involved with the use of spatial data as well. Because of the spectacular growth in the use of digital spatial data in all areas of decision making, Simmons (1994) predicted that most legal actions would involve the application of spatial analysis and the use of spatial databases in planning and public policy. This implies that the legal issues of spatial databases are not only confined to the conventional regimes of intellectual property rights, access to information and data protection, and liability of spatial data services as explained above, but they must also include the ability to withstand stringent reliability tests in a court of law. This is because if the reliability of spatial data is questionable as evidence in court, by extension the reliability of a decisions based on the data may also be questionable. Such vulnerability to error has the potential to trivialise the value of spatial databases as a decision support tool, and defeat the whole purpose of implementing spatial databases in the first place.

4.   SUMMARY

This chapter considered two important aspects of spatial database systems. First, the objectives and processes for establishing a user education and training support plan were considered as part of a spatial database implementation project. Second, the increasingly important legalities that involve virtually all aspects of enterprise-level database development, deployment and use were reviewed.

Several key methods for conducting a staff training needs assessment were discussed. Use of one or a combination of these methods on a continual basis within an organisation allows it to evolve the level of staff skills as the technology evolves, or at least to remain within a reasonable lag of new innovations. Failure to maintain skill levels with current technology may create an untenable time lag in staff skill sets that will hamper productivity and potentially create an insurmountable hurdle to overcome.

Failure to understand or 'know' the rights and wrongs of activities within a legal framework does not constitute a valid defence against stepping outside the bounds of 'legally' acceptable activities. Hence, the legal risks involved in not acquiring the requisite knowledge were identified and grouped into four general domains or areas including intellectual property and copyright, access to information legislation, the liabilities of supplying and using spatial data, and evidentiary standards of spatial data in court.

5.   REFERENCES

Altchuld, J.W. and Witkin, B.R. (2000) *From Needs Assessment to Action*, Thousand Oaks, CA: Sage Publications, Inc.

AIRTF (Access to Information Review Task Force) (2002) *Access to Information: Making it Work for Canadians*, Ottawa, ON: Public Works and Government Services Canada.

Cho, G. (1998) *Geographic Information Systems and the Law*, London: John Wiley & Sons.

CIPO (Canadian Intellectual Property Office) (2002) *A Guide to Copyrights*, Hull, Quebec: Publication Centre, Industry Canada.

Coleman, J. (2003) "Theories of Tort Law", in *Stanford Encyclopedia of Philosophy* by Zalta, E.N. (Principal Ed.), Stanford, CA: The Metaphysics Research Laboratory, Center for the Study of Language and Information, Stanford University (Http://www.plato.stanford.edu).

CSA (Canadian Standards Association) (1996) *Model Code for the Protection of Personal Information*, Ottawa, ON: Technical Committee of Privacy, Canadian Standards Association.

Davison, M.J. (2003) *The Legal Protection of Databases*, Cambridge, UK: Cambridge University Press.

Dobson, J. (1998) "Is GIS a Privacy Threat?", *GIS World*, Vol. 11, No. 7, pp. 34-35.

Epstein, E.F. (1987) "Litigation over Information: The Use and Misuse of Maps", *Proceedings, International Geographic Information Systems (IGIS) Symposium: The Research Agenda*, VA, Vol. 1, pp. 177-184, Washington, DC: National Aeronautics and Space Administration.

Epstein, E.F., Hunter, G.J. and Agumya, A. (1998) "Liability Insurance and the Use of Geographic Information", *International Journal of Geographical Information Science*, Vol. 12, No. 3, pp. 203-214.

FGDC (Federal Geographic Data Committee) (1998) "FGDC Policy on Access to Public Information and the Protection of Personal Information Privacy in Federal Government Databases", Washington, DC: Federal Geographic Data Committee.

Gaudet, C.H., Annulis, H.M. and Carr, J.C. (2001) *Workforce Development Model for the Geospatial Technology*, The University of Southern Mississippi, Harriesburg, MS.

Gaudet, C.H., Annulis, H.M. and Carr, J.C. (2003) "Building the Geospatial Workforce", *URISA Journal*, Vol. 15, No. 1, pp. 21-30.

Karjala, D.S. "Copyright in Electronic Maps", *Jurimetrics Journal*, Vol. 35, No. 4, pp. 395-416.

Lehman, B.A. (1995) *Intellectual Property and the National Information Infrastructure*, Washington, DC: U.S. Patent and Trademark Office.

Longhorn, R.A., Henson-Apollonio, V. and White, J.W. (2002) *Legal Issues in the Use of Geospatial Data and Tools for Agriculture and Natural Resource Management: A Primer*, Mexico, D.F.: International Maize and Wheat Improvement Center (CIMMYT).

McGlamery, P. (2000) "Issues of Authenticity of Spatial Data", Proceedings, 66th IFLA (International Federation of Library Associations and Institutions) Council and General Conference, Jerusalem, Israel.

Onsrud, H.J. (1999) "Liability in the Use of Geographic Information Systems and Geographic Datasets", in *Geographic Information Systems, Vol. 2: Management Issues and Applications*, by Longley, P., Goodchild, M.F., Maguire, D. and Rhind, D. (Eds.), New York, NY: John Wiley and Sons, Inc.

Onsrud, H.J., Johnson, J. and Lopez, X. (1994), "Protecting Personal Privacy in Using Geographic Information Systems", *Photogrammetric Engineering and Remote Sensing*, Vol. 60, No. 9, pp. 1083-1095

Onsrud, H.J. and Lopez, X. (1998) "Intellectual Property Rights in Disseminating Digital Geographic Data, Products, and Services: Conflicts and Commonalities among European Union and United States Approaches" in *European Geographic Information Infrastructure: Opportunities and Pitfalls* by Masser, I. and Salge, F. (Eds.), London, Taylor and Francis, pp. 153-167.

Onsrud, H. and Reis, R. (1994) "Introduction: Law and information Policy for Spatial Databases", *Proceedings of the Conference on Law and Information Policy for Spatial Databases*, Tempe, AZ.

Oxford English Dictionary, Second Edition, (1989) Oxford University Press, London.

Phillips, J.L. (1996) "Information Liability: The Possible Chilling Effect of Tort Claims against Producers of Geographic Information Systems Data", *Florida State University Law Review*, Vol. 26, No. 3.

Rose, N. and Radcliffe, J. (2003) "Death of Copyright – Long Live Patents and Database Right", *Proceedings of Cambridge Conference 2003*, Southampton, UK: Ordnance Survey.

Samuelson, P. (1976) "Legal Protection of Database Contents", *Communications of the Association of Computing Machinery*, Vol. 39, No. 12.

Simmons, T. (1994) "Spatial Data Analysis in the Formation of Public Policy and its Acceptance as Evidence: A Litigator's Perspective on Geographic Information and Analysis", *Proceedings of the Conference on Law and Information Policy for Spatial Databases*, Tempe, AZ.

UCGIS (2003) *Development of Model Undergraduate Curricula for Geographic Information Science & Technology: The Strawman Re*port, Task Force on the Development of Model Undergraduate Curricula, University Consortium for Geographic Information Science.

WIPO (2001) *WIPO Intellectual Property Handbook: Policy, Law and Use*, Geneva, Switzerland: World Intellectual Property Organisation.

Witkin, B.R. and Altschuld, J.W. (1995) *Planning and Conducting Needs Assessments: A Practical Guide*, Thousand Oaks, CA: Sage Publications, Inc.

Chapter 8

USER NEEDS ASSESSMENT AND MULTI-USER
SPATIAL SOLUTIONS

1.   INTRODUCTION

Successful implementation of spatial database technology demands that systems designers and developers carefully consider and plan for the needs of end users before embarking on hardware/software acquisition, data collection, database creation, and application development. This process embraces a systems development strategy that aims to bridge the gap between the initial scope of the database envisioned by the project sponsor and the detailed architecture of the database conceptualised by the systems designer.

This chapter explains the concepts and methods of user needs assessment in the context of spatial database design and implementation. The approach adopted to user needs assessment is based on several fundamental principles of best practice in information system development. These principles include data orientation, user-centred design, continuous quality assurance, iterative development, detailed documentation of activities, focus on architecture and infrastructure, and adherence to standards. Section 2 explains the nature and importance of user needs assessment in the context of spatial database design and implementation. Section 3 describes in detail the process of collecting user needs information using the method of Joint Application Development (JAD). The process of translating the results of JAD sessions into user needs specifications for database design and application development is then demonstrated in Section 4. Section 5 examines the ways by which different and possibly conflicting user needs can be satisfied in an enterprise database environment by means of multi-user database models and solutions. Finally, Section 6 summarises the discussion by highlighting the current status of user needs assessment and future development directions.

2.   CONCEPTS AND METHODS OF USER NEEDS ASSESSMENT

Systems developers generally agree that the hardest part of building a database system is to decide precisely what to build, rather than how to build it. Hence, the objective of user needs assessment is to help systems designers find out as precisely as possible what to build in a database system. The following discussion explains what user needs are, and how they should be assessed in order to produce the specifications for implementing a database system that precisely matches what the users want.

2.1   Nature and Definition of User Needs Assessment

In the context of database development, a *user need* defines what a typical user requires of a database in terms of the data to be stored and the applications to be performed on the data. Simply put, a user need is what the user expects to be able to do with the data in the database. Although the concept of user needs is apparently relatively straightforward, understanding user needs in database design is by no means a trivial task. Moreover, it is an extremely important step as failure to meet user needs during both the design and implementation stages of development will result in a database that will quickly fall into abeyance after deployment. Faulk (1997) has identified several common inherent difficulties with understanding user needs, including:

- *Comprehension.* Users do not always know exactly what they want. They may have some ideas of their own information needs but such ideas are often either too general or too narrowly focused to be useful to systems designers. User's ideas of their own information needs are also commonly limited to those pertaining to daily routines only. This implies that they frequently fail to appreciate and anticipate special cases and exceptions that are critical information to the system designer.
- *Communication.* Database systems are complex artefact of people's concepts and decisions, which are inherently difficult to describe and visualise. The difficulty of communicating concepts and decisions is compounded by the diversity of the backgrounds of the users. Different

users may describe and understand the same concept and description in different ways, using different language and terminology, and at different levels of detail or abstraction.

- *Arbitrary systems structure and continuous changes.* The arbitrary and invisible nature of data and databases makes it difficult to anticipate which needs can be met easily and which will potentially decimate the project's budget and schedule. The continuous changes of user needs over time and across functional areas often aggravate the difficulty of developing stable specifications of these needs.
- *Interdependency of needs.* In many database projects, it is commonplace for different users and/or user groups to have interrelated but potentially conflicting needs and priorities. Addressing the needs of one user may adversely affect the needs of many other users. The interdependency of concerns and interests amongst users increases the complexity of any effort aiming to satisfy individual and collective user needs.

In order to overcome these difficulties, it is necessary to assess user needs using a systematic and analytical approach involving the project sponsor, the systems analyst, and end users (Figure 8.1). Various methods enable the system designer to understand and capture user needs from the users' perspective (as opposed to the system designers' perspective). These methods, which are collectively called *user needs assessment*, include document analysis, job observations, questionnaires, interviews and focus group discussions. User needs assessment represents a proactive and collaborative approach to database development. During a needs assessment, the systems analyst and the users acquire progressively and intuitively a comparable understanding of the business functions of the organisation and the database functions required to serve them. By working together, they can turn unstructured and sometimes unclear user needs into formally structured database specifications.

Typically, users are represented in needs assessments by those who have in-depth knowledge of an organisation's business operations or the particular application domain the database system is intended to address. Such users are commonly called the *subject matter* or *domain experts* in project documentation in order to distinguish them from the end users who will use the system after it is fully implemented. However, the views of end users should not be left out of the process of user needs assessment. Often end-users are invited to validate and verify the needs identified by the subject matter experts, and to participate in systems testing at various stages of the systems development life cycle (SDLC). The immediate outcome of a user needs analysis, which is generally called the *requirements specifications*, include descriptions of spatial data and business processes, as well as guidelines, constraints and rules for database applications. These collectively constitute an agreement among key stakeholders on the system's design characteristics and functions. They are the primary source of information for database design, hardware and software selection, application and interface specifications, and project implementation planning.

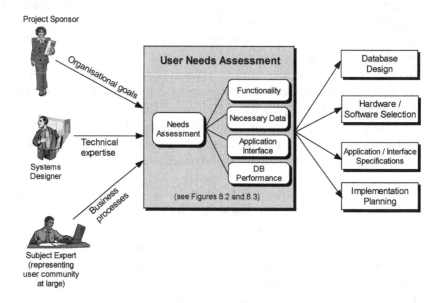

Figure 8-1. Elements of user needs assessment

The process of assessing user needs in database design is variably called *systems analysis, functional requirements study,* or *business function analysis* in the database literature (Batini et al., 1992). It is now also commonly referred to as *requirements engineering* as a result of the emerging concept of using a software or systems engineering approach to computer application development (Pressman, 2005; Thayer and Dorfman, 1997). The term "user needs assessment" is preferred because of its implication of generality with respect to the understanding of business processes, data and applications. The term "assessment" also implies the use of analytical techniques, structured approaches and commonly accepted practices in the identification, verification and documentation of user needs.

User needs assessment is an integral component of the initial database study phase of the SDLC, as discussed in Chapter 3. It provides the information necessary for the design phase of the life cycle and, therefore, can be regarded as an essential part of the conceptual and, to a lesser extent, logical modelling phases of database design.

2.2 Objectives of User Needs Assessment

The scope of user needs assessment as defined in the previous section is much broader than identifying the requirements of the database system and turning them into design guidelines and specifications. User needs assessment also addresses organisational issues including changes in management structure, business process re-engineering, and re-allocation of human and physical resources necessitated by the introduction or upgrading of database technology. The user needs assessment associated with a spatial database implementation project typically serves multiple purposes. These include:

- To provide a holistic, information science perspective of the database project before committing to specific software and hardware architectures. User needs assessment gives the project sponsor and the systems analyst the opportunity to evaluate critically the functional feasibility of the proposed spatial database and its relationships with corporate systems information technology policies and strategies. It reveals the organisational impacts of upgrading existing and introducing new spatial database technologies, and possible implications of the proposed spatial database on internal business operations and external client services.
- To provide a structured and systematic approach to identify desired systems functions and, equally importantly, to ensure that no major functions are overlooked and no unwarranted functions are included when designing and implementing the database system.
- To provide a coherent framework that relates business objects, business activities, and the associations among business objects and business activities as they pertain to the acquisition, structure, and use of spatial data in the organisation.
- To enable the solution of complex business problems by breaking them down into manageable pieces and, where appropriate, prioritising them in a systems development and technology migration plan according to the business goals and operational needs of the organisation.
- To provide a solid foundation and framework for database and application development, by focusing on reconciling and resolving the different user needs of multiple information sources and for multiple purposes.
- To facilitate knowledge transfer and integration between database users and the systems designers, and to support a channel of communication among all parties concerned throughout the entire SDLC.

- To provide a conceptual framework for sharable information services and interoperability, by approaching the database design and implementation process from a multi-user perspective that takes into consideration the needs across different business areas and application domains.
- To create an atmosphere of trust and respect between the systems development staff and the user community right from the beginning of the database project by having all parties concerned work together in the development of database specifications.
- To produce a blueprint for initiating organisational change and transformation necessitated by the introduction of database technology, for example, the re-engineering of business processes and practices, as well as the re-allocation of human, fiscal, and technical resources.

A poor understanding of user needs is generally recognised as one of the most important causes of failure in computer-based systems (Davis, 1993; McConnell, 1996). The complexity, and hence the cost, of correcting a database or application error caused by an oversight or deficiency in user needs assessment can be enormous. Based on the results of studies in the systems development industry, Figure 8.2 shows how quickly the relative cost of fixing a systems error can escalate at progressively advanced stages of the SDLC (Boehm, 1981; Faulk, 1997). An error detected in a user needs assessment that costs a dollar to correct may cost hundreds of dollars to correct if it is not discovered and fixed until after the system is fully implemented and is in the maintenance phase.

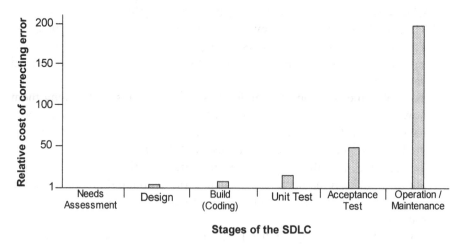

Figure 8-2. Relative cost for correcting systems errors in progressively advanced stages of the SDLC (Source: Faulk, 1997)

Experience has shown that most systems errors and deficiencies are due to the lack of understanding of user needs. Many of these errors and deficiencies do not manifest themselves until the system is operationalised in the production environment when different components and processes begin to interact and affect one another in ways that were not anticipated beforehand. This means that the relative cost of fixing errors can be easily translated into absolute costs, often amounting to tens or even hundreds of thousands of dollars. Frequent systems errors and deficiencies lead to loss of productivity, missed business opportunities, user frustration, and possibly the eventual demise of the database system itself. All these mishaps can be minimised by completing a properly executed user needs assessment prior to the actual design and construction of the database system and related applications.

From a project management perspective, therefore, the importance of user needs assessment can never be stressed enough. Good understanding of user needs encourages the design of spatial database systems that are more robust in functionality, faster and less costly to build, easier to maintain, more readily accepted by users, and closer to interoperability. Sponsors and managers of spatial database systems should never allow any database project to proceed unless they are satisfied that a user needs assessment has been adequately completed and documented. Failure to do so will inevitably lead to serious or irreparable problems in the design, implementation, and maintenance of the database system in the long run.

2.3 An Integrated Approach to User Needs Assessment

User needs assessment is an essential aspect of systems development. As such, it is always conducted by applying the principles of best practice in database system design and implementation. These include:

- *Data orientation*, to emphasise the fact that data rather than people, procedures or technology are among the most stable components of a database system. Data orientation also implies needs assessment of the complete *data life cycle* that includes identification, collection, structuring, database loading, management and eventual retirement from the database for archiving.
- *User-centred design*, which aims to provide a system that best reflects users' needs, rather than one that is built with the latest and most sophisticated hardware and software technologies that are available. User-centred design relies heavily on user participation in every phase of the SDLC, and promotes the accommodation of differing user

requirements by mediation, consensus building, and the development of multi-user solutions.

- *Continuous quality assurance*, which means using the problem solving/problem prevention cycle used in modern manufacturing that emphasises the detection and correction of errors before they occur and the prevention of problems both before and after delivery of the system to users.

- *Iterative development*, which encourages continuous evaluation of user needs and, where necessary, refinement and modification according to the findings of the evaluation by users at progressively advanced phases of the SDLC.

- *Managing change*, which refers to the ability to articulate the amendment, modification, and deletion of user needs resulting from an iterative development process, so changes that occur in one aspect of the system under development will be reflected in all related and affected places to maintain consistency and integrity of the documentation of the needs assessment.

- *Detailed project documentation*, which requires that all activities, findings, resolutions and other records of project meetings be systematically documented and filed in a project repository to ensure accountability, provide guidance, and enhance productivity throughout the SDLC.

- *Focus on architecture and infrastructure*, where architecture defines the logical structure of the data, hardware, software, and human resources required to support an organisation's business operations, and infrastructure providing the physical and policy framework for developing, managing, and deploying these resources and underlying technology in support of organisational goals in a cost-effective and sustainable manner.

- *Adherence to standards*, which demands that design methodology, use of hardware and software, data collection and structure, systems administration and maintenance, as well as documentation pertaining to these, be in compliance with applicable laws, regulations and prevailing standards.

Figure 8.3 depicts the conceptual framework of a user needs assessment based on the above best practice principles. This is an integrated approach that models database structure, business processes, and data-process interactions in a single assessment exercise. The simultaneous consideration of structural, operational and interface aspects provides a more coherent and comprehensive understanding of user needs than considering these aspects separately as in conventional methods. It also helps shorten the SDLC by articulating system development efforts, streamlining the project workflow, and sharing of resources by the database and application development teams.

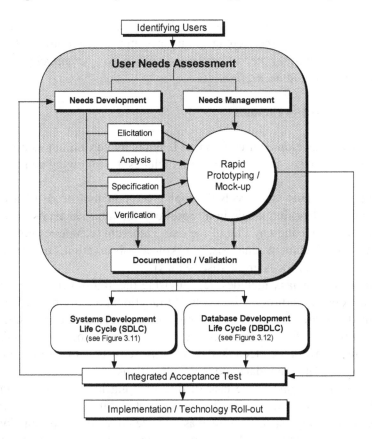

Figure 8-3. Steps and workflow of a typical user needs assessment as it related to the systems and database development life cycles

The first step of a user needs assessment is to identify who the users are. Depending on the purpose of the database system to be built, users can be internal or external, or both. Internal users in general are relatively easy to identify. However, external users are often hard to determine precisely, for example in the case of Web-based spatial database applications. Identification of the end users is usually provided by the project sponsor and front line managers whose business areas will be served by the proposed database system. The system analyst's experience and knowledge also play a significant role in many cases. Using information from all possible sources, the systems analyst estimates the number of potential users, how frequently they will access the database system, and what computer skills they possibly have. With the help of the project manager, the systems analyst also attempts to identify and get acquainted with potential subject experts who will be solicited to participate in needs assessment and subsequent stages of the SDLC.

The integrated approach to user needs assessment as illustrated in Figure 8.3 consists of four conceptually distinct but procedurally interrelated components, namely *needs development, needs management, rapid prototyping,* and *documentation.*

2.3.1   Needs Development

The needs development component is typically carried out as a four-step data gathering and analysis process that includes the following components:

- *Elicitation.* This task involves scoping the potential needs of users and filing the identified needs in a repository for future analysis. Elicitation can be done using any one of the data gathering techniques noted above (that is, document analysis, workplace observation, questionnaires, interviews and focus group discussions).
- *Analysis.* This task includes decomposing the high-level needs collected in the elicitation process into detailed functional requirements, which are called *features* in systems design terminology, and prioritising them according to factors such as the business goals of the organisation, resources available, technical feasibility, cost-benefit ratio, marketing force, legislated mandate and legal obligations.
- *Specification.* This is the task of documenting in detail the findings of elicitation and analysis in an accepted and structured format. Conventionally, specification was completed with the production of a systems requirements document. However, there is now a growing tendency toward automating specification by storing the needs in a database with the aid of a needs management tool, such as those noted in the following section.
- *Verification.* This is an on-going process that runs parallel with the previous three tasks and that may involve aspects of recursivity or feedback as needs are refined and verified. It involves evaluating the correctness and completeness of the elicited needs, the findings of the analysis as well as the resulting specifications during the iterative needs development process. Verification is a very tedious and time-consuming process that depends, to some extent, on the quality and quantity of documentation generated during the parallel tasks. As noted in Section 2.2, it is much cheaper to correct errors at the assessment stage. Hence, this process must be carried out as thoroughly and diligently as time and resources will allow.

2.3.2 Needs Management

The needs management component is designed to support needs development. As database systems are becoming increasingly sophisticated, user needs assessment is becoming equally complex and difficult to administer. Numerous needs management tools automate the process, either as standalone software applications or as a special module of a project management software package (Weigers, 1999).

A needs management tool contains functions for defining attributes of identified needs, such as version, stakeholder, origin and rationale, status, and priority. It also includes rules to specify how a proposed change will be submitted, evaluated, decided upon, and incorporated into the database of a needs specification exercise. Further, it has mechanisms to track the status of individual needs throughout the SDLC, to trace individual needs to completed database features and functions, and to facilitate the updating and refinement of the needs from one iteration to another during design and implementation. With the aid of a needs management tool, it is possible to store user needs and related information in a multi-user database. This enables project staff and systems developers to manipulate the specifications of particular needs, including importing and exporting them, and connecting them to related database features and functions completed at different stages of the SDLC.

2.3.3 Rapid Prototyping

Prototyping is the technique of constructing a partial implementation of a system so that both its designers and users can learn about a user need or proposed solution to a particular need prior to a complete implementation. The objective of *rapid prototyping* is to facilitate communication among the participants in a user needs assessment. As noted in the above discussion, user needs as they pertain to systems design concepts are inherently difficult to describe and visualise. This implies that decisions to accept or reject a candidate user need are not always easy to make. Rapid prototyping makes use of a variety of software development tools and computer graphics packages to augment the verbal description of user needs. It can be done with varying degrees of sophistication. In the simplest case, mock-ups of graphical user interfaces (GUIs) are produced to simulate the appearance of a completed system. More advanced prototyping produces functional systems that are very close to an operational system using the methods of throw-away, evolutionary and incremental prototyping. These methods are the basis of not only well conceived database development but in fact of most well specified systems development. Hence, they are explained in detail in the context of project management in Chapter 9.

2.3.4   Documentation

Documentation is a crucial, but commonly overlooked, aspect of user needs assessment. Poor documentation tends to cause confusion, errors, mistrust and loss of confidence in the development process. In contrast, proper documentation improves productivity, ensures accountability, facilitates communication, and keeps the needs assessment process transparent to all stakeholders. There are a number of measures of good quality documentation for user needs assessment. These include:

- *Completeness.* The assessment document should contain all information necessary to construct the database and code application programs acceptable to the project sponsor and users.
- *Freedom of ambiguity.* The documented needs can have only one possible interpretation.
- *Consistency.* The individual needs do not conflict with one another.
- *Correctness.* The descriptions of user needs are precisely what the users want.
- *Verifiability.* The ability to determine categorically that all documented needs are implemented in the final system.
- *Ease of reference.* The documentation must be organised and formatted to make it possible to answer questions about the needs quickly and easily.

The integrated approach to user needs assessment is an iterative process that goes on until the system is fully operationalised. It often happens that many user needs cannot be fully understood before the system is built. New ideas of using a system often emerge as users learn more about database technologies as the project implementation progresses. This implies that user needs have to be updated and refined continuously throughout the entire SDLC. Since it is practically impossible ever to produce a "perfect" user needs specification, some approaches to systems development do not attempt to produce a definitive user needs assessment. They advocate instead going directly from a problem model to design, or from a prototype implementation to coding.

Such an approach may be effective for some database implementation projects. However, most spatial database systems and applications are simply too complex to be understood without a thorough user needs assessment. Despite its inherent difficulties and the resources required, a carefully planned and conducted user needs assessment is still the best way of assuring the eventual success of a spatial database implementation project.

3.   COLLECTING USER NEEDS INFORMATION BY JOINT APPLICATION DEVELOPMENT

User needs information can be collected using different techniques, as noted in Section 2.1. For multi-user spatial database projects, the method of *Joint Application Development* (JAD) is perhaps the most widely used approach for eliciting user needs within the computer industry.

3.1   Definition of Joint Application Development

JAD was initially developed by IBM in the late 1970s (Wood and Silver, 1995). It was originally used simply as a means of bringing together systems developers and users with different backgrounds and opinions to explore the design requirements of new computer systems in a productive and creative environment. Over the years the concepts, techniques and scope of JAD have all changed considerably.

At present, JAD is no longer used merely for gathering information about user needs, but it has been formalised as a sequence of workshops designed to elicit, verify, decompose and prioritise these needs in an analytical and structured manner. It is used for developing new systems as well as for improving existing systems and building new business solutions for existing databases. In practice, it is common for JAD to continue at different intervals throughout the SDLC of a database development project. However, discussion in this chapter focuses particularly on needs assessment prior to systems design. Post-design needs assessment is different in terms of its objectives, but the concept and techniques used are essentially the same.

Technically, JAD can be considered as a variant of the focus group method of data gathering. It differs from conventional focus group discussions in its tight integration of behavioural and group dynamics techniques within the structure of a systematic information gathering and analysis methodology. JAD is conducted through a series of moderated workshops (see Section 3.3). However, it is also commonly supported by conventional data collection methods such as interviews, job analysis and questionnaire surveys.

One or more of these methods can be used to collect preliminary or supplementary information about user needs, which are then validated, refined and prioritised in JAD workshops. Although some JAD workshops are still conducted using flip charts and post-it notes, technology-based visual aids are now commonly used to improve efficiency and productivity in the data gathering and documentation process. Increasingly, Computer Assisted Software Engineering (CASE) and Rapid Application Development (RAD) tools are used to create data and process models in order to obtain immediate feedback from users regarding the accuracy and completeness of content and design-related discussion.

JAD is now popularly used in the information technology industry because of the numerous advantages that it offers over other approaches. These include, for example:

- *Robust methodological foundation.* Although JAD workshops can be organised and conducted in different ways, in essence they all follow the principles of best practice in systems development noted in Section 2.3. The popular use of JAD implies that considerable experience has been accumulated by systems development practitioners, and a great deal has been written about the concepts and techniques of JAD to give this particular approach a solid methodological foundation.
- *Simplicity and structure.* JAD provides a relatively simple but structured approach that enables user needs to be captured effectively. It provides a logical framework that is able to consolidate often months of application and development meetings, follow-up meetings, data acquisition and review meetings, individual and collective interviews, and teleconferences into individual workshops with highly structured agendas, clear objectives and well-defined deliverables.
- *User participation and commitment.* JAD helps identify people who have the knowledge to assist in database design and development. It gets users involved from the very beginning of a systems development project and encourages on-going involvement throughout the SDLC. Intimate user involvement minimises potential design errors and flaws. At the same time, active participation inspires users with a greater sense of belonging to the project, which in turn leads to a higher level of commitment and determination to make the project a success.
- *Communication and group dynamics.* Capturing and documenting user needs in JAD workshops improves communication between users and system developers, and increases their mutual understanding and respect. JAD helps system analysts understand what users face in the business environment and how the database approach can assist them. At the same time, it enables users to understand the rationale behind decisions pertaining to systems development priorities, interface design and database management. In short, JAD is an educational process for both

users and systems staff that generates the group dynamics for them to work together in a coherent project team.

- *Consensus building and conflict resolution.* By having representatives of different user groups sitting together, JAD helps identify important issues and the people responsible for resolving them. It offers a way to answer questions collectively and address issues that require decisions or further investigation.

- *Flexibility and transition management.* JAD supports the evolutionary process of defining business processes and their respective information needs. It can be used for both the centralised and view integration approaches of user needs assessment in the multi-user cross-functional database environment explained in Section 3.2. It provides a systematic way to track user requirements as the implementation project goes through the SDLC. In this way, JAD helps facilitate the transition between project phases, where outputs from one phase often constitute inputs to the next phase.

As explained in Section 2.1, understanding user needs is an inherently difficult task due to the common inability of users to express uncertain desires in unambiguous language on the one hand, and the frequent occurrence of conflicting or competing needs among users on the other. The JAD approach is a proven way of easing these difficulties. As noted above, it brings users and systems staff together so that users can learn the technology gradually and become comfortable with it. Conflicting interests and concerns among different user groups can be discussed openly, mediated and prioritised in face-to-face meetings moderated by a neutral facilitator. The accomplishment of these outcomes in practice is explained in the following section.

3.2 Joint Application Development in a Multi-user Database Environment

In a multi-user or cross-functional database environment, different users tend to have different views of what the database should and can do. One of the greatest challenges of JAD is to combine the views and needs of different users into one commonly accepted set of database design specifications.

Two approaches to dealing with multi-user views in JAD include:

- *Centralised approach.* This approach collates the views and needs of different users into a single set of requirements specifications (Figure 8.4a). Such a consolidated set of specifications is usually given a collective name that describes the business or functional area covered by all the merged views. A global conceptual database model representing the consolidated view is then created at the database design stage. Generally, this approach is best suited for situations where there are significant overlaps of needs for individual user views and the database applications are not overly complex.

- *View integration approach.* In contrast to the centralised approach, this approach leaves the needs for each user view as separate lists. A local conceptual database model is created for each list independently. These individual models are merged at a later stage of database design (Figure 8.4b). The view integration approach is generally preferred when there are significant differences among user views, and when database applications are so complex that it is more manageable to address them separately and individually rather than as a single total solution.

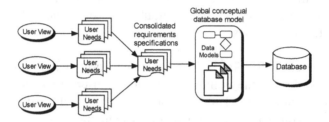

(a) The centralised approach to multi-user needs assessment

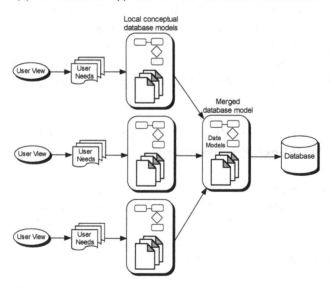

(b) The view-integration approach to multi-user needs assessment

Figure 8-4. Approaches to user needs assessment in a multi-user database environment

3.3 Key Participants in Joint Application Development

JAD requires participation, where possible, by database users as well as systems professionals. The number of participants is normally kept relatively small (between five and eight) to ensure active participation and to make the process more manageable in terms of scheduling, costs, and commitment. The typical ratio between users and systems staff is three to one. These participants form the *core team* and are required to attend every workshop from beginning to completion. Occasionally, people with special knowledge about a particular issue or business process can be invited to participate in workshops when such items are on the agenda.

The central figure of the JAD methodology is a facilitator who has had special training and experience in moderating group meetings and workshops. The facilitator can be chosen from among internal systems professionals with good moderation and human skills. However, some organisations prefer to hire a professional facilitator externally from a consulting firm. This is intended to avoid the potential perception of bias or favouritism toward particular user groups in the organisation. A good facilitator always remains neutral and impartial, in word and in behaviour, is willing to listen, recognises issues as they arise, and provides the leadership to help people work together. The objective of the facilitator is always to ensure that each participant is heard and has an equal opportunity to influence the decision, regardless of his or her position in the organisation.

Working closely with the facilitator is the scribe who is responsible for recording participants' questions, responses, comments and proposals verbally and by means of drawings. These are often projected in real time onto a large screen directly from the scribe's computer for users to verify for correctness and refine for clarity. The scribe edits the content of the transcription as it grows and is modified as the workshop progresses under the moderation of the facilitator.

Users are generally represented in JAD workshops by those who have an in-depth understanding or experience in particular business functions or processes of the organisation. As noted, these users are commonly called subject matter or domain experts to distinguish them from the end users. For the development of multi-user or cross-functional database systems, it is important for the facilitator and project manager to ensure that users in different parts of the organisation are represented. It is also desirable to have a mix of user representatives who will actually use the system (for example, front desk clerks), and those who look at the systems more from a managerial perspective (for example, front line supervisors). Occasionally, senior managers are invited to a workshop to show support for the project and to contribute their experience in areas covered by particular JAD workshops. It is also common for outside experts such as business and technology consultants to be contracted to provide advice that may not be available in-house.

The project manager usually takes part in JAD workshops as a logistical supporter of the facilitator and an observer rather than as an active participant in the discussion. The project manager helps the facilitator in pre-workshop preparation and assists in the editing of workshop documentation between workshops. The project manager also serves as the liaison between the JAD team and the project sponsor, and is responsible for signing off the JAD deliverables. When participating in workshop discussions, it is important for the project manager to remember that he or she is not speaking from the authority of his or her position. The facilitator has to ensure that the opinions of the project manager are not given special consideration in the decision making process as this would show unreasonable bias and may perhaps disincline others to participate fully in subsequent exercises.

3.4   The Process of Joint Application Development

As noted earlier, JAD is typically conducted as a series of moderated workshops, held in three distinct phases, away from the normal workplace of the participants, for example, in the conference facilities of a hotel. The objective is to keep the participants away from the distractions and interruptions that inevitably occur when people are meeting in their own workplace. Individual JAD workshops are generally held several days apart and are sequenced as follows: pre-workshop, workshop, and post-workshop activities (Figure 8.5). This allows participants to have sufficient time to review the results of a completed workshop, to prepare for the next one, and where necessary to carry out research on issues that have to be clarified and addressed in future workshops.

3.4.1   Pre-workshop Activities

Good preparation is essential for the success of JAD. Before the first workshop, the project manager and the facilitator should jointly identify its scope, gather necessary background information, and define the structure and goals of the workshops. They have also to identify and select participants, usually with the help of the project sponsor or senior managers in the organisation. It is common practice to talk to selected participants personally and, as a matter of courtesy, also to talk to their immediate managers. This will ensure that they are interested and available, and that they understand the commitment required before inviting them to participate at least two weeks but not more than one month prior to the first workshop.

The invitation should enclose background information on the project, a concise statement of the goals of the JAD workshops, the importance of the participation by users, as well as the time, place, and agenda of the first JAD workshop.

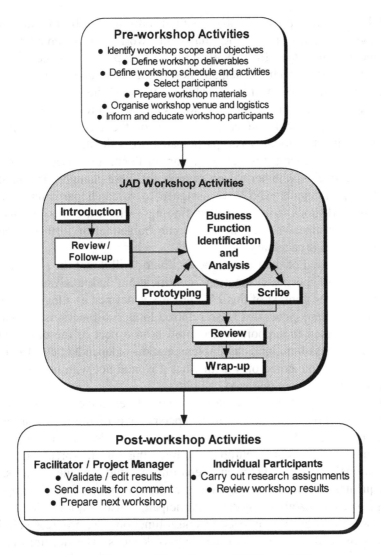

Figure 8-5. The three phases of JAD activities

At the same time, the project manager prepares and coordinates workshop logistics. These include, for example, booking of workshop venues, arranging accommodation for people who require overnight stay, ensuring the proper layout of the workshop room for optimal communication and interaction, as well as installing the computers, overhead projectors, screens, white boards and flip charts necessary for the workshop.

3.4.2   Workshop Activities

A JAD workshop usually starts with an introduction by the facilitator in which ground rules and the agenda of the workshop are announced. The introduction for the first workshop also typically includes a brief introduction from each participant describing who they are, where they work, and what they expect to contribute to the workshop.

With the exception of the first workshop, the introduction is normally followed by a critical and careful review of the results of the previous meeting. The facilitator asks participants to validate the findings and decisions, make changes to refine or correct them, and discuss all changes put forth to decide whether to accept or reject these changes. The facilitator also asks participants who were assigned to research particular issues in previous workshops to present their findings, which are then discussed in order to determine whether the issues can be settled or further discussion and/or research is necessary.

On completion of the review, the facilitator will direct the participants to the process of business function identification and analysis. The term "function" in the context of this JAD process is defined as a logical grouping of related *business processes* (also referred to as sub-functions, activities or tasks) undertaken to support or accomplish an aspect of the organisation's goals. A business function is an implementation-independent construct in the sense that it is concerned only with what it is, and not with how it is served by the database system that is to be built.

With the help of the facilitator, participants examine the organisation's business functions critically from the perspective of the organisation as a whole, rather than from the perspective of specific departments or business units individually. The purpose behind this is to break down the conventional barriers between departments and business units, and consequently to pave the way for the development of an interoperable enterprise-wide database system. Participants also explore the business processes that occur within each business function, and progressively break down individual activities into detailed work steps. The decomposition of individual processes and work steps continues until each one becomes a distinct and discrete work step that can no longer be decomposed any further. This process of analysing the business of an organisation starting from a high level and moving into smaller and more detailed work steps is called *business function decomposition* (BFD). The outcomes of BFD can be graphical, which is called a *function decomposition diagram* (FDD), or textual, which is called a *business function hierarchy* (BFH). The concept of BFD and the characteristics of the resulting FDD and BFD are explained in more detail in Section 4.1.

In order to help participants identify business functions and related processes, the facilitator directs the participants to think critically by answering the following questions:

- *What* triggers the function?
- *When* the function is performed?
- *Where* the function is performed?
- *How often* the function is performed?
- *What* work steps are required to complete the function?
- *What* opportunity exists to improve the function?
- *Who* owns the function?
- *Who* currently performs the function?

The facilitator should deliberately avoid asking any questions about data at this point in time. This is because people usually think of their work in terms of the business functions they perform rather than the data they use. The data used and generated by each function are derived in a separate phase of work after the identification of the business functions are completed, as explained in Section 4.1. When soliciting answers from participants, the facilitator usually starts by encouraging more knowledgeable participants to lead the discussion, and keeps on inspiring an atmosphere of collaboration among the rest of the participants. At the same time, the scribe transcribes the participants' discussion into a FDD or BFH, which is projected on to a large screen for real time review and editing, as explained in Section 3.3.

Occasionally, there are complex issues about which no conclusions can be drawn or no decisions can be made because participants do not understand them well enough. When such an issue arises, the facilitator should assign it to one or more of the participants to study in detail and report back to the overall group in future workshops. All assignments to undertake further research must be recorded so that their progress can be traced and follow-up actions can be taken until they are satisfactorily completed.

Prototypes are sometimes built to assist in workshop discussions at a later stage of JAD when the focus moves from business function identification to business function analysis (see Section 4.1). The throw-away approach of rapid prototyping (see Section 2.3) is commonly used at this stage, because the objective is to clarify concepts and demonstrate possible business solutions rather than actually to implement a lasting solution. Prototyping requires additional resources and expertise to build an interim product, however it is a very effective means of fostering communication among participants as they can visualise a tangible product rather than sometimes unclearly articulated and connected ideas. For complicated functions or features that are too difficult and time-consuming to prototype, simpler methods such as workflow diagrams, rough E/R or UML diagrams and screen mock-ups of GUIs, are sometimes created instead. All these allow participants to visualise potential solutions for desired user needs, to verify them, and make suggestions for improvement right on the spot.

Before concluding a workshop, it is essential for participants to review the results of the workshop collectively for accuracy and clarity. It is also advisable for the facilitator to go through all the assignments that were allocated to individuals between workshops once more. The facilitator should ensure that participants who are given the assignments have sufficient resources and time to carry out the research, and that they fully understand on what and when they are expected to report back in the future. Finally, the facilitator reminds the participants of the date of the next workshop. If for any reason some participants are unable to attend, a new date that is mutually convenient must be set by consensus.

3.4.3   Post-workshop Activities

Post-workshop JAD activities are the tasks that are undertaken between one workshop and another by the project manager, facilitator, and participants. On the part of the project manager and facilitator, these taks include the validation and proper documentation of the results, and sending the results to all workshop participants for review and comment well in advance of the next meeting. It is common practice to send a copy of the results to senior managers and people knowledgeable about the organisation's business, but who were unable to attend the workshops. In this way, their input and comments can be solicited. The facilitator also uses the results to set guidelines regarding what needs to be done to refine and acquire further specifications during the next workshop.

The primary post-workshop responsibility of the participants is to review critically the results of the previous workshop and to prepare as fully as possible for the next meeting according to the agenda they receive from the facilitator. For those participants who were assigned to undertake specific research on certain issues or topics, they have to make good use of the time in between workshops to investigate the issues by talking to their colleagues, conducting on-line literature searches, and consulting external subject matter experts if necessary.

The above three-phase workshop activities are repeated for all workshops until the business functions of the organisation are exhaustively examined and a relatively stable FDD or BFH is obtained and accepted by the participants. At this point the focus of subsequent workshops turns to the analysis of the FDD and BFH to identify possible user needs associated with individual work steps in each business function, as explained in Section 4.

4. TRANSLATING BUSINESS FUNCTIONS INTO USER NEEDS SPECIFICATIONS

The business functions and processes identified in JAD workshops are used in two ways. One is to identify what data are required by each of the business functions and processes within the organisation. The other is to identify how data required by individual business functions and processes are used and related to one another. The following discussion reviews the methods of translating business functions and processes into specifications, detailing what data are used and how they are used. The discussion begins with an examination of the underlying concept and technique of BFD and the characteristics of the resulting FDD and BFH.

4.1 Business Function Decomposition

Section 3.3 briefly introduced the method of BFD as a top-down approach of decomposing the business functions of an organisation progressively into discrete processes or work steps. Figure 8.6 is a FDD that illustrates the concept of BFD by showing how a typical business function of "change information in the parcel register" of a Land Records Office is progressively decomposed into finer business processes and work steps. Space does not allow presentation of the entire FDD but the portion shown is adequate to explain how the business function is decomposed first into business processes of "Change mailing address", "Change ownership information", "Register an Interest", and so on. The "Change ownership information" business process is then decomposed further into finer processes until each one of these processes becomes discrete work steps that can no longer be decomposed any more. Other business processes can be decomposed similarly.

A FDD is a very effective way of visually presenting the relationships between functions and processes. However, there are several critical limitations when using this graphical method to document the results of a BFD. For example:

- A FDD will grow very quickly to become too large to be effectively depicted on a single sheet of paper.
- It is difficult to relate one process to another if a FDD has to be recorded on several sheets of paper.
- The size of the text boxes severely restricts the number of words that can be used to describe a process.
- It is difficult to edit a FDD that is constructed manually, for example, when processes have to be moved around to reflect their sequence and when processes have to be added or deleted at an advanced stage of the development of the diagram.

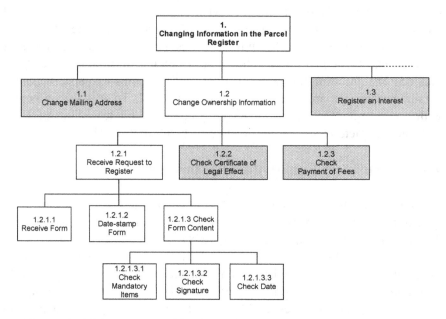

Figure 8-6. The concept of business function decomposition (BFD) as illustrated by a function decomposition diagram (FDD)

A more flexible alternative to a FDD is the BFH, a text-based method of documenting the results of a BFD. Figures 8.7a and 8.7b illustrate how the same business functions and processes in Figure 8.6 are documented in the form of a BFH. In a BFH, the relationships between functions and processes are denoted by the hierarchical identification of the functions, processes, and work steps. The business decomposition process starts with a high level of decomposition as shown in Figure 8.7a, where the four level-one processes are identified by single sets of digits. Each of these processes are then further decomposed progressively to lower levels, which are identified by two or more sets of digits separated by periods.

The number of sets of digits within an identification number denotes the levels of function decomposition and the position of individual functions and processes in the BFH. During the process of decomposition, the JAD team revisits identified functions and processes from time to time and amends the BFH where necessary. The idea is always to keep a global or enterprise-wide view of the organisation's business functions and to ensure no functions or processes are overlooked or excluded. This also helps to verify that that newly captured functions and processes are documented in the proper positions relative to those already in existence in the BFH.

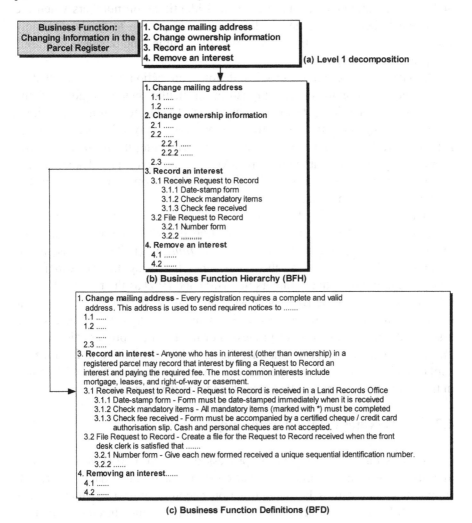

Figure 8-7. Example of a business function hierarchy (BFH) and its associated business function definitions (BFD)

A BFH is notably easier to edit than a FDD when it is documented using a word processor on a computer. New functions and processes can be added and unwanted ones removed in real time during the discussion in JAD workshops, as described in Section 3.4.2. Business functions and related processes can be moved around freely during post-workshop editing and refined to reflect more precisely their sequence relative to one another. By using the hierarchical identification numbers as a guide, it is relatively easy to comprehend the sequence of and relationships among business functions and processes. A special word processing template can be set up to change and manage automatically the hierarchical identification numbers when the document is being edited and updated. This has proven to be an extremely useful way of enhancing the productivity of documenting business functions and processes in JAD.

A business function or process denotes an action or activity that is performed to support or accomplish a particular business goal of the organisation. Therefore, it is customary to start each item in a BFH with a verb specifying an action or activity. However, many of these actions or activities are known only to users who are performing them directly. Other users and systems staff attending the JAD may not have any idea what they are, why they are necessary, and what a specific term really means in the context of the action or activity.

In order to assist in the understanding of the business functions and processes contained in a BFH by workshop participants not familiar with them, an accompanying *business function definitions* (BFD) table as shown in Figure 8.7c is created at the same time. By using the corresponding hierarchical identification numbers in a BFH and BFD as the means of cross-referencing, users of the BFH can easily find out the rationale of a particular business function or process, as well as the meanings of specific terms associated with the function or process. For the business process "Record an interest" in Figure 8.7b, for example, the associated definition in Figure 8.7c explains when this process is performed, and what the term "interest" means in the context of this particular business process using common examples.

4.2   Business Data Modelling

Business data modelling is the process by which a BFH is analysed to identify those business processes that consume or produce data. Figure 8.8 shows the relationships business function decomposition (in which a BFH and a BFD are produced), business data modelling (through which a sequence of data-related tables are generated systematically from the BFH), and database design (where the resulting tables of business data modelling are used as input for database modelling and creation).

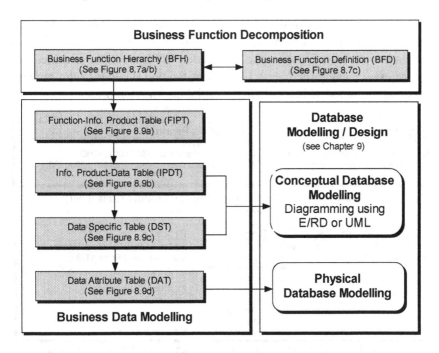

Figure 8-8. Using a BFH in business data modelling to generate information products and related data tables

In practice, business data modelling is typically carried out as a continuation of JAD by the same core team that participated in the initial JAD sessions. However, as the focus of the JAD workshops move from business functions and processes to data and information products, the system participants begin to play a more important role than before. It is also common to augment the JAD core team at this stage by inviting one or more data specialists to provide expert advice on data-specific issues (such as availability, sources, quality, and standards) if such expertise is not available from the original core team members.

The first task in business data modelling is to go through the BFH functions and processes them one by one to those that use or generate an *information product*. In this context, an information product is defined as any document containing information pertaining to a business function or process. These include text-based data forms, statistical tables and reports, maps, aerial photographs, remote sensing images, and existing databases and data warehouses. Led by the facilitator, participants examine each process carefully to determine whether it is associated with an information product and, if it is, they will mark it clearly for reference. On completion of the examination, the scribe should extract all marked processes to create a *Function-Information Product Table* (FIPT), as illustrated in Figure 8.9a.

The next step is to produce an *Information Product Data Table* (IPDT) from the FIPT by identifying the data required by or associated with each information product. This table, as illustrated in Figure 8.9b, also includes a cross-reference using the hierarchical identification number to the processes pertaining to the individual data items of the information product.

Function	Information Product
1.1 Receive notice of change	Notice from owner
2.1 Check Change Ownership form	Change of Ownership form MBS-0012-CHG
3.1.1 Obtain Req to Rec form	Request to Record an Interest form MBS-0009-REC
....	

(a) Partial listing of a **Function-Information Product Table** (FIPT) derived from Figure 8.8c

Information Product	Data Required	Process
Change of Ownership Form MBS-0012-CHG	Parcel ID	2.1 Verify parcel ID Land Records DB
	Vendor name	2.2 Check Land Record DB
	Vendor mailing address	
	Parcel location address	
	Date of transaction	2.2.1 Check date stamp
	Buyer name	
	Buyer mailing address	
	Reference to deed	
	Reference to survey	5.1 Load parcel map DB

(b) Partial listing of an **Information Product Data Table** (IPDT) showing data in an information product called "Change of Ownership Form"

Data Set	Source	Type	Attribute
Parcel info	Land Records Office	Text	Parcel ID
			Parcel location address
			Reference to deed
			Reference to survey
Owner info	Land Records Office	Text	Parcel ID
			Vendor name
			Vendor mailing address
			Buyer name
			Buyer mailing address
			Date of transaction

(c) Partial listing of a **Data Specific Table** (DST) showing data sets and associated attributes derived from (b)

Attribute	Type	Domain	Description
parcel_id	C, 20	PIN	Parcel Identification Number
vendor_name	C, 35		Name of vendor
vendor_address_1	C, 100		Mailing address of vendor line 1
vendor_address_2	C, 100		Mailing address of vendor line 2
buyer_name	C, 50		Name of buyer
buyer_address_1	C, 100		Mailing address of buyer line 1
buyer_address_2	C, 100		Mailing address of buyer line 2
date_transaction	Date		Transaction date yyyy/mm/dd

(d) Partial listing of a **Data Attribute Table** (DAT) showing characteristics of attributes associated with the "Owner info" data set in (c)

Figure 8-9. Illustration of data-related tables resulting from business data modelling

The data items identified with each information product are then organised into data sets by grouping related data items in a *Data Specific Table* (DST) under a descriptive name, for example, "parcel information" and "owner information", as shown in Figure 8.9c. In this table, participants also identify the sources of the data sets, their types such as text and numerical values (for attribute data), vector or raster (for spatial data), as well as the required qualities of the data (for example, currency, accuracy, scale, and standards).

Finally, a *Data Attribute Table* (DAT) as illustrated in Figure 8.9d is produced from the DST. A DAT provides characteristics of the attributes of each data set in the DST. These characteristics include the data type of the attribute (character string, numerical value, date, and Boolean expression), display spaces, domain and permissible values, and so on.

Although the nature of business data modelling is somewhat similar to conceptual database modelling as defined in Chapter 3, the primary objective of business data modelling is to assess user needs. The resulting tables represent only the users' views of the database to be developed. As such, they do not contain any reference to the systems designer's views of the database that include, for example, relationships between data sets and the cardinality of these relationships. In this regard, business data modelling is more appropriately viewed as a prerequisite or preparation for database design, rather than as part of database design.

In practice, there are systems designers who have attempted to integrate JAD into their database design process, however, the two processes should be regarded as separate in spatial database projects. This is because JAD and database design have relatively distinct objectives, require different expertise, and can be done better in different working environments.

4.3   Business Process Modelling

The method of business data modelling described in the previous section identifies what data users need to complete their work tasks. Such knowledge is essential for database design and construction. User needs assessment also entails an understanding of how data are used in the database. The method of identifying the various ways of using data is generally referred to as *business process modelling*. Understanding of business processes with respect to their sequence and interdependence is crucial for systems designers. In order to develop applications that allow users to access the database precisely in the ways the users want, systems designers have to know precisely how individual business functions and processes are related to the database and to one another. Such knowledge includes, for example, which process requires data from the database, what information will be produced, and how data from one process are used as input to another process.

It is obvious, therefore, that business process modelling is still very much concerned with data or, to be exact, data as they pertain to processes. There are three general approaches to business process modelling. These include (Pressman, 2005):

- *Data flow-oriented approach*, which has its origin in earlier design concepts that stressed modularity, top-down design and structured programming. Data flow-oriented process modelling is relatively simple as it centres around the construction of a *data flow diagram* (DFD) that depicts the movement of data between processes, data stores (that is, data in a database or in temporary memory) and interfaces (that is, sources and destinations of data). In theory, this approach can be applied to every application development effort. However, it is particularly useful in applications where data are processed or transformed sequentially in one or more transactions, and where no formal hierarchical data structure exists.

- *Data structure-oriented approach*, which focuses on the identification of information entities and the actions that are applied to them. The end product of data structure-oriented process modelling is a *structure chart* (STC) that depicts the procedural logic of application programs in a basic format consistent with the layout of structured computer programs. This approach of process modelling is best suited to applications that have a well-defined, hierarchical structure of information typical of the scientific and engineering domains.

- *Object-oriented approach*, which is modelling methodology that emerged as a result of the growing popularity of object-oriented technology in the computer industry. This approach of process modelling creates a representation of the real-world problem domain and maps it into a solution domain that is the software. Unlike the data flow- and structure-oriented approaches, object-oriented modelling results in a design that interconnects data (data objects or items) and processes (business activities or operations) in a way that modularises information and processing, rather than processing alone. This particular approach is now used in conjunction with object-oriented design and object-oriented programming to form a set of software engineering activities for the construction of an object-oriented system.

Although the above three process modelling approaches differ considerably in concept and technique, they share several common characteristics. These include, for example, that they are all data-driven, they attempt to transform data into an application representation, and they are based on sound systems design principles. All three approaches begin with an information collection phase by which user needs are systematically captured and documented. This implies that the method of business function decomposition and the resulting BFH and related tables can be used as input to any one of these approaches of business process modelling. This also implies that the choice of which approach to use depends more on the nature of the application, as noted in the explanation above, rather than the merits of one approach over another.

The following discussion briefly explains business process modelling using an object-oriented approach because this is fast becoming the industry standard. Object-oriented process modelling using a BFH as the source of information is made up of two steps, namely identification of objects, operations and messages, and diagramming the findings.

4.3.1   Identification of Objects, Operations and Messages

The first step is for the JAD core team to go through the BFH carefully and identify all key words that are considered to be an object, an operation or a message. In the context of object-oriented process modelling, an *object* is a component of the real world that is mapped into a software solution. An object is typically a consumer or producer of data or an information product, for example, front-desk clerks, managers, data files, interface displays, and printed reports. When an object is mapped into its software representation, it consists of a private data structure and processes, called *operations*. An operation contains control and procedural constructs that may be invoked by a *message*, which is a request that triggers the object to perform one of its operations.

4.3.2   Diagramming of Business Practices

The second step of business process modelling is to depict graphically the objects, operations, and messages identified in the previous step (Figures 8.10 and 8.11). This is an iterative process during which graphical elements have to be moved around, added or deleted, as the diagram is continuously refined and expanded. Therefore, it is essential to use a CASE tool or a general-purpose graphics software package to improve productivity in the diagramming process. UML provides a rich set of diagrams that can be used to depict the results of business process modelling relative to the object-oriented approach. Two diagrams are particularly useful at this stage.

The UML activity diagram shown in Figure 8.10 is an enhanced version of a conventional flowchart used for representing the workflow or functional view of a business process. The figure shows the activities involved in the business process of handling a Request to Record an Interest application in a Land Records Office. Closely related to the activity diagram, a UML sequence diagram provides a dynamic view of business processes by illustrating the interactions between the objects. These interactions show how objects interact with one another. Each time an object interacts with another object it invokes an operation. This implies that by modelling the interaction, the operations that the object requires and the messages that invoke the operations can be established.

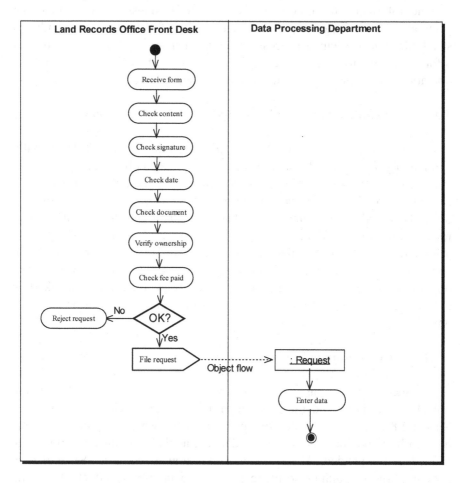

Figure 8-10. Example of a UML activity diagram showing the workflow of the business process of handling a Request to Record an Interest application

Figure 8.11 is an example of a sequence diagram that depicts the same activities involved in the business process of handling a Request to Record an Interest application in a Land Records Office. Note how the focus of this diagram has evolved from a relatively simple functional view of workflow in Figure 8.10 to a detailed representation of objects (that is, front-desk clerks, data entry clerk, Land Records database), operations (for example, verify existing ownership record and create new record), and messages (for example, filing of the application by the front-desk clerk to start the creation of new record by the data entry clerk).

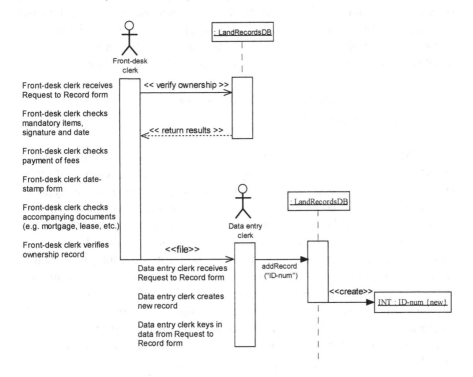

Figure 8-11. Example of a UML sequence diagram showing objects, operations and messages involved in the business process of handling a Request to Record an Interest application

Depending on the nature and complexity of the database and its applications that are to be constructed, other UML diagrams such as Use Case, Collaboration, and State Chart diagrams can also be used. However, in order not to overwhelm JAD participants who are not familiar with the technology, it is preferable to keep the process modelling method simple at this point. If the characteristics and behaviour of the business processes can be captured precisely using Activity and Sequence diagrams, it will not be difficult for systems designers to develop other UML diagrams at a later stage of the systems design process.

5.   MULTI-USER DATABASE MODELS AND SPATIAL SOLUTIONS

Database systems in general and spatial database systems in particular are typically set up to serve the needs of multiple users. The objective of user needs assessment is therefore helpful to develop a database system that all users can access equally and equitably with respect to data resources and systems functionality. The identification of user needs in a multi-user database environment using a JAD approach was discussed in previous sections. The following discussion examines how database models and solutions can be constructed to satisfy the needs of multiple users.

5.1   Characteristics of a Multi-user Database Environment

As the name implies, a multi-user database system is one that is designed to support multiple users concurrently. There are two broad classes of multi-user database systems. If a multi-user database supports a relatively small number of users (usually fewer than 50) or a specific department within an organisation, it is called a *workgroup database*. If a database is used by an entire organisation and supports more than 50, and possibly hundreds, of users across many departments or business units, it is commonly known as an *enterprise database*. Web-based database systems, which commonly serve tens of thousands of users inside and outside of an organisation also fall into this particular class of systems. This discussion focuses on the enterprise class of database systems because the design and implementation of a workgroup database system can only be simpler, not more complex, than those for an enterprise-level database system.

A multi-user database environment is characterised by several salient features that allow it to satisfy different needs for different users. These include, for example:

- *Multi-representation of data.* Data in a multi-user database typically represent a compromise or reconciliation of the needs of different users with respect to storage format, content and quality. Experience has shown that multi-representation is by far the most difficult task in the design and implementation of database systems. The problem is particularly acute for spatial database systems because of the additional considerations pertaining to spatial data such as map scale and projection, geodetic datum, data format, and cartographic symbology.

- *Common user interface*. The objective of a common user interface is to provide the same look and feel for applications in different domains and for different user groups. Having a common user interface minimises development and maintenance costs because computer code developed for one application is reusable for other applications. It also reduces the costs and need for additional user training because users familiar with one application can relatively easily apply the experience of using it to other applications.
- *Concurrent access and security*. A multi-user database system is built for access by different users of different applications simultaneously. In order to maintain database integrity and data consistency, the database system uses sophisticated algorithms, strict business rules and transaction control mechanisms to facilitate multiple access to the database concurrently without compromising its integrity (see Chapter 2).
- *Standards*. Adherence to standards is important for assuring that resources are sharable and applications are interoperable in a multi-user database. As was noted in Chapters 5 and 6, data standards such as structure and semantics are as critical as technical standards for hardware, software, and communications protocols in creating a harmonious working environment for different users.
- *Metadata support*. In order to facilitate multi-user applications, it is critical to set up the necessary metadata support so that all users can access the database system with full knowledge of the characteristics of the data that they use. At the same time, new users are able to determine the fitness of use of the database for newly planned applications (see Chapter 5).
- *Systems architecture*. Multi-user database systems are invariably set up using the data warehouse or federated database architectures discussed in Chapter 6. Both of these systems architectures allow users to access database resources seamlessly no matter where the resources and users are located geographically.

Since the characteristics of a multi-user database environment noted above have already been explained in previous chapters, the following discussion focuses on two particular issues from data and process modelling perspectives. Section 5.2 first considers the concept of multi-representation of spatial data. Section 5.3 builds upon this discussion considering multiple representation and multi-user database models. The chapter concludes with a discussion of multi-user application development.

5.2   The Concept of Multiple Representation of Spatial Data

Ideally, users in a multi-user spatial database environment are able to employ data from different sources for different applications seamlessly without the need for data conversion (for example, from one classification scheme to another for attribute data) and transformation (for example, from one map projection to another for graphical data). In practice, however, conversion and transformation are more a rule than an exception when an application requires spatial data from different sources. Diversity of spatial data representation arises because different people or organisations that share the same interest in the real world often have different needs for accuracy of real world features, and require data of different characteristics or specifications pertaining to these features for different applications and with different focuses. These needs call for different methods and standards of data collection and cartographic production, which inevitably result in the same real world features being represented in different data sets or databases with various levels of abstraction, detail of description, accuracy of measurement, frequency of observation, and cartographic symbology, among many other factors that govern the geometric and semantic characteristics of data.

Conventional database systems are generally weak in supporting different representations. Relational database systems, for example, support different representations by means of a view mechanism that allows the creation of new, virtual relations (called *views*) from any number of relations already defined in the database using a rich set of relational operators (see Chapter 2). However, since relational views are not derived from the underlying schema, it is not possible to use them to update the database, or to correct the geometric, syntactic and semantic inconsistencies that may exist among the representations. Further, since individual views are generated independently of, and are kept logically separate from, each other, the global view of the real world is somewhat fragmented, and navigating among views is not easy.

These reasons have restricted the possibility of using views to handle multiple representations in databases. Object-oriented systems follow the same idea of using views but they support simple filtering or selection defined by projection operations only. More complex view generation in this class of systems is constrained by the rule of inheritance, and requires sophisticated view creation mechanisms that are not generally available in contemporary commercial object-oriented database products.

The limitations of conventional database models and their derivatives (for example, object-relational, geo-relational and geodatabase models) to manage different representations also apply when they are used for representing spatial data. This implies that spatial database systems, as they currently exist, are incapable of managing consistently different representations of a particular spatial object as it is perceived by different people (for example, an engineer's and a driver's perceptions of highways) and represented from different perspectives (for example, depicting topography in two-dimensions for general-purpose topographic mapping, and in thee-dimensions for a detailed engineering design). In current spatial database technology, each of the data sets pertaining to the perceptions and representations in the above example must be modelled separately and represented as such in the database.

Without suitable database tools to model different representations of spatial data as a whole, organisations have little choice but to model individual representations separately and store them as discrete and independent databases. Applications working at different scales or for different purposes can use a particular database as appropriate. Mapping organisations in both the public and private sector typically have traditionally used this approach to address different representations. They keep data for different map series logically and physically separate from one another according to purpose and scale. When producing different types of maps, they use data in particular representations that are supposed to suit the best purpose of the maps or their clients' intended applications. While such an approach of matching data to applications provides an apparently convenient working environment for many end users of database systems, it is a work-around rather than a real solution to handle different representations for three important reasons:

- *Lack of flexibility.* Since it is practically impossible to anticipate the needs of all applications and maintain a large number of data sets in different representations, there is always the chance that some users are unable to find data compatible with their needs. When specific applications arise for which none of the existing representations is suitable, the user has either to use a representation that does not match application needs precisely, which will cause dubious results, or to carry out costly and time-consuming data conversion or transformation to derive a new data set that meets the required specifications.
- *Complexity in maintaining consistency among representations.* Maintaining several databases and trying to keep them consistent with each another is a costly and labour-intensive undertaking. As many real world features are dynamic in nature, frequent database updating is required in order to keep up with their evolving states. The absence of logical connections among different representations means that the process of database update cannot be automated. The problem of

database update is compounded by the requirements of long transactions, and the stringent rules for database replication in a distributed systems architecture, as explained in Chapter 2.

- *Uncertainty resulting from spatial analyses using different data sets for decision making.* Since there are no logical and physical connections among databases using different representations, it is not possible to know exactly how well the results of analyses using different representations are comparable with one another. This lack of understanding creates uncertainty that discourages the use of the database for decision support by senior managers and executives of the organisation.

The growing interest in *multiple representation* in spatial database research seeks to address the inability of current database technology to manage different representations consistently. Multiple representation is an evolving concept and, therefore, means different things to different people. In this chapter, multiple representation is defined from two interrelated aspects, namely (i) the coexistence of data in a database pertaining to the same real world features represented by different database models, geometries, descriptions and other characteristics, and (ii) the ability of database systems to abstract and manage such data consistently and without human intervention.

Figure 8.12 illustrates the concept of multiple representation of spatial data. In this diagram the topography of the real world is abstracted and stored by four data models, namely randomly sampled elevation points, contours, a digital elevation model (DEM) and a triangulated irregular network (TIN).

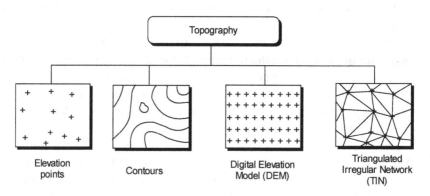

Figure 8-12. Depicting the same real world feature using different representations

In each one of these four representations, topographic data can be collected using different scales, with different accuracies, and according to different standards specified by different professional bodies. In a spatial database that supports multiple representation, data in individual representations are correlated spatially and logically so that change in one representation can be automatically propagated to all others. For example, when an error is detected in a local DEM collected at a 100-metre interval, related elevation values in regional and national DEMs collected at larger intervals, as well as related contour and TIN data sets, must also be amended to keep all topographic data sets consistent. The ability of a database system to change the values in different data sets simultaneously and consistently without human intervention is called *automated update propagation*.

In addition to representation by different data models, the concept of multiple representation can include geometric, graphical, semantic, and temporal dimensions. Figure 8.13 shows the concept of representing the same real world feature (a school) using different geometries, graphical symbols, textual descriptions and time frames. In this diagram, the school assumes different geometric shapes at different display scales. It is portrayed by different cartographic symbols on different types of maps. The school can be described by different textual descriptions according to the perceptions of individual data collectors or the purposes of data collection (that is, as a multi-storey building according to its form, an educational institution according to its function, and a built-up area according to a high-level land use classification scheme). At different points in time, the shape of the school building can change (for example, as the result of constructing an extension), and so can its function (for example, from a school to a community centre).

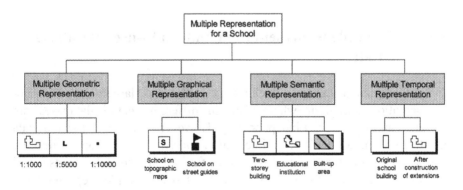

Figure 8-13. Different representations of the same real world features using different geometries, cartographic symbols and textual descriptions

As a multi-faceted concept involving database models, scales, cartographic symbology, classifications, and change over time, multiple representation provides a framework for different users to evaluate their respective needs globally. Once evaluated, the needs are then implemented logically and physically in database design to produce a *multiple representation spatial database* (MRSDB).

It is important to note that multiple representation is different from an apparently similar concept known as *versioning* in the geodatabase model of ArcGIS (ESRI, 2004). Whereas versioning is concerned with different "versions" or "states" of a spatial data item in a particular representation in multiple database transactions, multiple representation addresses the issues of consistency among corresponding data items representing the same real world features within a single database. Multiple representation and versioning are complementary concepts because a specific spatial data item in a particular representation can have different versions when it is accessed by multiple users simultaneously. In this regard, versioning can be seen as a partial implementation of multiple representation in spatial database systems.

Multiple representation is fundamental to the multi-user database environment. Successful implementation of a multiple representation strategy in spatial database implementation reduces the cost and resources for spatial data collection and management by enabling data sharing and database interoperability. Multiple representation allows all users to access a spatial database equally regardless of their respective needs and, more importantly it frees users from the burden of having to deal with incompatible data so that they can focus on developing useful solutions. The next section explains how multiple representation can be accomplished in practice by extending current spatial database models and data management principles.

5.3 Multiple Representation and Multi-user Database Models

Current spatial database technology still relies largely on a mono-representation paradigm where data in different spatial databases are modelled and represented independently. However, as the use of spatial data proliferates, and as spatial data integration and database interoperability are fast becoming the fundamental principles for building spatial databases and solutions, the demand for systems to have multiple representation functionality is more pressing than ever before for spatial data providers and users alike. This is an important driving force behind the surge of the interest in multiple representation research in spatial database systems in recent years.

The concept and techniques of multiple representation are rapidly changing as the limitations of current database concepts and techniques are better understood and technological advances allow existing methodologies to be refined and new methodologies to be developed. Earlier research on multiple representation in spatial database systems focused mainly on the geometries of spatial objects. This was inspired partly by the hierarchical nature of the transition of the geometries of spatial objects from one scale to another, and partly by the simplicity of treating this as a technical problem of changing geometries at different scales.

Much of the effort in earlier research focused on the use of cartographic generalisation, as explained in the following section. However, the growing realisation of the inadequacies of addressing geometries alone led to several propositions that sought to tackle the roots of the problem of multiple representation at the conceptual schema level. As an illustration of the evolution of multiple representation from a geometry-oriented to a schema-based approach, the theoretical frameworks of two emerging methodologies, namely multiple representation data warehousing modelling and MADS modelling, are introduced in Sections 5.3.2 and 5.3.3.

5.3.1 Multiple Representation by Cartographic Generalisation

Cartographic generalisation is broadly defined as the process of reducing the detail of a map as a result of reducing its scale. This process is one of the most intensively researched areas in cartography and GIS (Müller, et al., 1995). One of the primary objectives of cartographic generalisation is to enable the use of a single set of detailed cartographic data at a large scale as the basis for the automated production of less detailed maps at a range of successively smaller scales. Cartographic generalisation has traditionally focused on the use of algorithms for line and polygon simplification (Douglas and Peucker, 1973) to produce multiple representations of the geometries of spatial objects. Since multiple representation by cartographic generalisation is scale-dependent, this approach is commonly called *multi-scale representation*.

Several methods use the concept and techniques of cartographic generalisation in MRSDB (Kilpeläinen and Sarjakoski, 1995; Dunkars, 2004). Research results show that while these methods might have worked in proof-of-concept implementations, cartographic generalisation in a MRSDB is much more complicated than for map production and spatial visualisation. This is because generalisation of a spatial database is concerned not only with geometries, but also with attribute data. Several spatial database software packages, notably ArcGIS, now offer cartographic generalisation functionality, however the techniques that are used are too computationally intensive and resource-demanding to be deployed efficiently in an MRSDB. Cartographic generalisation is particularly inefficient in situations where a large number of spatial data sets of various representations have to be drawn from different sources, where database update requirements are frequent, and where data must be used frequently and freely at any user-defined scale such as in Web-based maps-on-demand applications.

5.3.2   Modelling of Multiple Representation in Spatial Data Warehouses

A data warehouse, as discussed in Chapter 6, is a centralised repository of legacy data that is used primarily for decision support purposes. It can also be set up as a central database that provides the framework data on which domain-specific applications are developed. Bédard et al. (2002) propose a two-tier data warehousing architecture where a multiple representation data warehouse is fed by independent operational database systems and, where possible, by cartographic generalisation processes. The same real world objects can be represented in such a data warehouse at different scales by different geometries (for example, a small town is represented by a detailed polygon at 1:20000, a simplified polygon at 1:100000, and a point at 1:500000), and then queried to provide only the appropriate geometry for the map being printed or for on-line database queries.

In order to facilitate and accelerate the process of multiple representation for data warehouse users, a three-tier infrastructure using data marts can be constructed around selected pivot themes (for example a road network, drainage, buildings) at pivot scales (for example, 1:5000, 1:10000, 1:50000, 1:100000). These pivot representations are intended for producing maps more quickly at compatible scales for domain-specific applications by automated generalisation. Since the production is based on a pivot theme and scale close to the user's requirements, the amount of 'on-the-fly' generalisation processing can be reduced considerably.

Using a CASE tool called Perceptory, Bédard et al. (2002) developed a solution for modelling a spatial data warehouse that supports multiple representation through generalisation using the approach described above. Perceptory is a visual modelling tool based on a UML-derived object-oriented formalism. It includes several UML basic components (for example, package, class, attribute, operation, association, among many others), as well as formal extension components that allow the modelling of spatial characteristics of spatial objects at the class name, attribute and operation levels. The modelling process is carried out by using two sets of modelling rules and a set of graphic notation including:

- *General modelling rules*, which model generalisation operations in the same way any other operations in the object-oriented paradigm are modeled. These rules apply only to data and operations that must be coded in the database. The modelling process is iterative in order to identify progressively those operations that can be fully automated.
- *Specific modelling rules*, which are applied to (i) manual, semi-manual or fully automated generalisation operations in an object class; (ii) all multiple representations as defined by Plug-in Visual Language (PVL), which is a font in which pictograms for modelling are embedded, and (iii) geometries that have to be stored in the system.
- *Graphic notation*, which is used to depict multiple representation (in object classes from large-scale detail level to small-scale more general level) and generalisation operations with source data scales and target data scales, as shown in Figure 8.14 to represent a HOUSE (upper box) and to denote how spatial objects are generalized from source scales to target scales (lower box).

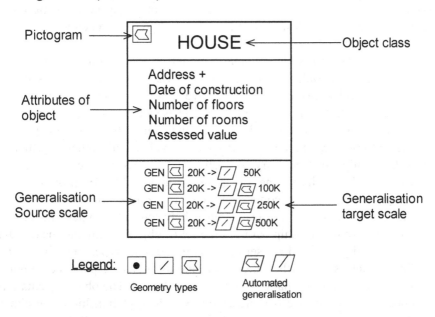

Figure 8-14. Modelling of the object class "HOUSE" using PVL graphic notation

The end product of modelling is a conceptual database schema that corresponds to the underlying spatial data warehouse. Bédard et al. (2002) note that the modelling method they present is just a subset of their intended solution as it addresses the problem of multiple representation and cartographic generalisation only. Research is in progress to develop a full solution that will support semantic and temporal multiple representation with generalisation.

5.3.3   MADS Modelling

Modélisation d'Applications á Données Spatiales (MADS) was developed in the late 1990s for the conceptual modelling of spatio-temporal databases (Parent et al., 1998). In essence, MADS is an entity-relationship modelling technique extended to support the main concepts of object orientation, including object type, relationship type, simple and complex attributes, spatial data type equivalent to those specified by the OGC (see Chapter 4), temporary data type, specific data type (for a raster view of space), and binary spatial or temporal relationships.

The characteristics of MADS can be understood from its three dimensions that address the thematic, spatial and temporal aspects of real world features. These three aspects are said to be "orthogonal" because each thematic concept (entity, relationship and attribute) may be thematic only, spatial, temporal, or both spatial and temporal, as explained below.

- *Thematic dimension.* The thematic model of MADS supports the usual concepts of extended entity relationship (EER) models such as entity types and relationships, as noted above. Entities or objects in a MADS model may or may not have attributes. These attributes may be simple or complex (that is, composed of more than one attribute), and may be derived by an expression from the values of other attributes. Relationships may be n-ary or cyclic (that is, relating back to an entity itself) and can have one or more attributes. MADS provides two categories of links that connect entities: (i) *is-a* link, which is used when there is a one-to-one correspondence between entity types and also a containment relationship between the set of object identifications (OIDs) in one entity type and the set of OIDs in the other type, and (ii) *may-be* link, which is used when there is a one-to-one correspondence among instances but no set relationship is to be enforced. The object orientation concept of inheritance is represented in an *is-a* relationship using either the concept of refinement or redefinition. Refinement allows a generic spatial data type to be represented in subtypes (for example, a generic spatial data type "settlement" is refined to subtypes "city" represented by an area geometry and "village" represented by a point geometry). On the other hand, redefinition enables the creation of a new attribute in the subtype with the same name, thus making it possible to associate several geometries to the same object. In order to support multiple representation, MADS offers three specific kinds of relationships. These include (i) aggregation, which models binary relationships that are directed from the composite to the component objects (for example, the Region of Waterloo includes the cities of Waterloo, Kitchener and Cambridge); (ii) generation, which links newly created entities to their parent entities; and (iii) transition, which allows applications to record the changing classification of database objects as states of the real world features that they represent.

- *Spatial dimension.* MADS supports the discrete (vector) and continuous (raster or field) spatial representations of real world features. In a discrete representation, a hierarchy of spatial abstract data types (SADT) contains generic simple data types (point, line, simple area and oriented line) and complex types (point set, line set, complex area, oriented line sets). Entity types may be spatial, which means that they have a specific attribute, called geometry, with a SADT for the domain of values. Attributes may also be spatial, in which case they have an SADT for the domain of values. MADS supports two predefined categories of spatial relationships, namely topological and spatial aggregation, corresponding to the usual requirements of spatial applications. Spatial relationships are used to force geometries of the linked entities to obey the spatial integrity constraints associated with the relationship type. Continuous representation in MADS is supported by *space-varying attributes.* In this context, an attribute is said to be space-varying if its value is defined by a function whose domain is the set of geometrical elements (points, lines or areas) in which the geometry of the corresponding object or attribute is decomposed (for example, the elevation of an area is represented by a set of elevation values measured at randomly or regularly spaced points).

- *Time dimension.* MADS supports discrete (instant) and continuous (interval) spatial representations of real world features by assuming that entities and relationship types have a life cycle. Time attributes take their values from generic simple (instant, interval) and composite (instant set, interval set) temporal abstract data types (TADT). *Time stamping* is an important feature of the MADS temporal model. Entities, attributes and relationships can all be time stamped to record when they are created and deleted, from which *transition relationships* (the behaviour of objects change classification), *generation relationships* (creating one object from another), *timing relationships* (describing temporal relationship between entities) and *time-based aggregations* (linking temporal entities to their snapshots) can be established. There are also synchronisation relationships that force the life cycles of the linked entities to obey the temporal integrity constraints associated with the relationship type.

MADS provides a rich set of visual modelling tools called SUPER-G that allow the graphical depiction of entities, attributes and relationships in the modelling process for better understanding and communication between database users and systems designers. Figure 8.15 is a simple example of conceptual modelling using MADS visual notation. It illustrates several of the key features noted in the explanation above. These include, for example, an is-a link between parcel and owner, a complex attribute "address" that is made up of "street" and "city", an attribute "landowner.land total" derived from other attributes, the ability of the relationship "own" to have an attribute, and a specialisation from "landowner" to "person".

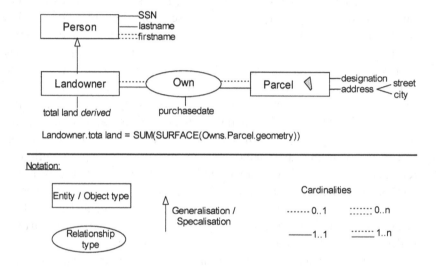

Figure 8-15. An example of conceptual modelling using MADS

As a conceptual modelling methodology for a mulit-user database, MADS provides a rigid theoretical framework that takes into account the thematic, spatial, and temporal user needs in an integrated modelling environment. The major drawback of MADS is that it is not based on UML, and translation from a MADS conceptual schema to a UML logical data schema may be problematic. However, Balley et al. (2004) have demonstrated that MADS can be used to support the unification of several mono-representation databases into a single multi-representation database. Through a MADS-to-Oracle translation module, they successfully implement a multiple representation database in an Oracle DBMS at the French national mapping agency (IGN). At the time of writing this book, MADS and similar research efforts toward multiple representation in spatial database systems are still largely at the proof-of-concept stage. However, results so far have shown that attacking the problem of multiple representation at the conceptual schema level provides a practical way of establishing database models capable of satisfying different users in a multi-user database environment.

5.4   Multi-user Application Development

A complete explanation of database application development is a complex and technical subject that is beyond the scope of this book. In keeping with the theme of the book, this section considers the application development environment from the perspectives of project management and systems implementation. The discussion focuses on how application development in a spatial database project is organised in a multi-user environment, rather than how application software is actually designed and coded. In particular the relationship between business process modelling, as explained in Section 4.3, and application development activities is stressed within the conceptual framework of an emerging systems development paradigm called *model-driven architecture* (MDA).

Multi-user application development is both a user- and systems developer-oriented methodology. For users, multi-user application development seeks to provide a consistent look-and-feel interface such that users familiar with one application are able to use another one with minimum training and adjustment of work procedures. It also seeks to maximise the level of information sharing by ensuring that output from one application can be transparently used as input to another, and that outputs from different applications can be seamlessly integrated for a high level of analysis and intelligence abstraction. From the systems developer's perspective, multi-user application development provides a working environment that allows delivery of software products that are portable (usable in different platforms), interoperable (able to communicate with other systems), and easy to document and maintain.

Central to multi-user application development is the use of standards. There are numerous advantages of using a standards-based approach in software development. These include, for example:

- Large-scale application development projects can be handled by dividing the tasks into manageable components, each of which is the responsibility of a specific team, with a standardised framework for coding as well as for communicating between component teams.
- Individual programmers or programming teams are freed from the need to develop coding styles and specifications, which allows them to focus on the actual production of efficient and robust solutions.

- Programs are easier to maintain and support because programmers are able to read and understand code written by others.
- Design specifications, code, user interfaces and documentation developed for one application can be easily reused for other applications with minimum modification, which leads to considerable savings in both cost and time of software development.

Standards in multi-user application development can be generally classified into four categories, as shown in Table 8.1.

Table 8-1. Categories of software application development standards (Source: Lo and Yeung, 2006)

Category of standards	Contents
Coding or scripting	o Templates for new programs o Program header o Syntax including comments o Parameters and options o Exit status o User and global variable names o Error handling
Data naming	o Entity names o Attribute names o Data name suffices o Table names o File names
Application and GUI	o Window and menu-naming conventions o Window and menu design and resources o Display management o Function keys o Tool bars and icons o Screen and hard copy input / output forms o Design documentation o User documentation o On-line help
Development and Version management	o Directory and file structure o Revision control of codes o Revision control of applications o Access control for collaborative application development o Component test and acceptance procedure

In addition to the contents noted in the table, these categories have the following general characteristics:

- *Coding or scripting standards*, which specify the format and style of program code so that a program written by one programmer can be readily understood, used and supported by other programmers.

- *Data-naming standards*, which specify conventions for naming data items to ensure that databases can be shared across organisational boundaries and application domains.
- *Application and GUI standards*, which aim to provide a consistent approach to design and build software products from the end user's point of view as well as from the systems developer's view.
- *Development and version management standards*, which are used to provide a consistent directory and file structure of the applications, and to control the revision of program coding in a systematic manner.

Over the years, the method of application development has changed from the traditional waterfall development strategy, through the spiral model and iterative development using prototyping and 4th generation non-procedural languages, to a combined paradigm that takes advantage of the merits of all of the above approaches (Pressman, 2005). For large-scale complex developments such spatial database applications, MDA is emerging as a useful methodology originally proposed by the Object Management Group (OMG) (Miller and Mukerji, 2003). Key to MDA is the role played by models in the application development process. As its name implies, within MDA the entire application development process is driven by the activities of modelling the software system. Therefore, MDA is fully compatible with business process modelling discussed in Section 4.2.

Figure 8.16 shows the MDA development life cycle (MDLC). In terms of the life cycle steps, this development life cycle is not much different from the traditional SDLC explained in Chapter 3. What distinguishes the MDLC and the SDLC is the nature of the artefacts that are developed during the requirements, analysis, and design phases of the development process. Whereas the artefacts of the conventional SDLC are diagrams and textual descriptions, those of the MDLC are two formal models called the platform independent model (PIM) and the platform specific model (PSM). These two models and the platforms that are associated with them are described as follows:

- *The platform independent model.* This is a model with a high level of abstraction that describes only business functions, processes, behaviours and rules in UML. Within PIM, the system is modeled from the perspective of how it best supports business functions and processes with respect to workflow and interfaces, in the manner explained in Section 4.2. Implementation details, such as hardware platform and types of database systems (relational or object-oriented), are not considered in a PIM.

- *The platform specific model.* This type of model can be transformed into one or more platform-specific models tailored to specify the systems in terms of implementation constructs. PSMs are described using specialisations and extensions to UML, called *UML profiles*, that contain terms tailored to particular platforms. For example, a relational database PSM includes terms such as "table", "column", "permissible values", "primary keys", "normalisation", and so on. Therefore, PSM is only understandable by systems developers who are familiar with the implementation technology.

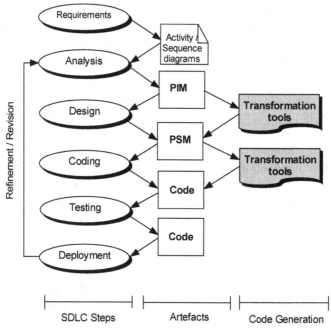

Figure 8-16. The MDA development life cycle

MDA defines PIM and PSM. It also defines how these are related to one another and to the code. The process starts with the creation of a PIM from business process models obtained from user needs analysis, followed by the transformation of PIM into one or more further PSMs, and ends with the transformation of the PSMs into programming code. The transformation from business process models, PIM and PSM to code represents different abstraction levels in the system specifications. By using a stepwise approach to identify implementation details progressively, systems developers can handle complex applications with less effort.

In traditional application development environments, transformation from model to model, or from model to code, was a manual process.

Attempts have sought to automate the generation of code from models but these seldom go beyond the production of basic template code. A considerable amount of manual work is still required to complete the final code. Because both PIM and PSM are expressed in UML, transformations from PIM to PSM and from PSM to code (in a language specified by the developer and supported by the selected platform) are always carried out by automated tools. The creation of PIM from business process modelling is the only manual task in the process. At the time of writing, an MDA approach is still relatively new and available transformation tools are not yet able to automate the process completely. Hence, it is often necessary for developers to enhance manually the transformed models and code.

Despite these limitations, an MDA approach is able to expedite application development considerably. MDA is particularly helpful in the development of multi-user solutions for several reasons, including:

- *Focus on PIM.* Since the transformations from PIM to PSM and from PSM to code can be largely automated, application developers can focus more on the development of PIM, and pay more attention to solving the solutions at hand. This will potentially result in a system that is less error-prone and is able to meet user needs more precisely.

- *Portability.* PIM is platform-independent and can be transformed automatically into multiple PSMs for implementation of different platforms. This allows the same applications to be implemented on different hardware and software systems. Further, platform-independence also means that when an existing system is upgraded using new technologies and platforms, the same PIM can be largely used to implement the new system quickly.

- *Interoperability.* MDA is able to generate bridges that define the relationships between different PSMs transformed from the same PIM, and between different code transformed from PSMs originating from the same PIM, thus enabling cross-platform interoperability in an intuitive manner.

- *Documentation.* PIM is used to generate PSM, which is in turn used to generate programming code. This means that PIM is an exact representation of the code and, as a result, fulfils the function of high-level documentation is essential for application development and software maintenance in multi-user environment.

In summary, MDA provides a systematic approach for developing multi-user solutions in current enterprise database environments. Automated systems development tools are not yet fully available for the use of MDA in software application development in general and spatial database applications in particular. However, many software developers have already started using approximations to MDA by automating the development process with their own transformation tools. It is strongly suggested that the conceptual framework, if not the full implementation, of MDA be adopted in spatial database projects to ascertain the production of quality application software for spatial database projects.

6.   SUMMARY

This chapter explained the concepts and methods of user needs assessment in the context of multi-user spatial database design and implementation. Multiple representations of spatial data and multi-user spatial databases were emphasised as they characterise the majority of current workplace implementations, especially at the enterprise level of database management. The chapter highlighted salient aspects of the major approaches to user needs assessment that have emerged in the context of a database-centred approach in recent years. Consistent with earlier chapters, the approach and discussion adopted utilised several fundamental principles of best practice in information system development.

The nature and importance of conducting intermittent user needs assessment within organisations was emphasised in Section 2, as this important aspect of spatial database management is often overlooked due to busy daily schedules that tend to relegate the important roles of backward and forward looking in management of enterprise spatial databases down the priority task list. Section 3 built upon the need to review user needs intermittently by describing commonly used methods for actually conducting a user needs assessment. The translation of identified needs gathered through an assessment process into new specifications, both technical and operational, was outlined in Section 4. The task of satisfying diverse and often divergent and conflicting user needs through multi-user data modelling was reviewed in Section 5.

7.   REFERENCES

Balley, S., Parent, C. and Spaccapietra, S. (2004) "Modelling Geographic Data with Multiple Representations", *International Journal of Geographic Information Science*, Vol. 18, No. 4, pp. 309-326.

Batini, C., Ceri, S. and Navathe, S.B. (1992) *Conceptual Database Design: An Entity-Relationship Approach*, Redwood City, CA: The Benjamin/Cummings Publishing Company, Inc.

Bédard, Y. (1999) "Principles of Spatial Database Analysis and Design", in *Geographical Information Systems, Volume 1: Principles and Technical Issues*, 2nd Ed., by Longley, P.A., Goodchild, M.F., Maguire, D.J. and Rhind, D.W. (eds.), New York, NY: John Wiley & Sons.

Bédard, Y., Proulx, M.-J., Larrivée, S. and Bernier, E. (2002) "Modeling Multiple Representation into Spatial Data Warehouses: A UML-based approach", *Proceedings, Symposium on Geospatial Theory, Processing and Applications*, Ottawa, ON.

Boehm, B. (1981) *Software Engineering Economics*, Englewood Cliffs, NJ: Prentice-Hall.

Brethour, C. and Yeung, A. (1994) "GIS Implementation in a Multi-participant Environment: the Example of Central and Northeast Regions, MNR", *Proceedings*, Decision Support – 2001, Toronto, ON.

Davis, A.M. (1993) *Software Requirements: Objectives, Functions and States*, Englewood Cliffs, NJ: Prentice-Hall.

De Floriani, L., Marzano, p. and Puppo, E. (1996) "Multi-resolution Modelling in Geographical Information Systems", in *Innovations in GIS 3* by Parker, D. (Ed.), London, UK: Taylor & Francis.

Devogele, T., Parent, C. and Spaccapietra, S. (1998) "On Spatial Database Integration", *International Journal of Geographical Information Systems*, Vol. 12, No. 4, pp. 335-352.

Douglas, D. and Peucker, T. (1973) "Algorithms for the Reduction of the Number of Points Required to Represent a Digitized Line or Its Caricature", *The Canadian Cartographer*, Vol. 10, No. 2, pp. 112-122.

Dunkars, M. (2004) "Automated Generalisation in a Multiple Representation Database", *Proceedings*, 12th International Conference on Geoinformatics, University of Gävle, Sweden.

ESRI (2004) *Versioning* (ESRI Technical Paper), Redlands, CA: Environmental Systems Research Institute, Inc.

Faulk, S.R. (1997) "Software Requirements: A Tutorial", in *Software Requirements Engineering*, 2nd Ed. by Thayer, R. and Dorfman, M. (Eds.), Los Alamitos, CA: IEEE Computer Society Press.

Faulk, S.R., Brackett, J., Ward, P. and Kirby, Jr. J. (1992) "The Core Model for Real-time Requirements", *IEEE Software*, Vol. 9, No. 5.

Frankel, D.S. (2003) *Model Driven Architecture: Applying MDA to Enterprise Computing*, New York, NY: John Wiley & Sons and OMG Press.

Kilpeläinen, T. and Sarjakoski, T. (1995) "Incremental Generalisation for Multiple Representations of Geographic Objects", in *GIS and Generalization* by Müller, J.-C., Lagrange, J.-P. and Weibel, R. (Eds.), London, UK: Taylor & Francis.

Kleppe, A., Warmer, J. and Bast, W. (2003) *MDA Explained: The Model Driven Architecture Practice and Promises*, Reading, MA: Addison Wesley.

Lo. C.P. and Yeung, A.K.W. (2006) *Concepts and Techniques of Geographic Information Systems*, 2nd ed., Upper Saddle River, NJ: Prentice-Hall.

Martel, C. (1999) *Dévelopment d'un cadre théorique pour la gestion des represéntations multiples dans les bases de données spatiales*, M.Sc. thesis, Laval University, Laval, QB, Canada (Quoted in Vangenot et al, 2002).

McConnell, S. (1996) *Rapid Development: Taming Wild Software Schedules*, Redmond, WA: Microsoft Press.

Miller, J. and Mukerji, J. (Eds.) (2003) *MDA Guide, Version 1.1*, Needham, MA: Open Management Group.

Müller, J.-C., Weibel, R., Lagrange, J.-P. and Slagé, F. (1995) "Generalization: State of the Art and Issues", in *GIS and Generalization* by Müller, J.-C., Lagrange, J.-P. and Weibel, R. (Eds.), London, UK: Taylor & Francis.

Parent, C., Spaccapietra, S., Zimanyi, E., Donini, P., Plazanet, C. and Vangenot, C. (1998) "Modeling Spatial Data in MADS Conceptual Model" in *Proceedings*, International Conference on Spatial Data Handling, Vancouver, BC.

Pressman, R.S. (2005) *Software Engineering: A Practitioner's Approach*, 6th Ed., New York, NY: McGraw-Hill.

Vangenot, C. (2001) "Supporting Decision-Making with Alternative Data Representations'" *Journal of Geographic Information and Decision Analysis*, Vol. 5, No. 2, pp. 66-82.

Vangenot, C., Parent, C. and Spaccapietra, S. (2002) "Modelling and Manipulating Multiple Representations of Spatial Data", *Proceedings*, Symposium on Geospatial Theory, Processing and Applications, Ottawa, ON.

Weigers, K. (1999) "Automating Requirements Management", *Software Development*, Vol. 7, No. 7, pp. 1-5.

Wood, J. and Silver, D. (1995) *Joint Application Development*, 2nd ed., New York, NY: John Wiley & Sons.

Chapter 9

PROJECT MANAGEMENT FOR SPATIAL DATABASE IMPLEMENTATION

1. INTRODUCTION

The implementation of database systems in general, and spatial database systems in particular, is a complex task. Despite the spectacular advances in database technologies in recent years, numerous implementation projects still fail to meet user needs, experience delays in delivery, suffer from cost over-runs, and even end up in complete failure. This situation can be attributed as much to sponsors and users of database systems as to systems development professionals. Many database sponsors and users have some idea of what they want, but seldom know how to turn expectations into reality. They commonly do not realise the possibilities and constraints of the implementation process and, therefore, are not sure what questions to ask at different stages of the systems development life cycle. On the other hand, many systems developers are often overly optimistic about the time and costs of technology implementation. They often focus too much on technical aspects such as computer architecture and database structure, at the expense of the financial, human resource and political factors that must be considered.

Clearly, advanced technologies alone do not always guarantee success in spatial database implementation. Non-technical factors also play a significant role in nurturing the eventual fruition of database products and services. This chapter considers spatial database implementation from a management perspective. The aim is to explain how basic principles and methods of project management can and should be used to avoid common pitfalls and mistakes experienced in spatial database projects. Section 2 provides a brief but concise explanation of the definition and objectives of project management, the concept of a *project management life cycle* (PMLC), as well as the competencies and skills required of project managers and systems development professionals. Following this, Sections 3 through 7 demonstrate how project management principles and methods can be usefully deployed to tie the technologies discussed in previous chapters together in a coherent manner through the five stages of the PMLC, namely initiation, planning, execution, monitoring and control, and closing and post-implementation evaluation.

2. PRINCIPLES OF PROJECT MANAGEMENT

Project management is a relatively complex field of study and professional practice. The following discussion explains what project management is, its objectives, and the competencies and the skills required by project managers to provide a proper perspective for the study of spatial database project management in the rest of the chapter.

2.1 Definition and Objectives of Project Management

Wysocki et al. (2003, p. 38) define a project as "a sequence of unique, complex and connected activities having one goal or purpose that must be completed by a specific time, within budget, and according to specification". Central to this definition is a logical sequence of activities that must be completed within a specific time frame. These activities are said to be unique in the sense that no two projects are exactly identical with respect to their objectives, processes and operating environment. The methods used to implement one project successfully are not necessarily applicable directly to another project of the same nature. Since each project is unique and prior experiences cannot be exactly duplicated, all projects involve risks and uncertainty that are essentially random in occurrence and hence difficult to predict precisely.

Project activities are complex because they rarely involve routine repetitive acts, but often require specific knowledge and skills to be used in their design, execution and management. The idea of connectivity between individual project activities is closely related to the notion of time frame and logical and systematic order noted above. It is helpful to think of the connection between project activities in terms of inputs and outputs where the output of one activity becomes the input for the next. Although overlapping activities are commonplace, the concept of sequence is important in the connection among project activities (see Section 2.2).

A project must have a well-defined goal with respect to the mission or mandate of an organisation. In many instances, a project may be too complicated to be carried out as a single undertaking. Hence, it is necessary to divide it into several sub- or part-projects according to the prevailing organisation structure (e.g., by departments or business functions) or geographical divisions (e.g., by regions, sales territories or watersheds). Under such circumstances, each sub-project is considered as a separate but interdependent undertaking in its own right. All sub-projects have their specific goals but when added together these goals collectively constitute the specific goal of the parent project.

Every project or sub-project is generally subject to three constraints regardless of its objective and scale. These are:

- *Time.* Projects have definitive milestones that specify when particular components (e.g., progress reports, prototypes) must be delivered, and a completion date when the database system being developed will become fully functional and operational.
- *Cost.* Projects have cost or budgetary limits, which will impact on the availability of human and technical resources.
- *Specification.* Deliverables of a project are required to meet a specific level of functionality and quality both independently and when working as a whole.

It is important to understand that the constraints of time, cost and specification are interdependent and, as a result, changes in one constraint always cause changes in the others. For example, a change in specification will inevitably lead to changes in the time and cost requirements. Similarly, delays in the delivery of intermediate and final products inevitably necessitate an extension of the project time frame, which will in turn increase the cost and resource requirements of the project. Clearly, the dynamic nature of the interplay among the constraints requires that projects must be properly managed in order to succeed. The principles and practice of management are often deployed in the context of managing tangible entities such as people, physical and financial assets, and the business operations of an organisation. These same principles, however, can be equally applied to the management of tangible resources and non-tangible activities that are required to complete a project.

The Project Management Institute (PMI) (2004, p. 8) defines *project management* "as the application of knowledge, skills, tools, and techniques to a broad range of activities in order to meet the requirements of a particular project". This definition is supplemented by five Project Management Process Groups (PMPG) that describe the life cycle of typical projects, and nine Knowledge Areas (KA) in which project managers must be competent. The five PMPG are initiating processes, planning processes, executing processes, monitoring and controlling processes, and closing processes. The nine KA, on the other hand, focus on management expertise in project integration management, project scope management, time management, cost management, quality management, human resources management, communications management, risk management and procurement management.

Project management has also been defined in many other ways in related literature. However, it is apparent that many authors have accepted the PMI proposition that project management is a special branch of management characterised by the application of management principles and best practices that seek to steer the initiation, planning, implementation, monitoring and closing of projects toward their ultimate success. It is also apparent that many authors have adopted the PMI's approach to group all project management activities into five sequential phases or levels, commonly called the *project management life cycle* (PMLC), as explained in Section 2.2.

2.2   The Project Management Life Cycle

The PMLC defines how a project is managed effectively and efficiently from its conceptualisation through implementation to its operationalisation. A further term, namely the *project life cycle* (PLC) is also used to describe this process. The terms PMLC and PLC are always used in conjunction in project management but they actually refer to two relatively distinct sets of concepts and processes. The purpose of a PLC is to describe the activities that must be completed in order to create a product or a service. The PLC of individual projects varies from one to another because of the uniqueness of the nature of each project, as noted in Section 2.1. The Systems Development Life Cycle (SDLC) and Database Development Life Cycle (DDLC) discussed in Chapter 3 are examples of the PLC for spatial database implementation projects.

The PLC focuses on the tasks that are necessary in a project. In contrast, the focus of the PMLC is on how these tasks can be managed. In this regard, the PMLC is more of a conceptual framework for the systematic application of managerial principles and best practices, rather than the actual steps of building a product or service. As such, the PMLC remains the same for all projects regardless of the PLC being employed. This means that while the PLC of a spatial database project is markedly different from that of, for example, a project to build a new highway, the PMLC for both projects is essentially the same, in terms of the project management cycle or phases. Throughout the course of every project, the PLC and PMLC work in conjunction with one another. It is the project manager's responsibility to ensure all PLC activities use the conceptual framework of the PMLC.

The PMI groups project activities generally into five phases. Figure 9.1 shows the relationships among the phases of the PMLC, as well as typical project management activities undertaken in each of these phases. This diagram is adapted from the PMBOK Guide (Project Management Institute, 2004) and is used as the basis for the discussion in this chapter. The five PMLC phases are essentially sequential in nature. However, there is a feedback loop from the monitoring and control phase to the planning phase, as noted in the figure. This loop can be followed enough times as is required until the project manager is satisfied that the project is sufficiently complete for it to be closed out and for the evaluation process to commence.

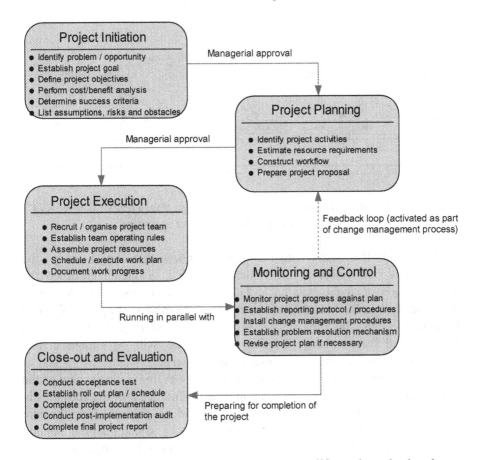

Figure 9-1. Phases of the project management life cycle and related activities (Modified from PMI, 2004)

It is important also to note that the sequential PMLC phases tend to overlap somewhat with one another. The duration of each phase and the amount of overlapping between sequential phases may vary considerably depending on the naure of the project, the complexity of the activities in, and hence the efforts and resources required by, individual phases as shown in Figure 9.2. Each phase of PMLC is discussed in detail in Sections 3 through 7.

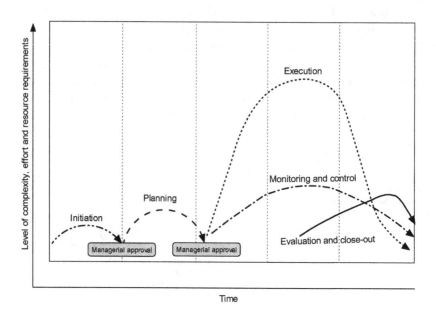

Figure 9-2. Relationships between phases of the PMLC, complexity, and resource requirements of project activities

The full PMLC starts with the initiation phase that aims to scope the project. The deliverable of the initiation phase is a document called the *project proposal* or *business case*. This is a three- to four-page document that describes, in the language of business, what problem or opportunity will be addressed by the project, the project goal and objectives, the costs incurred and the resulting benefits, how success will be measured, and potential risks and obstacles that may be encountered.

Once the project proposal has received approval from management of the organisation, the planning phase commences. While most of the activities in this phase are undertaken by one or a few individuals that will form the core of the project team, it is common practice to hold a formal planning session for all stakeholders who will affect or be affected by the project. The deliverable of the planning activities is a detailed *project plan* that provides a description of each project activity, the resources required to complete the activities, the project schedule, different milestones for the delivery of intermediate results, as well as the dates for acceptance tests and final delivery of the project products.

The project execution phase, also commonly called the project implementation phase, is probably the most labour- and resource-intensive phase of the PMLC. This phase usually starts with the organisation of the project team. Members of a typical project team include people transferred internally from other departments within the organisation, new staff hired specifically for the project, and contract staff from external consulting firms. The actual people assigned to work on each activity of the project are identified, and detailed descriptions of the activities are developed, reviewed and signed off. This signifies the actual design, building and testing of the spatial database system to be delivered by the project.

The monitoring and control phase begins as soon as the implementation activities have commenced. It runs in parallel with the project execution phase as a way of quality assurance and quality control. Generally speaking, change management is the most critical component of this particular phase, and very clear protocols and procedures must be established to ensure that when conflicts arise (between users and developers as well as between different users), the problem(s) can be addressed expediently and effectively. As change requests always have some impact on the initial time, cost and resource allocation, adjustment to the original project plan is necessary, and the feedback loop is then activated.

Preparation for the final phase of the project usually starts well ahead of the conclusion of the execution phase. In this phase, the close-out activities include the installation and testing of the deliverables, post-implementation audit or evaluation, and the compilation of the final project report summarising all project progress reports, acceptance of test results and a brief description of the lessons learned. The final project report may also include recommendations to enhance and refine the spatial database system in response to anticipated changing user needs and advancements in the technological environment.

2.3 Competencies and Skills of Project Management

Project management is the responsibility of a project manager. This individual seldom participates directly in the activities that produce the spatial database system, but rather strives to maintain the progress and productive interaction of various parties in order to minimise the overall risk of failure. A project manager is expected to have a specific level of competency and skills in technical management, financial management and people management in order to accomplish this objective. In many organisations, the project manager is required to possess a professional designation from an accreditation body such as the PMI noted above. Although a professional designation for a manager is by no means a panacea for the success of a project, such a requirement at least reflects the general realisation that a set of managerial competencies and skills is essential for the practice of effective project management.

The New York State Project Management Guidebook (Mulholland, 2003) identifies five core sets of competencies and skills for project managers. These include:

- *Communication.* Good communication skills are a critical requirement for project managers. Communication in project management is bi-directional in the sense that talking and listening are equally important. The project manager must be able to convey his or her messages clearly both verbally and in writing. At the same time, he or she must be willing to listen to suggestions and ideas put forth by project sponsors, project team members, and other stakeholders of the project both within and outside of the organisation.
- *Leadership of a changing organisation.* All projects inevitably involve changes in the organisational culture, institutional structure, business processes and people. The project manager assumes the role of a change manager who will steer the organisation through the transition with minimum disruption to the organisation's business, the least anxiety of its staff, at the lowest possible cost, and within the shortest realistic time frame.
- *Managing politics and conflicts.* While a project manager has considerable responsibility for the success of a project, he or she does not always have absolute authority and control over human, financial and technical resources allocated to the project. Therefore, it is essential for a project manager to be politically astute, and have good networking and negotiation skills to deal with senior managers and all stakeholders to ensure appropriate and sustainable support for the project. It is also important for a project manager to have good mediation skills to resolve disputes or conflicts among members of the project team as well as those between the project team and other stakeholders of the project.
- *Team leadership.* Experience has shown that project teams seldom become high-performing immediately after their formation. It takes time and effort to build an effective and coherent project team. As the team leader, the project manager should continuously motivate members by letting them know the benefits and potential opportunities of the experience and skills to be gained in the project. He or she must provide

team members the necessary training to perform their assigned tasks effectively, and recognise their efforts and accomplishments appropriately. Team leadership also means delegation and empowerment, where necessary, to increase the sense of belonging among team members. At the same time, team leadership never ignores accountability and discipline, which will be applied impartially, transparently and promptly when and if the situation warrants them.

• *Building trust and credibility.* Project management can be a very difficult task if the project manager is unable to gain the trust of the project sponsor, project staff and other stakeholders. Trust and credibility cannot be built overnight. They must be developed over time and can be inspired only if the project manager exhibits behaviours compatible with the competency and skill requirements described above, is willing to admit mistakes and accept responsibility for actions, values differences and diversity of opinions and cultures, and treats everyone equally and equitably.

Technical competencies and skills are absent from competencies and skills required of project managers noted above. Hence, it is useful to consider these requirements from another perspective by taking technical competencies and skills into consideration. Figure 9.3 shows the role of technical competencies and skills in project management. In this figure, project management competencies and skills are classified into three categories according to their respective nature, namely strategic, tactical, and technical. Strategic competencies and skills cover mainly the two areas of "communication" and "leadership of a changing organisation" described above. These two areas of competencies and skills are applied mainly in the initiation and closing phases of the PMLC. Tactical competencies and skills, on the other hand, transcend the areas of "communication", "managing politics and conflicts", "team building" and "building trust and credibility". These four areas of competencies and skills are needed most for the planning and monitoring and control phases of the PMLC.

Technical competencies and skills are essential throughout the entire course of the PMLC, but are particularly significant in the project execution phase. Such competencies and skills are project-specific because different types of projects require different technical skill sets. For spatial database implementation projects, the project manager is not normally expected to be an expert in geographic information science and database systems. However, it is essential for the project manager to have, as a minimum, a good understanding of the basic principles of geographic information systems (GIS), including the characteristics and sources of spatial data. He or she should be familiar with fundamental spatial database concepts and methods presented in the other eleven chapters of this book. He or she should also be up-to-date with the latest advances in spatial database technology. These include in particular Web-based systems and applications covered in Chapter 10, and the trends of spatial database technologies discussed in Chapter 12.

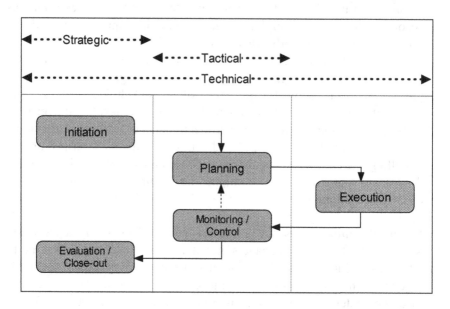

Figure 9-3. Competencies and skill requirements at different stages of the PMLC

Application-specific knowledge and skills pertaining to a particular spatial database project are not as important. The project manager can usually pick up relevant knowledge and skills through interaction with application specialists and user representatives as the project unfolds.

3. PROJECT INITIATION ACTIVITIES

Logically, project initiation is the first phase of the PMLC. In practice, however, project managers actually start at the end and work backwards mentally. Therefore, the discussion of project initiation starts first by defining the required outputs upon completion of the project, then makes preliminary estimates of the resources required, and finally develops the strategies required to construct and deliver the project output in time, within budget and according to specification.

3.1 Preliminary Project Organisation

Activities in the project initiation phase are conducted mostly by the project manager, possibly with the aid of one or more systems analysts and application specialists who are identified as potential members of the future project team. The very first task of project initiation is to define the goal of the project that is to be undertaken. The goal statements define the scope of the project in terms of what is included and what is excluded. Furthermore, the goal statements identify constraints of time, cost and specification.

Writing a well conceived goal statement requires considerable brainstorming and discussion with the project sponsor and representatives of potential users of the resulting spatial database system. It is sometimes also necessary to research the possible relationships between the project and legislated responsibilities and regulatory obligations of the organisation, as well as safety and quality standards of the spatial database system that will result from the project. Goal statements should be:

- *Specific*, so that any individuals with basic knowledge of the project can understand the goal(s).
- *Concise and clear*, so that there is no scope for ambiguity or project creep within the context of the project's activities.
- *Measurable*, to determine clearly whether particular goal statements were achieved.
- *Agreed upon*, by the sponsor and representatives of potential stakeholders.
- *Realistic*, or achievable, affordable and sustainable fiscally, technically and politically.

Some organisations prefer the use of a *pro forma* project proposal to standardise the submission and management of project initiation information (see Section 3.5). In this context, one of the most difficult challenges experienced in project management is the maintenance of documentation (see Section 6.1). Thus, it is advisable to use a *project binder* to keep all project-related documents in on one place so that they are easily accessible and filed in a systematic manner. Project documents are now often prepared electronically using word processing and project management software packages. In a computing environment, relevant document files can be stored and managed in a *project folder*. However, experience suggests that a hard copy of all relevant documents also be made and kept in a project binder for handy reference, particularly for people who do not have ready access to a computer.

3.2   Developing Project Activities

In general, a top-down approach that uses a hierarchy of the sort shown in Figure 9.4 is recommended for recording project activities. The hierarchy is constructed based on the previous experience of the project manager and the systems analysts and application specialists assisting him or her at the preliminary stage of the project's development.

Phase 1 - Initiate project

Activity 1A - Create preliminary project goal

    Task 1A1: Develop project goal statements
    Task 1A2: Create project org-chart
    Task 1A3: Complete project business case form
    Task 1A4: Create project folder
    Task 1A5: Create project binder

Activity 1B - Create a list of activities

Activity 1C - Identify project resources

    Task 1C1: Complete project resource requirements form
    Task 1C2: Conduct cost-benefit analysis

Phase 2 - Plan project

Activity 2A - Organise project team
Activity 2B - Meet with project team

    Task 2B1: Confirm commitment of team members
    Task 2B2: Seek approval from Human Resources Department
    Task 2B3: Determine date, time and place of first meeting
    Task 2B4: Prepare meeting ground rules and agenda

Activity 2C - Refine project plan

    Task 2C1: ....
    Task 2C2: ....

    ....
    ....

Phase 3 - Execute project

Activity 3A - ....
Activity 3B - ....

Figure 9-4. Partial listing of a project activity hierarchy

The top-down approach of building the hierarchy starts with the highest-level activities identified for each PMLC phase. These activities are then analysed progressively one by one for completeness and to identify individual tasks that are required to complete a particular activity. The information captured in this exercise is later used to determine the project cost, the human and technical resources requirements for the project, and the expected completion date. Contingencies in terms of time and resource requirements should also be estimated at each stage as rarely, if ever, do projects unfold without some form of unforeseen events occurring that require contingency planning.

3.3   Identifying Project Resources

All project activities require financial, human and technical resources to complete or support them. While the types of resources required by individual activities can be identified relatively easily using existing knowledge and past experience of the project manager, it is sometimes quite difficult to determine the precise amount of required resources. To assist in the documentation of resource requirements, a resource requirements form, such as that illustrated in Figure 9.5, can be used.

Organisation Name Spatial Database Project - Resource Requirements					
Activity	*Personnel*	*Facilities*	*Equipment*	*Materials*	*Other*
	Person(s): Time: Cost:				
	Person(s): Time: Cost:				
	Person(s): Time: Cost:				
	Person(s): Time: Cost:				
	Person(s): Time: Cost:				

Figure 9-5. Example of a resource requirements form

For each identified activity, the resources needed can be entered by determining the knowledge and skill levels required of project personnel, along with the necessary equipment, facility and space, materials, as well as the estimated time and cost of the resources. It is also useful to develop an *organisation chart* (org-chart) for all project participants (Figure 9.6), together with a brief description of their respective roles and responsibilities in the project. The org-chart can become quite complicated quickly and in the case of a large and complex organisation, specialised software can be used for this purpose. Once completed, the form and the org-chart should be saved in the project folder on a computer, and printed copies of these documents should be made for the project binder.

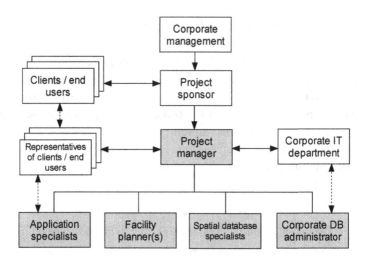

Figure 9-6. Example of an organisational chart (org-chart) of a typical spatial database implementation project

3.4   Cost-benefit Analysis

Many organisations require a *cost-benefit analysis* to be conducted as part of the project initiation process. Cost-benefit analysis, also called *investment analysis*, is based on the realisation that justification for investments are best made when accompanied by some level of analysis of the associated costs and benefits both of the investment itself and its returns over a relevant time frame. In addition to providing a framework for planning, a cost-benefit analysis provides some assurance of the prudence of the initial capital expenditures on the project and the long-term support and maintenance costs of the resulting spatial database system.

However, as a project management instrument, cost-benefit analyses are often imprecise because they do not account for many of the non-monetarised subtleties and complexities that surround spatial database projects. Cost-benefit analysis is particularly difficult in the case of large-scale spatial data infrastructure projects that take several years to complete. Although the costs of information technology are relatively easy to measure and account for, the benefits side of the equation is difficult to formulate. This is because many significant benefits are intangible (e.g., improving the integrity of land tenure systems, better stewardship of the environment, and increasing public awareness of and interest in participatory democracy), and hence cannot be quantified precisely in monetary terms.

Cost-benefit calculations are further complicated by three other factors. One of these includes what kinds of costs and benefits are measured and how they are measured, at what time the benefits are realised, and whether to include the values of external benefits or only those that are relate directly to the project being undertaken. Further complicating matters is the *principle of the time-value of money*, which stipulates that costs incurred today are "worth" more than the benefits of the same monetary values received in the future. The third factor is concerned with the variation and variability of the life cycles of different project components (e.g., the relatively short, perhaps three to five-year, technology life cycle and the longer term, and often indefinite, data life cycle).

Working within the above limitations, a typical cost-benefit analysis can be conducted as an "educated guess" to assist in project management decision making, using the following two steps (GeoAnalytics, 2003):

- *Assumptions*. There are four sets of assumed parameters or variables used for the calculation of cost and benefits. These include:

  - *Time frame* for project life cycle, payback period, initial investment period, and benefit accrual period.
  - *Costs* for personnel (wages and benefits), facilities, equipment, and materials obtained from the Resource Requirements Form (see Section 3.3).
  - *Benefits* from productivity gain, cost recovery (including royalty from sale of data, potential revenues and user fees), value-added services, economic spin-off internally and externally.
  - *Discount rates* for use as differential weightings in the calculation of future costs and benefits.

- *Calculation*. Cost-benefit analysis calculates annual benefits (Table 9.1) less annual costs (Table 9.2) relating to the creation and maintenance of the spatial database system resulting from the project. Two sets of values are normally calculated, namely:

  - *Sum of Flow of Net Benefit* (SFNB), which simply calculates a sum of all the payments and income/benefits over the same assumed payback period. An example of SFNB calculation based on productivity gain is shown in Table 9.3.

- *Net Present Value* (NPV), also called "discounted net present value", which calculates the sum of future payments (negative values) and income/benefits (positive values) over an assumed payback period (for example, 10 years) and reduces them to present value using an assumed discount rate (such as 3.95%). An example of NPV calculation based on productivity gain is shown in Table 9.4.

Table 9-1. Value of potential benefits by percentage of productivity gain

Productivity Gain (%)	FY 2003 (%)	FY 2004 (%)	FY 2005 (%)	FY 2006 (%)	FY 2007 (%)	FY 2008 (%)
1	0	0	0	0	284,704	284,641
3	0	0	0	0	854,112	862.624
5	0	0	0	0	1,423,968	1,438,207
10	0	0	0	0	2,847,935	2,876,414
15	0	0	0	0	4,271,903	4,314,622

Productivity Gain (%)	FY 2009 (%)	FY 2010 (%)	FY 2011 (%)	FY 2012 (%)	FY 2013 (%)	Total (%)
1	290,518	293,423	296,357	299,321	305,607	**2,057,660**
3	871,554	880,269	889,072	897,962	916,820	**6,172,981**
5	1,452,589	1,467,115	1,481,786	1,496,604	1,528,033	**10,288,302**
10	2,905,178	2,934,230	2,963,573	2,993,208	3,056,066	**20,567,605**
15	4,357,768	4,401,345	4,445,359	4,489,812	4,584,099	**30,864,907**

Table 9-2. Total capital investment and operating costs by year

FY 2003 ($)	FY 2004 ($)	FY 2005 ($)	FY 2006 ($)	FY 2007 ($)	FY 2008 ($)
10,000	618,000	932,000	583,654	182,214	125,880

FY 2009 ($)	FY 2010 ($)	FY 2011 ($)	FY 2012 ($)	FY 2013 ($)	Total ($)
309,656	133,546	137,553	141,679	145,929	**3,320,911**

Table 9-3. Sum of flow of net benefits by year based on percentage of productivity gain

Productivity Gain (%)	FY 2003 (%)	FY 2004 (%)	FY 2005 (%)	FY 2006 (%)	FY 2007 (%)	FY 2008 (%)
1	-10,000	-618,000	-932,800	-583,654	102,490	161,761
3	-10,000	-618,000	-932,800	-583,654	672,167	737,044
5	-10,000	-618,000	-932,800	-583,654	1,241,754	1,312,327
10	-10,000	-618,000	-932,800	-583,654	2,665,721	2,750,534
15	-10,000	-618,000	-932,800	-583,654	4,089,689	4,188,742

Productivity Gain (%)	FY 2009 (%)	FY 2010 (%)	FY 2011 (%)	FY 2012 (%)	FY 2013 (%)	Total (%)
1	-19,139	159,877	158,805	157,642	159,677	**-1,263,251**
3	561,897	746,723	751,519	756,283	770,890	**2,852,070**
5	1,142,933	1,333,569	1,344,234	1,354,925	1,382,103	**6,967,391**
10	2,592,522	2,800,684	2,826,020	2,851,529	2,910,136	**17,255,693**
15	4,048,111	4,267,799	4,307,806	4,348,133	4,438,169	**27,543,996**

Table 9-4. Discounted net present value (NPV) of costs and benefits and pay period based on productivity gain

Productivity Gain (%)	Net Present Value ($)	Payback Period (Years)
1	-1,321,386	9
2	235,859	6.75
3	1,793,103	5.75
4	3,350,347	5.25
5	4,907,591	5
6	6,464,836	4.75
7	8,022,080	4.75
8	9,579,324	4.5
9	11,136,568	4.25
10	12,693,813	4.25
11	14,251,057	4.25
12	15,808,301	4.25
13	17,385,545	4.25
14	18,922,790	4.25
15	20,480,034	4.25

(Source for all tables: GeoAnalytics, 2003)

Cost-benefit analysis involves a substantial amount of relatively simple but repetitive calculations because for each combination of the identified costs and benefits, NPV and SFNB tables are required using different payback periods and discount rates. While the tedium of the calculation can be reduced by using spreadsheet software, the interpretation of the resulting multi-dimensional tables still requires a great deal of manual effort on the part of the project manager.

3.5　　Writing a Project Proposal

A project proposal, also referred to as a project overview statement or project definition form, is a document that summarises the findings of the project initiation phase for approval by senior management of the organisation. A Project Business Case (PBC) form (Figure 9.7) can be developed for this purpose using a *template*. This approach is often pre-ferred because a template is easy to create, edit, read and standardise so that all projects within the same organisation can be presented for management consideration in the same format.

Alternatively a *Proposed Solution* (PS) form (Figure 9.8) can be used for this purpose. The PBC form contains information that aims to help senior managers understand the justifications of the project. The information items in the form should be written in plain and concise business language. The PS form, on the other hand, provides a clear description of the proposed solutions. If alternative solutions are proposed to address a particular problem, a clear explanation of the pros and cons of different approaches as well as the reasons for picking the final choice must be given. This form also contains two tables listing the budget and personnel requirements of the project. In practice, it is advisable to keep the entire proposal to three or four pages.

Name of Organisation
Project Business Case

- -

Project Identification
Project name: ————————————— Date: ——————————
Division/Department/Unit: ———————————————————————
Project sponsor: ————————————— Project manager: ——————————

Business needs / problems and project goal statements

Solutions *(Summarised from and cross-referenced with Proposed Solutions)*

Consistency with organisational mission, legislated and regulatory requirements

Resource requirements *(Summarised from Resource Requirements Form and org-chart)*

Anticipated benefits and opportunities *(Both quantitative and qualitative)*

Cost estimate *(Abstracted from and cross-referenced with Proposed Solutions)*

Cost/Benefit analysis *(Summarised results only)*

Risk and organisational impact

Sources of funding *(Include both internal and external funding where applicable)*

Figure 9-7. Example of a project business case form

<table>
<tr><td colspan="4" align="center">**Name of Organisation**
Proposed Solutions</td></tr>
</table>

Project Identification

Project name: _____ Date: _____

Division/Department/Unit: _____

Project sponsor: _____ Project manager: _____

Summary of business needs / problems *(From Business Case)*

Proposed solutions / Project approaches

Solution / Approach / Alternative	Why chosen / not chosen

Budget / Estimate costs

Item	Initial (Development)	Annual (Recurring)	Remarks
Hardware			
Software licences			
Supplies			
Data			
Consulting services			
Other			
Total			

Personnel / Estimate resources

Item	Initial (Development)	Annual (Recurring)	Remarks
Design / development	------- hours	------- hours	
QA / QC / Testing	------- hours	------- hours	
Training	------- hours	------- hours	
Legal services	------- hours	------- hours	
Outsourcing	------- hours	------- hours	
Total	------- hours	------- hours	

Figure 9-8. Example of a proposed solutions form

Although filling out the PBC and PS forms is a relatively straightforward task, it is always advisable to read the forms several times, from the reader's perspective rather than from the writer's perspective, before submitting them for management approval. Moreover, it is important to file the forms in the project binder (hardcopy) and project folder (digital copy), along with all other important information, for possible use in subsequent projects where the same or similar issues need to be reviewed. Such procedures may seem mundane and trivial, however experience has shown that good organisation and documentation is a important factor for the success of a project.

4.   PROJECT PLANNING PROCESSES

The objective of the project planning phase is relatively simple, namely to pave the way for turning ideas (project initiation) into realities (project execution) by establishing the project team and developing a detailed project plan on the basis of the identified project activities in the previous phase of the PMLC. At this stage of the project, it is also advisable to consider seriously potential risks that the project may encounter, as well as the possible opportunities that the project may bring to the organisation.

4.1   Recruiting and Organising the Project Team

Unlike the project initiation phase, which is largely the work of the project manager, project planning is dependent on team work because of the increasing amount of effort and level of technical skills required. The first job of the project manager in the planning phase, therefore, is to form the project team.

A typical project team includes the project sponsor, the project manager and a number of project team members. The project sponsor is the champion and promoter of the project. He or she also serves as the immediate supervisor of the project manager. The project sponsor is usually a senior manager of the department or business unit that is home to the spatial database system that will result from the project. However, it is not uncommon for a senior manager who has no direct vested interest in the project to be appointed project sponsor. The project sponsor does not take part in the daily administration of the project. His or her primary role is to represent the project to the management of an organisation, and to secure necessary resources for the project from its beginning through to its completion. The project sponsor has the final authority to sign off intermediate and final delivery of the resulting spatial database system on the recommendation of the project manager.

The project manager is the key person in the project team. In many large organisations, such as government agencies, banks and engineering consulting firms, there are professional project managers on the permanent staff whose job is to lead corporate projects. However, it is commonplace for organisations to appoint project managers from existing unit managers or senior IT personnel who have the training and experience to assume such a job function. As noted in Section 2.3, all project managers are expected to be strategic in their thinking, tactful in dealing with people, and technically knowledgeable in the business area served by the project. The project manager serves as the chief executive officer of a project. He or she is the technical advisor to the project sponsor, mentor and supervisor of members of the project team, and representative of the project when dealing with internal and external stakeholders. In the case of spatial database implementation projects, the project manager usually has to work closely with members of the organisation's IT department to ensure adherence to corporate standards, protocols and resource sharing policies.

It is the responsibility of the project manager to recruit members for the project team and organise them into a coherent working group. Team members can be co-opted from internal staff, or hired externally on a contractual or permanent basis if internal expertise is not available. The number of members of a project team varies according to the nature of a particular project. However, it is essential to recruit and choose team members with the understanding that they have collectively all of the necessary skills that are required to complete the project successfully. For large-scale spatial database implementation projects, it is helpful to divide the team into small working groups, each headed by an experienced technical lead, such as map data conversion and acquisition, database design and system development, quality assurance and control, and end-user training. This makes the team more manageable and creates a clear sense of accountability and responsibility.

Experience has shown that while recruiting is seldom a problem, organising members into a coherent high-performance team can be problematic. It is a real challenge for the project manager to use his or her people skills to ensure the commitment of the members to the project (i.e., their other competing duties or tasks will not negatively impact on the project). The project manager must also keep motivating team members continuously throughout the course of the project by mentoring team members, practicing open communication, giving mutual understanding and respect, recognising achievements, as well as using disciplinary actions where and if necessary, as noted in Section 2.3.

4.2 Refining the Project Plan

The project activity hierarchy prepared during the project initiation phase (see Section 3.2) contains only very general high-level information about the tasks to be completed at various phases of the PMLC. It is not detailed enough for use as a blueprint to design and develop the spatial database system of the project. As well, the sequence of the activities may not be as logical as was originally conceived. The objective of refining the project plan is to revisit the original activities, individually and holistically, with a view to identifying the actions, cost, start and end dates, order of tasks relative to each other, and the project team responsible for every task associated with each of the activities. The intent is for the project team to develop the plan to the point where it is very clear how each task will be executed at the project execution phase.

The refinement of the project plan involves four basic aspects of work:

- *Project tasks.* The aim here is to identify the details of all tasks within each activity, and sort them into a logical order for execution. The project activity hierarchy is the natural starting point of the refinement process, which is usually carried out in a series of workshop sessions attended by the project team. Subject experts who are not regular members of the project team are sometimes invited to attend the workshops to provide input to specific issues. During the workshop, project team members review the items in the task hierarchy one by one. Tasks deemed to be clear and definite can be left untouched. Ill-defined and vague tasks must be broken down using an hierarchical approach into definite sub-tasks (e.g., "Administer contract" can be broken into four sub-tasks that include "Contact Legal Services Department to get a sample contract", "Amend sample contract", "Submit amended sample contract to Legal Services Department for approval", and "Incorporate amended contract in information package for vendors"). Some of these processes may require several iterations before passing comfortably to the next task. On completion of refining the tasks, the entire project activity hierarchy must be reviewed globally to ensure that the logical ordering of the tasks makes sense for their execution. The deliverable of the refinement process is a detailed and logical task hierarchy to be accepted as the formal project plan.
- *Resources.* The aim for this aspect of the project plan is to quantify the resources required for each task. The project team should first identify the type of resources needed, and then estimate the cost for that particular task. Cost can be estimated in one of two ways, namely a fixed cost (e.g., $5000.00 per year for a multi-user software license) or a variable cost (e.g., 50 hours of initial VB.NET programming and 20 hours or less of additional programming work per year for maintenance and update). The project team should carefully consider the matching of available resources and estimated resources. If, for example, a particular task requires skills that are not found among project team members, then arrangements must be made for outside assistance. This may necessitate

amendments to estimated time and costs, and the project plan must be adjusted accordingly.

- *Project schedule and task precedence.* The aim is to determine the most efficient order for the tasks by maximising the use of resources. This is usually accomplished by having more than one task in progress at the same time and by preventing the delay of any given task from holding up the start of another task. The concept of "precedence relationships" between individual tasks is the governing principle for project scheduling. This concept identifies tasks either as "dependent" or "independent". Dependent tasks are those that cannot proceed until another task is completed, whereas independent tasks are those that can be carried out any time during the course of project execution. As noted earlier, an effective way of project scheduling is to consider the project from the end point, and work backwards to the starting point. In this way, it is relatively easy to identify those tasks that must be completed prior to the commencement of their respective predecessors. There are two deliverables for project scheduling. One of these is the *Gantt Chart*, which depicts graphically the relationship between time lines and individual tasks (Figure 9.9). The other is the *Project Network Diagram,* which provides a visual representation of the overall project schedule, including sequence and relationships among tasks, their respective start dates and duration (i.e., the number of days to complete the task) (Figure 9.10). Tools exist in many software products to produce Gantt Charts and project network diagrams relatively easily.
- *Project member's roles and responsibility.* Of all the project resource requirements, human resources are the most precarious and hence the most difficult to estimate in advance and manage after the fact. In order to make the best use of the expertise and time of individual team members, it is important for the project team to establish a Responsibility Chart as part of the planning process. This is a simple table which lists all the tasks along the left hand side and the names of project team members on the top. This forms a matrix of tasks against names in the body of the table in which a cross or check mark is placed under each member's name relative to the task(s) that he or she is responsible for. Tasks are permanent in the sense that they persevere and must be completed, while personnel responsible for one or more tasks may change over time.

As with all such project management tools, there should be consistency in cross referencing, for example between a Gantt Chart and a Responsibility Chart. While a general rule of thumb suggests that visual tools help to ensure transparency and efficiency in project tasks, it is never a good idea to use more tools than are necessary for a specific project.

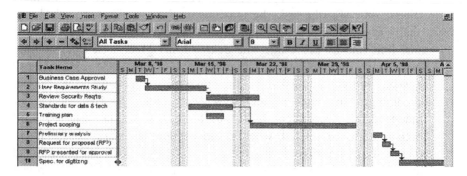

Figure 9-9. Example of a Gantt chart

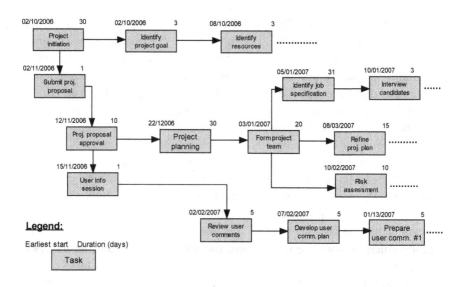

Figure 9-10. Example of a project network diagram

On completion of the above project planning activities, the project team should revisit the project plan, Gantt Chart, Project Network Diagram, and Responsibility Chart carefully, and refine them further if and where necessary. Since considerable amendments and expansions were made to the original project activity and resource requirements documents, it is necessary to submit all of the above project planning documents to the project sponsor for approval. Once these documents are approved, they should be filed properly in the project folder.

4.3 Potential Risk and Opportunity Analysis

Despite the best of intentions and careful thought invested in the project initiation and planning processes, the possibility of unexpected problems and benefits occurring during the course of the PLC is real. Risk and opportunity analysis are not always seen as integral and critical components of project planning, however a prudent project manager should realise that there is no perfect approach to project planning, and if something can go wrong it usually will. The idea of potential risk and opportunity analysis is based on the belief that an experienced project team does not and cannot know exactly what problems will occur and when they will occur, but they are able to anticipate the types of potential problems and have the capability of dealing with them if they materialise.

In essence, potential risk analysis is a preventive measure rather than a regular element of a project. Table 9.5 lists common risks and their corresponding preventive measures in spatial database implementation projects. The idea of potential risk analysis is closely related to the principles and methods of project monitoring and control explained in Section 6.

Table 9-5. Common risks and preventative measures in spatial database implementation projects

Possible Risks	*Preventative Measures*	
Loss of project personnel (due to resignation, reassignment of duty and/or long term absence such as illness, maternity leave, etc.)	O	Always have a back-up person for all critical areas, who can quickly be reassigned with minimum delay or productivity loss.
	O	Ensure detailed and up-to-date documentation of work done by all project team members.
Late delivery and delayed availability of materials	O	Adjust project schedule by launching independent project tasks when dependent tasks are being put on hold.
	O	Keep a record of alternative source materials and services.
Processing errors	O	Establish stringent backup policies and set up properly managed back-up facilities to ensure all completed work and work in progress are properly backed up.
Data errors	O	Avoid the assumption of "clean" data.
	O	Exercise stringent quality assurance and quality control when evaluating, acquiring, storing and archiving data.
	O	Keep an up-to-date record of metadata to enable the tracking of errors if and when they occur.

Risk analysis essentially aims to help the project team develop contingency plans so that it can respond quickly and correctly to problematic situations before irreparable damage is done. One of the most common problems is the resignation or reassignment of project team members to another project. This problem is particularly acute in the case of large-scale multi-year infrastructure projects or for projects in organisations where staff resources are limited and priorities change within short time frames. Another very common problem is the late delivery or delays in the availability of source materials. Computer-related projects are particularly susceptible to errors in processing, including inadvertent deletion or over-writing of application and data files. A problem that is specific to spatial database projects is related to errors in the source data, corrupt files, incompatible formats, and out-of-date data. Many project managers tend to plan their projects as if all spatial data are "clean", and get caught when data errors begin to surface as the project unfolds.

In addition to risks, the project manager should also look at the possibility of maximising the benefits from the project. They can do this by, for example, exploring the potential of extending the functionality of the spatial database system, providing value-added sales of the data to other users, and using their experience gained in the project being undertaken to offer consulting services to other organisations. All these will potentially bring additional revenue to the project, thus increasing its return on investment and long term sustainability.

5.   PROJECT EXECUTION ACTIVITIES AND METHODS

Project execution is the most challenging phase of a spatial database project in terms of resource and technical requirements during the PMLC. This section discusses the major project execution activities only in general terms, because the tasks involved in individual projects vary considerably from one to another. These activities are typically determined by a combination of project-specific factors that are practically impossible to cover within the space of a single chapter. The following discussion focuses on the primary project execution tasks of procurement for services through a request for a proposal, spatial data acquisition, selection and acquisition of technology, and application development.

5.1   Request for a Proposal

Spatial database projects can be implemented by internal staff or by external consultants either in total or in part (a process commonly referred to as *outsourcing*). There are three common approaches to outsourcing, namely *sole source, invitation to tender* (ITT) and *request for a proposal* (RFP). As the name implies, outsourcing by sole source means that the entire contract for a project is awarded to a single contractor without going through a competitive procurement process. When compared with the other two approaches, sole sourcing is less flexible and not as rigourous a procedure, although it takes much less time to pick a contractor from a list of *contractors of record* and start the project (Table 9.6). For public organisations, sole sourcing can be seen as a form of favouritism, and the decision may be challenged by consultants or companies who feel that they have not been fairly treated. Hence, sole sourcing should be avoided except in the cases where the chosen consultant is the only possible candidate who can supply the material or service required.

Table 9-6. Comparison of approaches to procurement

Approach	Flexibility	Selection Time	Rigour
Sole Source	Low	Short	Low
Invitation to Tender	Low	Moderate	Moderate
Request for a Proposal	High	High	High

An ITT and a RFP are competitive processes with different intents and purposes. An ITT is used when the organisation is absolutely clear what it wants, and how it wants things to be done. Suppliers or vendors are openly invited to bid for the contract on the basis of factors such as price and ability to meet the specified requirements. An ITT is most suitable in situations where the supply of materials (e.g., computers and peripherals) and the type of services involved are relatively well-defined (e.g., facility maintenance and security services). A RFP, on the other hand, is used when the organisation knows generally or exactly what it wants, but prefers to seek solutions from the consulting community. In spatial database implementation projects, a RFP is the more prevalent approach than an ITT in terms of soliciting external consulting services.

A RFP is a relatively complex procurement process that demands considerable effort and expenditure on both the part of the requesting organisation and the responding consultants. The complexity of a RFP is dependent on its objective and scope, which may cover all or a substantial part of the project execution activities shown in Figure 9.11.

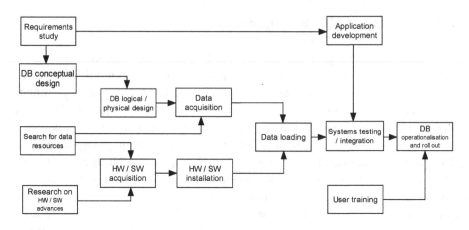

Figure 9-11. Major activities and workflow of the project execution phase

There is no standard procedure for conducting a RFP but the workflow of a typical RFP exercise starts with approval of the recommendation in a project proposal to elicit external work (Figure 9.12). The project team then prepares the RFP information package for approval. Since a RFP always ends up with a contract between the organisation and the selected consultant, legal advice must be sought in advance so that the wording and general content can be reviewed from a legal standpoint, and a sample of the contract to be signed can be prepared. Any changes that are made to the RFP or the contract should, of course, be re-scrutinised by legal counsel before either document is issued and signed off on.

The workflow required for developing a RFP comprises mandatory and optional components, as shown in Figure 9.12. The extent to which individual components are included in the final RFP document and the steps that are followed in the process of compiling the RFP document is to some extent a function of the scale of the work that is being called for. Some tasks are relatively straightforward and have minimal risk associated with them. However, other tasks and indeed overall projects have considerable risks, especially where subcontractors may be involved or where the actual configuration of outputs is not clear at the time of developing the RFP document. In the latter case care should be taken not to rush the preparation of the RFP document as errors or omissions that are made at the stage of the call for work will likely be compounded into the work that is produced. In the former case the lines of responsibility for the completion of any subcontracted work must be made clear in the RFP document and all legal issues concerning subcontracting must be accounted for. Hence, it is advisable to approach developing a RFP and producing the RFP package gradually making sure that all angles are considered and taken into account in the configuration of the work requirements.

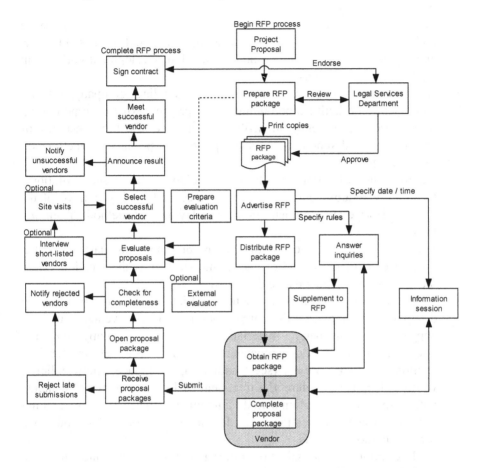

Figure 9-12. Workflow of a request for a proposal (RFP) exercise

The content of the RFP package varies according to the nature and purpose of the procurement. It normally includes but is not limited to the following items:

- The background of the project, including the business mission and mandate of the organisation.
- The scope of the project, which is commonly defined and presented in terms of the results of a preliminary user requirement study and conceptual database modelling.

- Objectives and all mandatory, desirable and optional outcomes of the project, including the performance standards of all spatial database products and services.
- Specification of the organisation's present hardware and software architecture, internal and industry standards for spatial database products and services, resource sharing and application development partnership with other organisations.
- Request for respondent's information, including company history, equipment and facilities pertaining to the project, experience pertaining to the project, professional affiliations and certification, resumés of project staff and sub-contractors (if applicable), primary contact person and references.
- Descriptions of proposal format, including proposed solutions, technology requirements, quality assurance and quality control plans, requested price and compensation, page layout and method of binding the proposal.
- Specification of the method of submission, including data and time, number of copies, method of packaging, as well as the location where the proposal package must be deposited.
- Supplementary information about conflict resolution in case of disagreement between the organisation and the vendor, use of copyrighted third-party materials, ownership of copyright and intellectual property of components of all database products and services resulting from the project (see Chapter 7), workplace safety policies, working language, preferential treatment of local and national contents with respect to international trade agreements, the protocol of making inquiries where clarification is needed, as well as time, date and place for information sessions.
- A description of the selection criteria and procedure, including the organisation's discretion to terminate the project with no obligation to compensate the responding companies for the cost incurred.
- Request for certification of insurance and redemption coverage of a specified amount, evidence from government revenue department and, where relevant, statement from a worker compensation board clearing the responding company of any outstanding tax and payments.
- Requirement for the signature of the principal officer of the responding company certifying the truthfulness of the submission, together with the date and time of signing.
- A sample of the contract to be signed.
- A glossary of terms pertaining to the project.

When the RFP package is ready for distribution, the project manager announces it through advertisements in local and national newspapers, and professional and trade journals. Some organisations also place the announcement in its corporate Web site and if relevant on government contract procurement Web sites. The announcement usually highlights the time, date and location of the information session, if any, and the rules for submitting inquiries in case of questions. It is common practice to give six to ten weeks for the interested parties to respond because the preparation of a proposal is quite a daunting and time-consuming task.

No matter how carefully a RFP package is prepared, there will always be questions from potential respondents. As a rule, all inquiries must be submitted in writing and no telephone inquiries are allowed. Answers to inquiries are not addressed to the inquirer alone. Instead, questions are collected and answered in the form of one or more supplements to the RFP that are distributed to all those who have requested the RFP package. This provides all interested parties with equal access to the information. The supplements are considered an integral part of the RFP process.

Sealed proposals are received at a specified location, and must be deposited personally by the submitting parties into a sealed box after they are date-stamped by the reception staff. The proposals are opened in public, usually immediately after the specified closing date and time. The project manager reviews each proposal carefully to ensure completeness. Proposals that violate any of the mandatory requirements (e.g., signature of the principal officer, inadequate insurance and redemption) are rejected out of hand (i.e., the respondent is not given the opportunity to re-submit the missing information). Arrangements are then made to convene a proposal evaluation session in which all qualified proposals are critically assessed according to pre-defined evaluation criteria (see Section 5.3). It is common practice to invite an independent evaluator, for example, from a university or an accounting firm, to sit in the session to assure the respondents that the evaluation process is carried out in a fair and impartial manner.

For large-scale projects, it is desirable to short-list and interview respondents deemed to have the most feasible solutions as measured by their respective scores in the evaluation. This gives the evaluators an opportunity to ask questions about various aspects of the proposal. At the same time, it allows respondents to demonstrate and clarify proposed concepts, processes and techniques that are not easy to comprehend when presented in writing. Optionally, the project team can visit the short-listed companies if and when on-site equipment and facilities are critical components to be considered in the evaluation (e.g., a specialised micron scanner or mass storage equipment in the case of large-scale aerial photography and spatial data conversion projects).

On completion of the interviews, the short-listed respondents are ranked on the merit of their proposal, results of the interviews and site visits, as well as price and other proposed terms of service. Before the contract is signed, it is common practice to meet with the top-ranking responding company to confirm that it has the intention to proceed. In many instances, this meeting is the last opportunity for the two parties to negotiate and make minor adjustments to requested project deliverables (e.g., enhancement to business processes not anticipated in the original RFP) and corresponding amendment to the overall price of the contract.

It is rare but not uncommon for companies to withdraw from the process at this stage if they realise that they may not be able to make any profit from the project after carefully considering the price of the contract and the expenditures involved. If this occurs, the second top-ranking company will be invited to a meeting to see if they are willing to accept the offer of a contract. In the very exceptional case when no company is willing to accept the project contract for one reason or another, the project team must return to the start of the process to review the project requirements and its feasibility. This will often result in changes in the objective and scope of the project, or in the worst case scenario to abandon the project completely.

5.2   Spatial Data Acquisition and Evaluation

Data acquisition is often regarded as the most challenging and expensive part of spatial database implementation. However, maintaining data quality through the life cycle of a database system can, in total, create greater challenges and be more expensive (Chrisman, 1983). Conventionally, organisations contemplating the implementation of a spatial database system usually have to convert existing paper-based spatial information into digital form either internally or by outsourcing. No matter which of these approaches is used, the process is invariably costly, time-consuming and prone to error.

Spatial data are much more readily available now than ever before. With the construction and inception of global and national geospatial data infrastructures, data warehouses and Web-based information dissemination technologies (see Chapter 6), spatial database projects now rely more on third-party data than on internal data conversion. New data collection technologies are now able to capture spatial data directly in digital form quickly and with a higher degree of accuracy than in the past. As a result, the focus of data in spatial database projects has moved from acquisition to quality or usability. However, up to now greater data availability has not necessarily been translated into higher usability and, therefore, data and data quality remain mission-critical issues in spatial database implementation projects.

The issue of data quality in spatial database implementation cannot be considered simply from a project management perspective alone. Instead, it must be considered holistically within the context of an organisation's vision, policy and strategy for quality data (Chapman, 2005). These three terms include the following considerations:

- *Vision.* This provides a conceptual framework for assuring and controlling data quality within an organisation. In developing a vision for quality data, project managers should focus on achieving an integrated management framework in which leadership, people, computer hardware, software applications, quality control and data are all brought together with appropriate tools, guidelines and standards to maintain the data, and turn them into quality information products.
- *Policy.* This refers to the formal guiding principles for implementing the vision of organisational data quality. Specifically, a sound data quality policy should force the organisation to think broadly about quality and to re-examine their day-to-day business practices. It should aim to formalise the processes of data management by focusing on reducing costs, improving data quality, improving customer service and relations, and improving the decision-making process. At the same time, it should provide users with confidence when accessing and using data originating from the organisation. Furthermore, it should improve the relations and communication with the organisation's own staff and external clients, and thus enhance the integrity and credibility of the organisation within the broader community.
- *Strategy.* There are different strategies to implement the policy for quality data. The dominant strategy is based on the realisation that prevention is a far more effective way, both technically and economically, to correct potential errors than to fix them when problems surface at a later time. The costs of data collection and database construction can be substantial but are only a fraction of the cost of checking and correcting errors occurring in data already in a database. Making corrections retrospectively not only causes uncertainty in analyses already done using the data before they were corrected, but it may also lead to legal liabilities and political responsibility for decisions and policies made using the erroneous data.

A preventative approach to data quality uses a four-tier *Total Data Quality Management* (TDQM) model to define, measure, analyse and improve data quality (Wang, 1998). Figure 9.13 shows how this particular approach is conceptualised and used in practice when data obtained from data suppliers are validated and cleaned before they are loaded into a spatial database. This approach differs from conventional methods in two ways. First is the emphasis on the use of pre-defined rules to validate the data before they are entered into the database. Second is the systematic capture and analysis of detected errors as well as the subsequent feedback of information to develop preventative measures at the data collection and supply end of the database construction process.

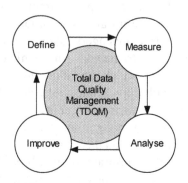

(a) The conceptual TDQM model

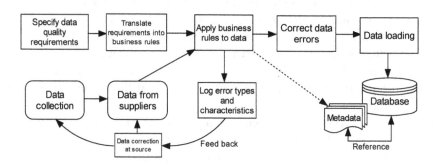

(b) Workflow of applying TDQM in practice

Figure 9-13. The concept and application of total data quality management (TDQM)

In practice, errors in graphical spatial data are much easier to detect and correct than attribute data. GIS software packages such as ArcGIS have map editing tools that allow errors to be detected and corrected automatically using the concepts of tolerances, topological relationships and business rules. Errors in attribute data, on the other hand, can only be detected by careful query and inspection and then corrected manually. In spatial analysis and decision making, the quality of graphical data is as important as the quality of attribute data. Thus, both types of data must be given the same level of attention in spatial database project management.

5.3   Technology Acquisition and Evaluation

Technology acquisition and evaluation is a relatively straightforward process when compared with data acquisition and evaluation. Technology in this case refers to computer hardware, software, and local and wide area networks, together with related peripherals and supplies. In many organisations, spatial database systems must be implemented using corporate architectures and standards. Therefore, a spatial database project team is often more concerned with the evaluation of the suitability of corporate resources specifically for spatial applications, than with their acquisition.

The common exception is spatial database or GIS software if the organisation has not determined a spatial software standard. In this case, a carefully planned software evaluation exercise must be conducted before the issue of a RFP. This is because the project team has to equip itself with sufficient knowledge of the relative merits and usefulness of the possible choices for the project to be undertaken. The actual acquisition usually takes place after a specific solution is selected from the RFP process.

Developing a technology evaluation plan is a relatively complex task (Berdusco et al., 1999). It normally starts with a review of a user requirements study that seeks to identify both the hardware, software and network needs of the deliverables of the project. The project team then identifies the key features of these relative to the generation of the deliverables. These features are progressively broken down into further levels of detail and recorded in a software evaluation manual that is regularly updated during the process of development (Figure 9.14). The features to be evaluated are assigned either as a "must" or "should" requirement to indicate their relative significance in the evaluation process. In this context, a "must" requirement means mandatory functionality (i.e. hardware equipment or software programs) that must work in the manner specified in the evaluation process. On the other hand, a "should" requirement is functionality that is expected and, if it does not already exist, a work-around solution may be acceptable.

In the evaluation process, potential vendors are required to supply in detail specific information about each of the features of their products that are to be evaluated. The project team then reviews the completed product evaluations and short lists the four or five best submissions that are deemed to meet the requirements of the project. A rigourous technical evaluation is then conducted for each of the short listed vendors. The evaluation can take place at a site within the organisation where the vendors will set up the equipment or software, and demonstrate to the evaluation team the functionality of each and every feature they claim it can achieve at the specified performance level. Alternatively, the evaluation can be conducted at the vendor's site if the equipment is too bulky to be transported and/or too complicated to set up at another location. While the latter option may be mote convenient, care should be taken that what works well outside of the workplace will also work well within it. On completion of the evaluation sessions, the project team tabulates the evaluations of all evaluators, compares notes, ranks the vendors' submissions, and then selects the vendor that satisfies all aspects of the evaluation process.

Product name: _____   Page F-001

7.1 Database implementation 7.`1 Spatial data structure and organisation		
Ref #	Requirement	Comments / Observations
7001	**MUST** represent point, line, area and surface spatial entities and topological relationships using vector data models • Describe the specific data models in each case • Describe the data structures used to implement the models • Describe how these entities can be used to construct higher order entities, i.e. user-defined objects	
7002	**MUST** represent areas and surfaces using a grid data model • Describe the data models and the data structure used to implement them	
7003	**MUST** support appended images for point, line and area entities • Describe how images are retrieved through attribute data queries and through selection of spatial entities • Describe data format used and access/retrieval methods	
7004	**SHOULD** support full UTM Grid Reference values • Identify the maximum and minimum coordinate values the system can handle, and the internal storage format and data type	
7005	**SHOULD** support storage of elevation values with each planimetric coordinate pair used to represent point, line and area entities • Describe format used.	
7006	**MUST** support data transformation from vector-to-grid and grid-to-vector for area, line and point entities • Describe the process used and any limitations on data set size.	

Figure 9-14. Example of a page of a software evaluation manual

5.4 Application Development Strategies and Techniques

In addition to the construction of a database, a spatial database project involves a substantial amount of effort and resources to be focused on application development. This includes the design, programming, testing and integration of software modules for the user interface, database connectivity, information retrieval and analysis, generation of reports, presentation of graphics, and multimedia information products.

Spatial database applications are now developed predominantly using the methods of *software engineering* (Pressman, 2005; Summerville, 2005). One of the prevailing software engineering approaches is called *rapid prototyping* (Figure 9.15). This approach aims to provide the users with a working prototype for an application within a relatively short span of time, so that users can see operationally whether or not the end product meets their needs.

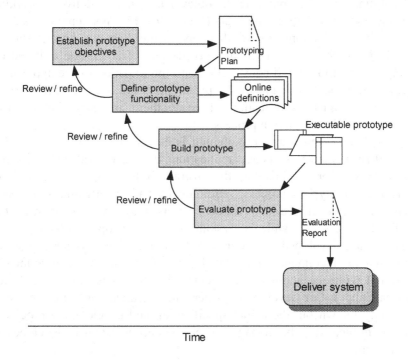

Figure 9-15. Workflow and processes of application development by (rapid) prototyping (see also Figures 9.16 and 9.17)

Prototyping generally starts with a rudimentary definition and objective of the application, which is then progressively refined and enhanced as the development process continues. When compared with conventional application development strategies, rapid prototyping has numerous advantages over other approaches. These include, for example, improved communication between the users as represented by the project team and the application developer, early detection and correction of missing functionality, reduction of risks by progressive testing, and opportunity for users to educate and train themselves during the process of development.

There are two strategies to prototyping, namely *throw-away prototyping* and *evolutionary prototyping*. In the throw-away strategy, a practical prototype is developed to help discover requirement problems, after which the prototype is discarded. A fully functional application is then constructed, usually with software modules or components developed at the prototyping stage in order to cut down the cost and time of the development process (Figure 9.16a). This approach is best suited for situations where the application is new to the users and hence their requirements are hard to understand and capture in advance.

Evolutionary prototyping, on the other hand, aims to deliver a working application progressively to the users (Figure 9.16b). This approach is most suitable for applications where the specification cannot be determined precisely in advance (e.g., user interface design). The method is based on techniques that allow rapid construction and modification of the application through iterations. Validation of the prototype is not done by checking its functionality against specifications, because there are none, but by user participation in the evaluation process.

A variant of evolutionary prototyping is called *incremental prototyping* (Figure 9.16c). In this strategy, applications are developed and delivered incrementally after establishing the overall architecture of the system. During the prototyping process, requirements and specifications for each increment are developed and refined. Users may experiment with the completed increments while others are being developed. When all the increments are developed, a rigourous integration process is undertaken to ensure that all increments and related software modules are functionally operational with one another. Incremental prototyping is most suitable for large-scale application development where the overall architecture is more or less exactly known but the actual specification and development processes are too complicated to be managed when the entire system is developed at the same time.

Techniques used for system programming can be classified into three categories, namely high-level language programming, database programming, and visual programming. In practice, these techniques are not exclusive to one another but are often used together. For example, dynamic programming is a coding method that uses a high-level language with powerful database management functionality (for example, Smalltalk, Java and Prolog). It is particularly suitable for small-scale applications and for the rapid prototyping of small program modules that seek to prove a design concept prior to full implementation. For large- and medium-scale application development, the complexity of the implementation often necessitates the use of database programming and visual programming as part of the prototyping process.

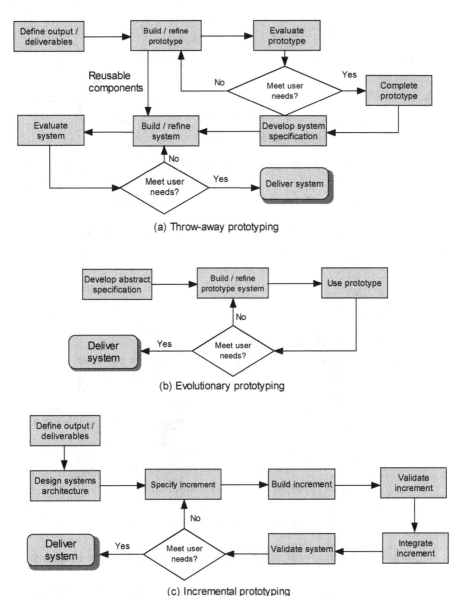

Figure 9-16. Types of prototyping

Database programming makes use of domain-specific language for business applications built around a database management system (Figure 9.17a). It normally includes a database query language (e.g., SQL), a screen or interface generator, a spreadsheet, and a report generator. It may also be integrated with a CASE tool.

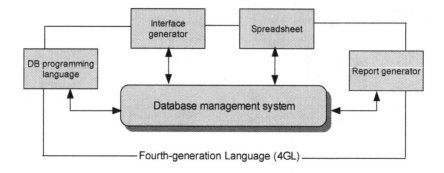

(a) The Fourth-generation Language programming environment

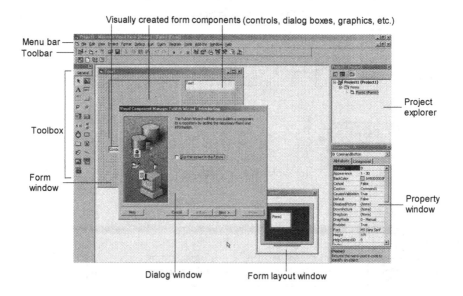

(b) The visual programming environment of Visual Basic

Figure 9-17. Methods of prototyping by 4GL and visual programming

Many spatial database projects have taken advantage of advances in visual programming to create rapid prototypes. Visual programming is supported by languages such as Visual Basic and Visual C++ that contain software components for the rapid creation of interfaces and associating these with the database (Figure 9.17b). In fact, visual programming can be relatively easily tailored to meet the needs of specific application environments. However, team-based development is difficult to coordinate because there is often no explicit system architecture. Further, the complex dependencies among different program parts can be difficult to maintain as the complexity of the application grows. Thus, careful documentation of the prototyping process is absolutely necessary to enable easy tracking of problems when they arise and *revision control software* is often used to track coding changes.

6. PROJECT MONITORING AND CONTROL

Some authors consider project monitoring and control to be a part of project execution in the PMLC. As Figures 9.1 and 9.2 showed, there are indeed considerable overlaps between the activities of these phases. However, project monitoring and control are best treated as separate and distinct phases due to the fact that they should be done independently of project execution in order to prevent potential conflicts of interest from occurring when the individuals executing a project monitor and control their own activities.

6.1 Project Documentation and Record Keeping

Project monitoring and control starts with proper project documentation and record keeping. As noted in Section 3.1, the latter two activities are among the greatest challenges for a project manager. Project documentation in this context includes all documents resulting from the project initiation and planning phases as well as on-going amendments to project schedules, memoranda distributed to project team members, minutes of project team meetings, communication with the contractor and representatives of stakeholders, progress reports from the contractors, expenditure reports, and progress reports from internal project team members.

Project documentation and record keeping serves several important purposes in project management. One of these is to help the project manager maintain control of the project and keep track of its progress. The project manager monitors the progress of the project by establishing a reporting system using two types of progress reports:

- *Regular or periodic progress reports.* These reports are completed by the contractor as well as internal project staff, and are submitted to the project manager on a weekly, bi-weekly or monthly basis. Individual regular progress reports cover only those activities that have commenced or are in progress during the reporting period. They highlight activities completed, and the variance between scheduled and actual completion dates. There should be some discussion of the reason for the variance for any activities whose progress is falling behind the scheduled time of completion, as well as possible corrective measures to bring the activities back on schedule.

- *Cumulative reports*. These reports, usually completed by the project manager, contain the history of the project from its beginning up to a particular point that is identified as a critical milestone. These reports show the trends in project progress by summarising the regular progress reports. Cumulative reports are typically required by projects of long duration to provide the project manager and the project sponsor a brief but up-to-date status of the project as a whole rather than of individual project activities.

Project documentation and record keeping also allow the project manager to track the resources used against the activities completed and in progress. This can be done using the resource management functionality of a project management software package to generate reports periodically to show the progress of the project and the status of the resources for the project. Alternatively, use can be made of the project Gantt chart and the project network diagram (Section 4.2 and Figures 9.9 and 9.10) by highlighting with different colours the activities that are completed, modified and in progress. It is advisable to replicate and present the project network diagram in the form of a wall chart so that everyone involved in the project can have direct access to the up-to-date status of the progress of the project throughout its entire life cycle.

From the beginning of a project, its manager must ensure that all members of the project team keep detailed notes of their respective activities (e.g., where they store their data and application program files, and the methods used for as well as progress made in the development of a particular application software module). If a particular member is reassigned to a new role in the project or leaves the project altogether, these notes will allow the project manager quickly to provide the replacement with sufficient information to continue seamlessly with the work in progress. These notes can be kept by individual project staff themselves, but should be summarised in the regular progress reports submitted to the project manager.

No matter whether a computer-based or manual method is used for progress reporting and resource tracking, all documents generated must be carefully kept, indexed and, where appropriate, cross-referenced for easy access in the project folder and binder that were prepared at the project planning phase (see Section 4.2). Project documentation and record keeping are tedious and time-consuming tasks. However, it is important to keep a balance between the time and cost of record keeping on the one hand, and the usefulness of the documentation on the other hand. Obsession with detailed documentation and record keeping may be a waste of valuable resources, but keeping a collection of unorganised documents and project notes will not help either. The project manager has to use his or her discretion to set up a project documentation and record keeping system that is both economically sustainable to maintain and easy enough to access and use.

6.2　　Contract Management and Control

Contract management requires considerable people skills. The project manager not only deals on a day-to-day basis with the contractor but also serves as the primary liaison between the contractor and the project team on various aspects of the project.

A common pitfall in contract management is the use of more than one contractor in the same project. Although it is not uncommon for a project contract to be awarded to more than one consultant (e.g., one for the database, the other for the application program development), awarding multiple contracts for a single project can be the project manager's worst nightmare and, therefore, should be avoided if possible. In the case of large-scale projects that are difficult for one contractor to complete alone, the usual solution is to allow the contractor to sub-contract, subject to compliance with certain conditions (e.g., the principal contractor must be able to satisfy the project manager that the sub-contractors have the capabilities and resources to complete the assigned components to the required performance standard, or the contractor takes full responsibility for delivery of sub-contracted components). When this happens, the project manager typically only interacts directly with the principal contractor. He or she should avoid intervening in the working relationship between principal contractor and any sub-contractors (e.g., getting involved in dispute resolution between the principal contractor and sub-contractors on any specific aspect of the project).

During the course of the project, disagreements between the project team and the contractor occur from time to time. In order to facilitate the resolution of potential conflicts, it is important for the project manager to specify in the RFP the rules and protocols for communication between the two parties, and apply these rules strictly and consistently when problems occur. Typical rules and protocols often include communications channels (e.g., all communication must be done between the person designated by the principal contractor and the project manager only), the exclusive use of written communication by mail and fax, and the time limit for the contractor to respond when questions are raised, as well as the time limit for the project manager to indicate acceptance after a response is received from the contractor.

In order to keep track of the communication between the contractor and the project manager and the progress of the problem fixing process, it is advisable to use a *pro forma* problem reporting form as shown in Figure 9.18.

Problem Report Form

Problem # [□]	Date: _____ Time: _____
	Reported by:

Description of problem
New problem? [yes/no]
Data-related? [yes/no]
Priority? [Severe/Critical/annoyance]

Name of system components/functions where the problem occurs:
_____

Error messages:
_____
_____
_____

Notify contractor:
Date/Time: _____
By [Fax/e-mail] _____
Acknowledgement date/time: _____

Problem Fixing status
Clarify [y/n] _____ Date/Time _____
In progress [y/n] _____ Date/Time _____
Fixed _____ Date/Time _____
Distributed _____ Date/Time _____

Acceptance test
Performed by _____ Date/Time _____
Description of solution:
_____
_____
_____

Reports received from contractor (MUST be filed with is Problem Report Form)
1. _____ Date/Time _____
2. _____ Date/Time _____
3. _____ Date/Time _____
4. _____ Date/Time _____
5. _____ Date/Time _____

Additional information
Problem/error satisfactory resolved. Signed (Project Manager / Date)
_____
Problem/error not resolved. Possible further course of action/mediation:
_____
_____

Figure 9-18. Example of an error report form

Completed forms should be properly filed in the project folder and binder, with a copy given to the contractor. These forms will be used as a source of information for post-implementation evaluation (see Section 7.2), and can be used for record keeping in quality assurance and control to be discussed below, as well as in the post-implementation maintenance of the spatial database system.

A commonly overlooked aspect of contract management is change management. Often, no matter how carefully a project manager plans a project and develops project specifications for hardware, software and performance standards, unforeseen events may happen to necessitate a change in the plan and the specifications. These events may include, but are not limited to, advances in technology, additional user requirements, changing business practices, and unexpected difficulties in the coding of application programs. A good contract management strategy should have a change management process in place.

Change in project plans and project specifications can be generally classified into five categories, as shown in Table 9.7. Whenever a change is required, the person (e.g., the contractor or a particular project team member) initiating the change will start the change request process (Figure 9.19). This is usually done by completing a project change request form, which typically contains the following sections:

- Part(s) of the project that is affected (e.g. specific hardware and software components).
- Reason(s) for change.
- Proposed solutions, refinement and enhancement.
- Resource implications (e.g., additional/new hardware and software required, additional human resources or expertise required, needs for extension of dates of delivery).
- Impact analysis, including the benefits and risks relating to the change.

Table 9-7. Categories of change in project management

Categories	Characteristics
1	Can be accommodated within the project resources and timelines.
2	Can be accommodated but will require an extension of the deliverables schedule.
3	Can be accommodated but within the current deliverable schedule but additional resources will be needed.
4	Can be accommodated by additional resources and an extension of the deliverable schedule will be required.
5	Cannot be accommodated without a significant change to the project.

Upon receiving the request form, the project manager will assess the justification for the change as well as the resource implications and the feasibility of the change by classifying the change into one of the categories listed in Table 9.7.

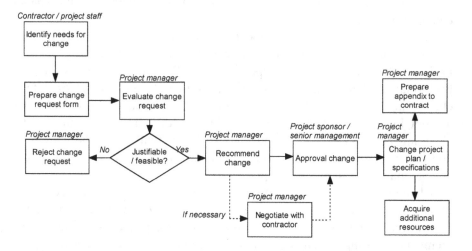

Figure 9-19. The change management process

If the change is initiated by the contractor, and if it involves an increase or decrease of the contractor's resources, the project manager has to start the negotiation process with the contractor for an adjustment to the compensation agreed upon in the original contract. When an agreement is reached, the project manager must submit the change request form to the project sponsor, or through the project manager to senior management for approval. Legal advice is sometimes required as well. When approval is granted, the project manager will change the project plan or the affected project specifications. He or she will also prepare an appendix detailing the changes and adjustment to compensation to be attached to the original contract. Finally, he or she must proceed to adjust the project schedule and acquire the necessary resources to complete the project accordingly.

6.3   Quality Assurance and Quality Control

The terms *quality assurance* (QA) and *quality control* (QC) are often used in conjunction with one another (QA/QC). In reality, however, they refer to two relatively distinct but closely interrelated sets of concepts and practices. QA and QC have their origin in manufacturing and engineering. The purpose is to safeguard the quality of products against possible imperfection in design, craftsmanship, material used, and production processes. Thus, QA/QC are preventive measures that are applied throughout the entire production or construction cycle of hardware manufacturing and engineering development projects. Spatial database projects, being a design and implementation undertaking, require stringent QA/QC measures to ensure that all activities, from design, development, production, installation, servicing and documentation, are carried out according to accepted practices and standards, and will lead to the delivery of a system that is useful for its intended purposes.

The relationship between QA and QC in spatial database projects can be easily understood from the different levels at which quality management (QM) activities are applied:

- *Organisational Level.* At the organisational level, QM activities are conducted to implement the infrastructure necessary for all projects undertaken by an organisation, and to make sure that all project activities have the resources they need to be completed successfully. This is usually conducted as a standard corporate policy and a regular audit. These QM activities are commonly called QA activities.
- *Project Level.* At the project level, QM activities support the success of individual projects by ensuring that accurate information is channeled to the right people at the right time so that decisions made during project implementation are defensible and cost-effective. These QM activities are more of an "assurance" rather than a "control" nature, and are therefore also QA activities.
- *Technical Level.* At the technical level, QM activities aim to ensure that individual project activities that design, build, test and install intermediate and final deliverables are performing within accepted limits or tolerances. Since these activities aim to "control" the quality of the project deliverables, they are aptly labeled as QC (also commonly called technical QA).

Figure 9.20 shows how the concepts and techniques of QA and QC are articulated in spatial database projects. In this figure, QA is largely perceived as the umbrella activities that provide the support infrastructure for the technical application of QC measures. The support infrastructure includes corporate information quality and standards, competencies of project staff, and performance standards of business procedures served by the spatial database system resulting from individual projects. Of particular interest here is ISO 9000 certification. ISO 9000 is a family of standards approved by the International Organisation for Standardisation (ISO) that defines a quality assurance program. Organisations that conform to these standards receive *ISO 9000 certification.* This does not necessarily mean that the organisation's products and services have a high quality, but rather that the organisation follows well-defined procedures for ensuring quality products. In many jurisdictions, particularly in Europe, government agencies increasingly use ISO 9000 certification as a mandatory requirement for data suppliers and application developers bidding for spatial database contracts.

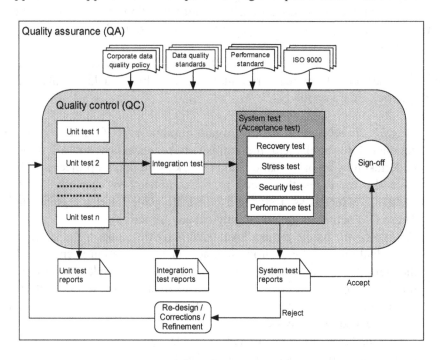

Figure 9-20. A conceptual framework of QA/QC in spatial database implementation

At the technical level, QC is applied by conducting a sequential series of software tests that include:

- *Unit tests.* These tests target the smallest unit of software construction, namely the functional modules, and are carried out immediately after the source-level code is developed, reviewed and checked for correct syntax.
- *Integration tests.* These tests aim at detecting potential errors associated with the interaction between individual functional modules. There are different approaches to integration testing. The most commonly used one is called regression testing, which is carried out incrementally each time a new module is completed and added to the software system.

- *Systems test*. This is the final stage of software testing conducted when the entire development process is complete. In practice, systems testing is actually a series of tests with different purposes that collectively seek to verify the functionality of the database system as a whole. The four commonly used tests are a *recovery test* (to verify the ability of the system to restore itself to the state before the failure), a *security test* (to verify the ability of the system to protect itself from unauthorised use), a *stress test* (to examine the response of the system to abnormal events such as excessive interruption, excessive number of users, and maximum memory use), and a *performance test* (to monitor the functioning of the system in run time). As the systems test is carried out to determine whether the final system is functionally acceptable for implementation, it is also commonly called the *acceptance test*.

7. PROJECT CLOSING AND POST-IMPLEMENTATION EVALUATION

The successful completion of the acceptance test signifies the completion of both the project execution and monitoring and control phases of the PMLC. It also marks the beginning of activities to close the project and put the resulting spatial database system into operation.

7.1 Technology Roll-out Activities

The actual activities required to implement the technology in a spatial database tend to vary from one project to another. This is because each project is undertaken under different situations, with different user profiles, and in different technical and organisational environments. Generally speaking, typical roll-out activities include but are not limited to the following:

- Site preparation for the implementation of the database system at one or more locations.
- Hardware and software installation and testing.
- Development of user education and training programs.
- Development of user support facilities, such as the training of corporate helpdesk personnel, documentation, and self-help materials.
- Development of an on-going error/problem reporting and response system.

- Ordering of supplies and materials required by the on-going operation of the problem.
- Announcing, through news release and brochures, to internal staff and external clients, information about the methods of accessing the database and ordering its services.

7.2   Post-implementation Evaluation

Spatial database implementation is seldom a one-time endeavour. In many cases when a project is completed, it is time to start thinking about its upgrade and enhancement. The purpose of post-implementation evaluation is to summarise the experiences gained in the project just completed, so that the same mistakes will not be repeated in the next version of the database system and in other similar projects.

In essence, a post-implementation evaluation is a comparison between the predicted events and the events that actually occurred during the entire course of the project's life cycle. Experience has shown that projects rarely run exactly as planned and scheduled. Variances between planned and actual events can often be generally traced to two primary causes, namely technical and personnel. Thus, a post-implementation evaluation report is best approached from these two perspectives. It is common practice for the project sponsor to analyse the performance of project personnel, while the project manager is responsible for the evaluation of the performance of equipment and applications. In conducting the technical portion of the evaluation, the project manager should pay special attention to the problems encountered and the solutions that were used to deal with the problems successfully.

Upon completion of the evaluation report, the project manager should submit copies to the project sponsor and senior management to signify the completion of the entire project. He or she should distribute copies of the report to all project staff. As a matter of courtesy, the principal contractor is usually provided with a copy of the report, which may be in an abridged form to avoid confidential information being released to people outside of the organisation. A copy of the evaluation should also be inserted into the project folder and binder which can then be closed for safe keeping by the project sponsor or a designated person or department in the organisation. Once the folder and binder are complete, they become an important resource of information for future projects of the same or a similar nature.

8. SUMMARY

This chapter presented a concise and comprehensive overview of project management in general and as it applies to spatial database implementation in particular. The placement of this chapter at this point in the book does not suggest that management should take a back seat in spatial database projects. On the contrary, data, technology and management play three equally important roles in turning the idea of implementing a spatial database system into reality. All people engaged in spatial database management are expected to be equally competent in these three aspects of spatial database implementation.

Project management was introduced from first principles. As project management requires a multi-faceted set of skills to deal with both technical and human resource-related issues, a brief description of the competencies and skill sets expected of a typical project manager was provided. The process of project management was then described by following sequentially through the major activities of each of the five phases of the PMLC.

Project management for spatial database implementation, just like project management in other application domains, has different variants of best practices. Thus, it is important to remember to approach project management with an open mind and to consider the relative merits and drawbacks of different options, with respect to the nature and objective of individual projects, before a particular course of action is chosen. It is also important to stress that project management requires technical skills that can be learned from books and training programs, as well as soft skills (i.e., managerial behaviours and styles) that can only be developed through practical experience over time.

9. REFERENCES

Berdusco, B., Trowell, N.F., Ayer, J., Madon, Z., van Haaften, S. and A. Yeung (1999) "From functional analysis to CD – Digital compilation of the Timmins map sheet, Abitibi Greenstone Belt, Ontario", *U.S. Geological Survey Open File # 99-386*, pp. 113-121, Reston, VA: USGS.

Chapman, A.D. (2005) *Principles of Data Quality*, Copenhagen, Denmark: Global Biodiversity Information Facility.

Chrisman, N.R. (1983) "The role of quality information in the long-term functioning of GIS", *Proceedings of AutoCarto 6*, Vol. 2, pp. 305-321, Falls Church, VA: ASPRS.

EPA (2003) *Guidance for Geospatial Data Quality Assurance Project Plan*, Washington, DC: Office of Environmental Information, US Environmental Protection Agency.

ESRI (2003) PLTS GIS Data ReViewer 4.2 (An ESRI White Paper), Redlands, CA: Environmental Systems Research Institute.

GeoAnalytics (2003) *A Cost-benefit Investment Analysis of an Enterprise Geographic Information System for Scott County, Iowa*, Madison, WI: GeoAnalytics, Inc.

Kaplan, R.B. and Norton, D.P. (2001) *The Strategy-focused Organisation: How Balanced Scorecard Companies Thrive in the New Business Environment*, Boston, MA: Harvard Business School Press.

Mulholland, N. (ed.) (2003) *New York State Project Management Guidebook*, Rel. 2., Albany, NY: New York State Office for Technology.

Peters, D. (2005) *System Design Strategies* (An ESRI Technical Reference Document), Rev. 1, Redlands, CA: Environmental Systems Research Institute, Inc.

Pressman, R. (2005) *Software Engineering: A Practitioner's Approach*, 6th ed., New York, NY: McGraw Hall.

Project Management Institute (PMI) (2004) *Guide of the Project Management Body of Knowledge* (commonly referred to as the *PMBOK Guide*), 3rd ed., Newtown Square, PA: Project Management Institute.

Summerville, I. (2005) *Software Engineering*, 7th ed., Boston, MA: Addison Wesley.

Tomlinson, R. (2005) *Thinking About GIS: Geographic Information System Planning for Managers* (Revised and updated edition), Redlands, CA: ESRI Press.

Wang, R.Y. (1998) "A product perspective on total data quality management", *Communications of ACM*, Vol. 41, No. 2, pp. 58-65.

Wysocki, R.K., Beck, R. Jr. and Crane, D.B. (2003) *Effective Project Management*, 3rd ed., New York, NY: John Wiley & Sons.

Chapter 10

WEB-ENABLED SPATIAL DATABASE SYSTEMS

1. INTRODUCTION

The Internet is the *de facto* standard of today's global communications. It is host to millions of corporate and organisational Web sites and *Web portals* where various types of information services are provided and accessed. Spatial database vendors and users alike have closely followed the advances of the Internet to develop and adopt new software tools and solutions that have basically revolutionised the ways spatial information is processed and used. The spectacular growth of spatial information technology in recent years is due more to the Internet than to any other factor. In most contemporary spatial database projects, access to the Internet is a necessary rather than optional requirement.

Aspects of using the Internet for spatial database applications are discussed at various points elsewhere in this book. This chapter consolidates this discussion by focusing on spatial database systems in the Internet environment. The objective is to explain how spatial database and Internet technologies are integrated within the framework of organisational, national and global spatial information infrastructures. The discussion covers both the conceptual design and technical implementation of Web-enabled spatial database systems.

Section 2 introduces the nature and characteristics of Web spatial database systems. Section 3 provides an overview of the development and standards of the Internet and the Web in general. The working principles of Web-enabled spatial database systems using different architectures are explained in Section 4. These initial sections provide the conceptual framework for the strategies of implementing Web-enabled spatial database systems that are discussed in Section 5. Section 6 summarises the state of the art in Web-enabled spatial database technology as well as the trends of systems and application development in the future.

2.   DEFINITION AND CHARACTERISTICS OF WEB-ENABLED SPATIAL DATABASE SYSTEMS

Web-enabled spatial database systems are database systems designed especially for access over the Internet using the World Wide Web (Web) protocol. These systems combine spatial database and Internet technologies to provide a distributed and network-centric approach to spatial information. They are increasingly used not only to enable external access to enterprise information by clients and business partners, but also to support internal business operations such as data sharing among project teams and work groups at the same or different locations of an organisation.

2.1   Characteristics of Web-enabled Spatial Database Systems

Technically and functionally, Web-enabled spatial database systems are not simply an extension of their conventional counterparts, nor are they just another one of the growing number of modern Internet applications. Rather, Web-enabled spatial database systems have several relatively unique characteristics and requirements in their design, implementation and application, for example:

- They have *distinct architectures* that integrate spatial database software tools, server-to-database middleware, application programming languages, and Internet markup languages. The operation of these systems requires expertise in Web server administration, telecommunications and data transfer protocols, computer network and systems security. Effective Web interfaces also require good design skills and standards.
- They are commonly *used in conjunction with new data collection, processing and dissemination technologies* (for example, Web cameras, weather stations, in-vehicle navigation systems) as an integral component of organisational, national or global information infrastructure, rather than as standalone monolithic database systems within a particular organisation.

- They are used to *serve new business functions* (for example, e-commerce, location-based services and self-served information kiosks, see Chapter 12) that require spatial information to be used in a more mobile and flexible manner than other contexts, as well as better able to share data and communicate among people and organisations.
- They are typically *directed toward the widest possible audience* with different types of end users whose knowledge, skill and ability of using a database may vary considerably from one user to another, which requires specific considerations for human-computer interaction, user interface design, and end user support and training.

Architecturally, Web-enabled spatial database systems use a client/server configuration with four tiers of components as shown in Figure 10.1:

- *The presentation tier,* which typically uses a Web browser as the GUI to interact with the database system and display the results of database queries and analysis.
- *The communication infrastructure tier,* which includes local area networks (LAN) and local telephone networks connected to the global telecommunication network that forms the backbone of the Internet.
- *The business logic tier,* which contains rules and protocols that govern the interaction between the client and server computers. There are two components, namely a *Web server* that is responsible for the Internet connection between the client and server computers, and a *Web mapping engine* (also called an *Internet mapping server*) that is responsible for spatial processing in response to client requests.
- *The data management tier,* which is made up of one or more spatial databases, data warehouses and off-line data inventory and archives, as well as operating and database system software. For systems security reasons, data are normally served from a Web server using a replica of the production server where the data are maintained and used by internal users.

All major vendors of spatial database systems have closely followed the general thrust toward the use of the Internet for the delivery of information and business services, by developing and promoting the use of Web-enabled solutions (Lake, 2001). There has also been considerable interest in the academic and *Open Source software* communities in developing Web-enabled spatial database tools. Many of these tools are freely available on the Internet (Lowe, 2002). The availability of a wide range of commercial and *free software tools* with various levels of functionality allows users to construct Web-enabled spatial database systems with different architectures and functionality. The working principles of Web-enabled spatial database systems of different architectures and their applicability in spatial information are explained in Section 4.

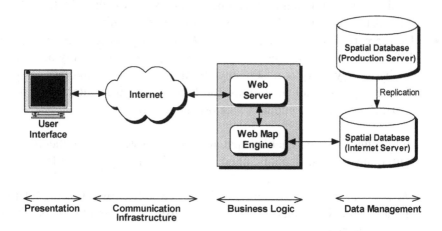

Figure 10-1. Generic Architecture of a web-enabled spatial database system

2.2   Advantages and Disadvantages of Web-enabled Spatial Database Systems

When compared with conventional spatial database systems, the Web-enabled approach to spatial information is characterised by numerous advantages. The more important ones include:

- *Interoperability in a distributed computing environment.* By using a client/server architecture, internationally accepted industry standards and the Web browser as the universal client on the Internet, Web-enabled spatial database systems provide one of the most interoperable environments in distributed computing, not only among spatial database systems themselves, but also between spatial database systems and other types of information systems.
- *Hardware and software independence.* Web-enabled spatial databases can be accessed transparently regardless of the types of computers that information providers and end users use. This eliminates the need for multiple-platform application development (that is, developing different versions of the same application for computers using different operating systems).
- *Rapid deployment and universal accessibility at manageable cost.* The technologies for Web-enabled spatial database systems are relatively

mature now. Because the Web browser is a standard application installed on all desktop computers applications can be developed relatively quickly using commercial or open-source software tools, and deployed to serve unlimited numbers of end users within a relatively short time frame at minimal cost.

- *Lowering the cost of using spatial information technology.* Web-enabled spatial database systems can be built and deployed without using expensive proprietary software products. The growing market of spatial information has also led to the introduction into the spatial database arena of a new business model of providing information services, known as application service providers (ASP), discussed in Section 4.4, in the information technology industry. ASP offer complete fee-based service packages, including data and application, to enable organisations to use spatial information in their business without incurring the capital costs for hardware/software acquisition, database creation and application development, and on-going costs for systems operation, maintenance and upgrading.

- *Improved spatial information customer services.* By taking advantage of the ubiquitous nature of the Internet as the medium of communication, organisations are able to provide spatial information services to their clients throughout the world twenty-four hours a day, seven days a week. At the same time, anyone with access to the Internet is able to use spatial information to enjoy, to learn and to make decisions whenever they want to. As data can be downloaded electronically over the Internet, there is no need for users to travel to information centres or points of sales to acquire data. Further, Web-enabled spatial databases can be linked directly to field data loggers or collectors (for example, GPS, Web cameras, Web sensors, weather stations and satellite imagery receiving stations) to provide spatial information to end users in real time. These features of Web-enabled spatial database systems provide a new business model of customer service that conventional over-the-counter information services can never match.

Web-enabled spatial databases also have several limitations. These include, for example:

- *Inherent limitations of the Web.* The Web was initially designed to facilitate the search, retrieval and presentation of text- and image-based information by means of Hypertext Markup Language (HTML) documents. It has limited capabilities for spatial database applications using vector data without enhancements that require development costs and effort (see Sections 4.2 to 4.4).

- *Inherent limitations of spatial databases.* As noted in Chapter 6 spatial databases are characteristically heterogeneous because they are constructed using different database models, represented at different scales and with different cartographic and attribute classification schemes. The structural and semantic differences among spatial databases, together with the inadequacy of spatial metadata information (see Chapter 6) have inhibited the potential applicability of Web-enabled spatial database systems.
- *Security and privacy concerns.* Spatial data transmitted over the Internet are vulnerable to abuse and can be spied on relatively easily. Although encryption methods for text-based data are well developed and reliable for online transaction processing, encryption technology has apparently not been as well developed for spatial data. This limits the usability of Web-enabled applications particularly when sensitive spatial data are involved or authenticity of the data being delivered is required.
- *Copyright control and liability of abuse and misuse.* Many organisations are not willing to adopt a Web-enabled approach to spatial information because it is hard to control and retain the copyright of the data disseminated to end users. Some organisations are also deterred by the increasing possibility of legal liability when spatial data are made easily obtainable by a large number of unknown end users who may abuse or misuse the data (see Chapter 7).

In recent years, concerted effort has been made and success achieved by various communities including stakeholders in industry, academia and research organisations, and standards organisations to address the above limitations. Web-enabled solutions are now an integral part of spatial database technology because of the general thrust toward the use of the Internet on the one hand, and the proven benefits of using the technology for the delivery of spatial information services on the other hand. The Web-based approach has fundamentally changed the ways spatial information is provided and used. The study of spatial database systems cannot be complete without a good understanding of the concepts and techniques of implementing these systems in the Internet environment.

3.   TECHNOLOGIES AND STANDARDS

This section first provides an overview of the technologies and standards that form the building blocks of Web-enabled spatial database technology. Subsequently, the focus of the discussion turns to how information is transmitted over the Internet, and how the Web has evolved to become the primary conduit for using spatial information today. *XML* is also introduced along with related technologies and standards with particular reference to their impacts on the development of Web-enabled spatial database technology.

3.1   Information Communication over the Internet

Computers on the Internet are usually referred to as *nodes* or *hosts*. They are connected using existing public telecommunications networks and a set of protocols called *Transmission Control Protocol/Internet Protocol* (TCP/IP). The current standard is that each node on the Internet is identified by a unique 32-bit binary number called the *Internet Protocol address* (IP address) that is written as four fields each representing an eight-bit number in the range 0 to 255 (called *octets*) separated by decimal points, for example, 215.45.7.159. IP addresses are configured by software and are not hardware-specific, and are made up of a *domain address* component and a *node address* component. All computers and network devices on the same local area network (LAN) have the same domain address. This means that when a new computer or network device is added to the LAN, it will be assigned a node address only. Its domain address is the same as all other computers or network devices on the LAN.

The domain address is assigned by a network of international, regional and national registration agencies coordinated by the *Internet Corporation for Assigned Names and Numbers* (ICANN) in the United States. ICANN is an internationally organised, non-profit corporation that has responsibility for IP address space allocation, protocol identifier assignment, generic and country code top-level domain name system management, and root server system management functions. The node address, on the other hand, is assigned locally by the systems administrator or network designer of a LAN. The class of an IP address determines which part of it represents the domain address and which part represents the node address. Classes can be distinguished by the first octet of the IP address as shown in Table 10.1.

As an example, 129.97.54.72 is a Class B address characteristic of a mid-sized network found in a typical University, with the first octet in the 128-191 range. Class B addresses include the second octet as part of the net identifier and the other two octets are used to identify each host or node. There are a possible 16,384 (2^{14}) class B networks, each with 65,534 (2^{16} -2) possible hosts/nodes and a total of 1,073,741,824 (2^{30}) unique IP addresses. If the number of computers and network devices on the LAN of an organisation exceeds the allowable number of nodes relative to its current class, then it is necessary for the organisation to apply to the appropriate international or national registration agency for, in the above example, the assignment of a Class A address that is capable of accommodating a maximum of 16,777,214 (as compared to 65,534 Class B) possible hosts.

Table 10-1. Classes of IP addresses

Class of IP Address	First Octet of an IP Address	Domain and Node Parts of an IP Address D = Domain Address; n = Node Address
A	1 to 126	DDD.nnn.nnn.nnn
	127	Reserved for internal testing on the local machine
B	128 to 191	DDD.DDD.nnn.nnn
C	192 to 223	DDD.DDD.DDD.nnn
D	224 to 239	Reserved for multi-casting
E	240 to 255	Reserved for future use

IP addresses are used in the same ways street addresses are used for mail delivery and are instrumental in the correct delivery of information to the targeted computer. Information is transmitted over the Internet in blocks called "*packets*". Each packet contains the IP address of the node for which it is destined. Only the particular computer represented by the IP address gets the packet. As the size of a packet is fixed, the information contained in a document being transmitted probably requires a large number of packets. These packets may or may not arrive in the proper sequence. The machine which is receiving the packets is responsible for assembling them in the correct order.

Although IP addresses as binary numerical values are perfect for computer data processing, they are hard to remember by users. In practice, therefore, *domain names* are used instead. These are the familiar character strings that are used for accessing Web sites and sending e-mails. When a user enters a domain name to access a Web site or send an e-mail, it is broken down into its corresponding IP address through a process called *name resolution* in a *domain name server* on the network.

Just like an IP address, a domain name is made up of different parts, separated by dots, that follow the *Domain Name System* (DNS) standard set out by the ICANN noted above. The top level part of a typical domain name, represented by the suffix, is a generic code (for example, .com, .org, .edu, .mil, .biz, and so on) or a country code (for example, .nz, .ca, .cl, .pe and so on). The secondary part is the node name that can be made up of one or more character strings that identify a particular computer uniquely. This node name is created by the systems administrator or network designer of an organisation. A node name normally contains the Internet protocol used by the node, for example, "www" for the World Wide Web and "ftp" for File Transfer Protocol.

Before a domain name can be used, it must be registered with a domain name registrar accredited by the ICANN in the United States, or a national domain name registrar in other countries, for example, the Canadian Association of Internet Providers (CAIP) in Canada. The ICANN maintains the InterNIC Web site (http://www.internic.org) that provides public information about domain name registration services including records of existing domain names in use and a directory of accredited domain name registrars worldwide. By using this Web site and by overseeing the assignment of unique IP addresses and domain names in cooperation with accredited and national domain name registrars, ICANN ensures that every domain address is unique and that all users of the Internet can find all valid addresses. It also ensures that each domain name maps to the correct IP address and, consequently, keeps the complex task of managing millions of domain address and names worldwide relatively simple and efficient.

3.2   Intranet and Extranet

An *intranet* is a private communications network within an organisation. A typical intranet consists of many interlinked local area networks (LAN) and also uses leased lines in the wide area network (WAN). The primary purpose of an intranet is to share an organisation's computing and data resources among the employees within the organisation. In practice, an intranet also includes connections through one or more computers, known as *gateways*, to the Internet. Therefore, an intranet functions somewhat like a private version of the Internet. Since both the Internet and the intranet use TCP/IP, users within an organisation can send private electronic mail and other text-based messages with special encryption/decryption and other security safeguards to connect one part of their intranet to another. In order to maintain systems security, a *firewall server* is often used to screen documents and files being transmitted in both directions.

When part of an intranet is made accessible to users external to the organisation, that part becomes an *extranet*. Whereas an intranet resides behind a firewall and is accessible only to users inside an organisation, an extranet provides various levels of accessibility to outsiders. Users can access an extranet only if they have a valid username and password, and their identities determine which parts of the extranet they can view and use. Extranets are nowadays becoming a very popular means for business partners to exchange information, including spatial information, with one another.

Since the intranet and the extranet use the same technologies and protocols of the Internet, the concepts and techniques of Web-enabled spatial database systems discussed in this chapter generally apply to both of these two network architectures as well. However, there are some significant differences between the Internet and intranet environments, for example the purpose of implementing a spatial database system, the knowledge and skills of users, the technologies used, the systems security requirements and other factors. These differences are examined further in Section 5.1.

3.3   Characteristics of the World Wide Web (Web)

The Web is probably the most widely used part of the Internet. The Web, which was developed by the European Laboratory of Particle Physics (CERN) in 1993, can be conceived as a gigantic collection of information resources linked together to form one global file server. An information store on the Web is called a *Web site*, *Web server* or *portal* and is identified by a unique IP address and domain name associated with an Internet node, as explained above. A Web site contains *Web pages* that are designed using HTML or one of the associated markup languages. In its most basic form, a Web page is a plain ASCII text file containing HTML commands, known as *tags*, that describe the structure and appearance of a Web page in an Internet browser. HTML tags also define *hyperlinks* that reference other Web pages or resources including text-based documents, graphics and images, downloadable files, application programs and services.

Each individual page and resource on a Web site has its own address, called a *Uniform Resource Identifier* (URI, more commonly known as a *Uniform Resource Locator* [URL]). A typical URL is a string of characters made up of the IP address or the domain name of the hosting Web site and a unique identifier with one or more components separated by a forward slash (/). In the URL "http://www.fes.uwaterloo.ca/crs/gp555/index.html", the fes.uwaterloo.ca component represents the domain name of the host computer and the /crs/gp555/index.html component uniquely identifies the location of a particular Web page on the host computer.

The Web uses the *Hypertext Transfer Protocol* (HTTP) on top of TCP/IP to enable access to and exchange of resources residing in computers anywhere on the Internet by using a *Web browser* as software interface to the Internet. HTTP can be regarded formally as a 'stateless' (that is, transient and temporary) *remote procedure call* (RCP) that works in the sequence illustrated in Figure 10.2. The process starts with the client computer establishing a TCP/IP connection to a remote server using a URL as the address of the server and location of the desired Web pages and issuing a *request* to the server. The server then processes the request, returns the results of the request to the client's browser software in the form of a hypertext page, and closes the connection.

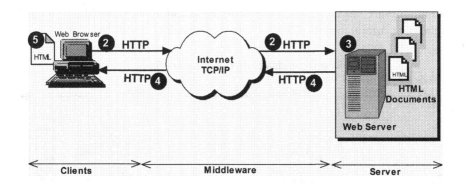

Figure 10-2. Web client/server interaction using HTTP

Once the client has received the hypertext page, it interprets the commands of the page and displays the content accordingly. Note that in the above process HTTP sets up a new connection for each request. The client computer must wait for a response before a new request can be sent out. This means that, for example, it takes six connections to access an HTML document containing five inline images, one to retrieve the document, and five to retrieve each of the images individually. Thus, although an HTTP request is exceedingly simple and easy to implement, it is not necessarily an efficient protocol. Obviously, in this context, the Web architecture has opted for simplicity over performance.

In October, 1994, the *World Wide Web Consortium* (W3C) was created to coordinate the development of common protocols and technical architecture that promote the Web's evolution and ensure its interoperability. The W3C has around 350 member organisations from all over the world. Since its inception, the W3C has produced numerous technical specifications, recommendations, open source software, and validation services that defined the evolution of the Web as shown in Figure 10.3. The W3C has also collaborated with various international and national standards and regulatory bodies. It works closely with the ISO in the development of technical Web standards (see Chapter 5), and plays a key role in the effort of the World Intellectual Property Organization (WIPO) to create and maintain protocols for the protection of privacy (see Chapter 7).

At present, the W3C is working aggressively to transform the architecture of the original Web, based essentially on HTTP, HTML and URLs as explained above, into the architecture of the next generation Web, built on top of a foundation provided by XML. The transition of Web-enabled database architectures is discussed in more detail in Section 4. In the meantime, the main features of XML and several important standards that are instrumental in nurturing the spectacular growth of Web-enabled spatial database technology are considered as they apply now and as they are likely to apply in the near future.

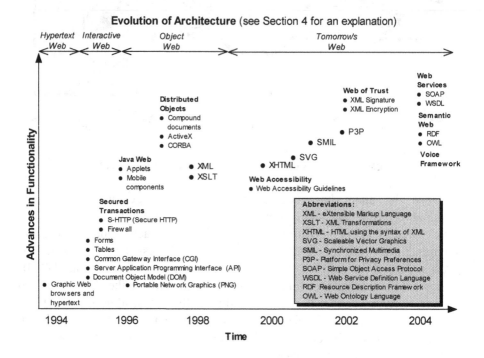

Figure 10-3. The evolution of web technologies

3.4   Extensible Markup Language (XML) and Related Standards

XML was created by the W3C to overcome the limitations of HTML which, as noted in Section 3.3, is the language used to format information for display in a Web browser. As the Web grew and its applications became more sophisticated, new functions were added to HTML to deal with the diverse needs of the users. This made HTML increasingly complex and difficult to use and it remained essentially only a static form of markup using an ever extending array of tags. XML was designed to curtail the growing complexity of the static, single and all-embracing compendium of HTML tags by permitting the creation of domain-specific dialects or grammars as well as the ability to dynamically update Web pages through XML style sheets, as discussed in the next section. This is achieved by allowing the users to create tags liberally that handle syntax and semantics specific to applications in particular domains. This ability to extend is the reason why XML carries the word "extensible" in its name.

Today, XML is seen not only as a single markup language simply for Web page documents, but also as a family of technologies and standards that includes the following (see also Figure 10.4):

- A meta-markup language, which is a collection of rules for defining elements and attributes, and tags enclosed within XML documents and documents in an XML dialect.
- A program library of tools that are used to describe and validate the structure of an XML document (that is, Document Type Definitions [DTD] and XML Schema), format data in XML documents for display (that is, XML Stylesheet Language [XSL]), link to a database (for example, XLL), query an XML database (for example, XSQ and XQuery), and convert XML documents from one dialect to another dialect or for export into a non-XML format (for example, XML Stylesheet Language Transform [XSLT]).
- Domain-specific dialects or grammars such as X3D for 3-dimensional graphics, Scalable Vector Graphics (SVG) for 2-dimensional vector graphics, Geography Markup Language (GML) for spatial data, Mathematics Markup Language (MathML) for mathematics equations, and so on. Some of these dialects were developed by the W3C itself (for example, SVG), others were created by industry consortia (for example, GML by the OGC) or individual organisations (for example, ArcXML by ESRI, Inc.).

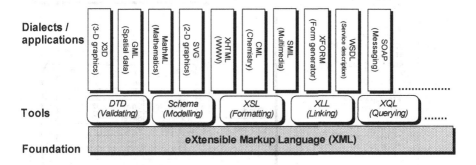

Figure 10-4. The XML family of technologies and standards

Figure 10.5 illustrates how typical components of the XML family of technologies and standards are used and interact with one another during data processing. The sample data in the XML data box show how the fundamental principle of separating content from presentation is applied in

XML. Unlike an HTML document, standard formatting tags are characteristically absent in an XML document. The structure of the data is described in this example by an accompanying DTD file. Different XML documents interoperate with one another by sharing the same DTD. The use of the relatively simple DTD is being gradually replaced by an emerging W3C XML Schema that provides a more sophisticated means of describing the structure and constraints on the content of XML documents.

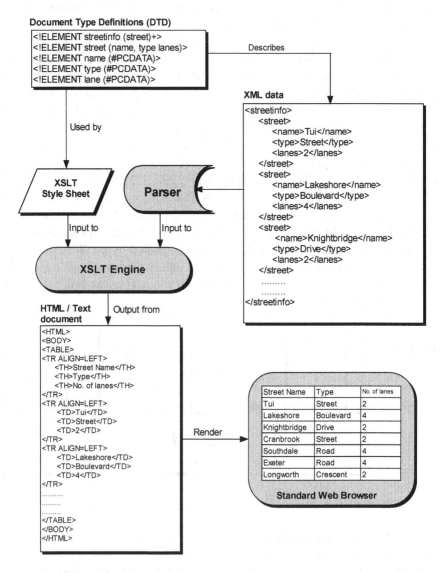

Figure 10-5. Storing, processing and displaying data using XML

The parser is an XML processor that separates data from XML tags during data processing. In order for the parsed data, which are formally called a *parse tree*, to be displayed on a Web browser, they must be translated first into an HTML file by the XSLT engine, which also takes information from the style sheet associated with the XML data file. The XML style sheet plays the same role of the cascading style sheet (CSS) in HTML. It provides the formatting information for the display of the data in the Web browser. Note how formatting tags (including header information, use of table, and alignment of displayed text) are added to the HTML document by the XSLT engine to generate the display of the original XML data in the standard browser.

For database applications, XML is used in one of two ways (Chaudhri et al., 2003). One is the *document-centric* model of XML where XML is used as a means to create semi-structured documents with irregular content. In essence, this is a functional extension of HTML and an XML dialect called XHTML was created to serve such a purpose. The other is the *data-centric* model of XML where XML is used as a storage or interchange format for data that are stored in a relational database or similar repository. In the data-centric model, data are transferred back and forth between XML and the database by a middleware application (for example, Microsoft's ADO.NET) or alternatively by XML-enabling the database. This is the strategy used by Oracle, which has completely integrated XML into its Oracle 9i database as well as the rest of its family of products. Microsoft's SQL Server 2000 also supports XML operations on relational data.

Both the document-centric and data-centric models have been usefully employed for Web-enabled spatial database systems. The former is now commonly used to create and deliver spatial metadata (see Chapter 5) whereas the latter is rapidly becoming a popular way of storing, transporting and sharing spatial information using the dialect GML.

The creation of GML is probably one of the most important milestones not only for Web-enabled spatial database systems in particular, but also for spatial database technology as a whole. Since GML is a dialect of XML, it makes possible the full integration of geographic information into daily business applications in the enterprise computing environment that is increasingly XML-based. Further, since GML is non-proprietary and based on the *OGC Abstract Specification* (a common model of geography developed and agreed to by the vast majority of spatial database software vendors in the world), it is an functional format for spatial data sharing using the open-source approach. In addition, data in GML can be displayed, queried, edited and analysed at the feature level using a Web browser. Except for highly advanced and sophisticated spatial applications, there is no longer any need for users to install expensive proprietary software in order to use spatial data in the desktop environment.

GML is now widely accepted by the spatial database industry. In the United Kingdom, for example, the British Ordnance Survey adopted GML as the spatial data format for its MasterMap program that contains over 400 million geographic features, making it one of the most important spatial data formats in the country. All key companies in the spatial information arena, such as ESRI, Intergraph, Laser-Scan and GE Network Solutions, among many others, support GML in their products.

Figure 10.6 illustrates how GML is used for Web-enabled spatial database applications. The GML data box in the figure shows how the same sample XML data in the XML data box in Figure 10.5 is represented in GML. Note how coordinates and descriptions of the individual spatial features are tagged. Tags are also available for capturing topology and other text-based characteristics of spatial data. Technical standards developed by the OGC also include a standard for automated coordinate transformation, geocoding, messaging, routing services, information searching and mining, style descriptors, and so on.

Writing a GML data file manually is an extremely time-consuming and error-prone process. GML data sets are therefore mostly created through an import procedure from existing spatial databases using a translation program. It is also possible to structure and store spatial data captured by on-screen heads-up digitising and field surveying directly in GML. For GML data to be displayed and queried, they must be first parsed to separate data from GML tags. The output data can be displayed using a GML viewer or more commonly in a Web browser. For display in a Web browser, it is necessary to convert the data to one of the several XML-based specifications for describing vector graphics including SVG, Vector Markup Language (VML) or X3D (for 3-dimensional data). The purpose of these specifications is to define how graphics elements (that is, points, lines, polygons and arcs) are presented with respect to position, colour, line weight, transparency and so on.

The most broadly supported and used output specification is SVG. This XML-based specification was created by the W3C specifically for describing 2-dimensional graphics. It supports three types of graphic objects, namely vector graphic shapes, images and text. All graphic objects can be grouped, styled, transformed and composed into previously rendered objects. For spatial data in SVG format to be displayed in a Web browser, a plug-in is required. SVG displays are interactive and dynamic. Interactivity between a user and the SVG display is achieved by a rich set of event handlers (for example, on-mouse-over and on-click) that generate attribute displays, pan and zoom, and data editing functions. For more advanced applications it is possible to animate the displays by embedding SVG animation elements in SVG content or by scripting using the SVG Document Object Model (DOM), which is a method of accessing and dynamically editing and manipulating elements, attributes and properties in XML documents. When a hard copy of the display is required, formatting information including map margin and symbology information can be imported from a cartographic style sheet to generate a fully custom-designed map.

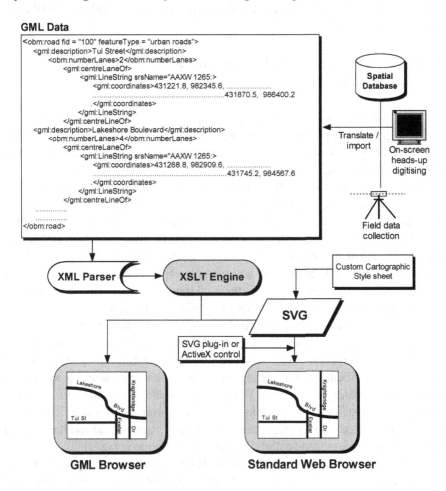

Figure 10-6. Using GML for web-enabled spatial database applications

In general, XML has become the dominant language in application development on the Web. Its impact is increasingly felt in all aspects of Web-enabled spatial database technology. Traditionally, the spatial information industry was dominated by a handful of key companies. These companies controlled not only the manner in which spatial data were structured and encoded, but also the way spatial applications were developed. The ability to store and use spatial data generically in GML, and to display the results of spatial queries and analysis in a regular Web browser, opens a whole new world of possibilities for spatial database solutions by creating and popularising an open source alternative to the dependence on expensive proprietary software. The use of XML for the delivery of Web-enabled spatial database applications is explained further in Section 4.4.

3.5   Open Geospatial Consortium Web Mapping Standards

The OGC was introduced in Chapter 5 as a spatial information technology industry consortium that promotes interoperability through standardisation. As a result of the rapid growth of Web-enabled spatial database technology, the OGC has turned much of its focus on interoperability to the Web mapping environment in recent years. The objective is to provide a set of standards that will enable client and server computers to communicate using a common software syntax and semantics in a Web-enabled spatial database environment. By using an OGC standards-compliant client to connect to one or more OGC standards-compliant servers, a user is able to request and visualise spatial data dynamically from multiple sources, regardless of their respective data models, formats, georeferencing systems and proprietary vendor technologies.

OGC specifications cover both data and systems architecture. GML as explained in the previous section is an OGC specification that is fast becoming the universal XML standard for encoding and transporting spatial data. The specification for setting up a Web-enabled mapping system is called the *Web Mapping Server* (WMS) *Implementation Specification* (OGC, 2001). This specification contains a set of common interfaces for client computers (called a *Web map viewer*) to query, request and display spatial information from remote spatial databases (called *map servers*).

Figure 10.7 shows how a WMS works. When a user requires spatial data for an application, a query is sent to a WMS-compliant map server via standard Web protocols (that is, TCP/IP, HTTP, HTML). A WMS-compliant web server is one that implements this particular specification. This request is used to obtain service-level metadata, which describe the content and acceptable request parameters of the request (for example, layer names, bounding box, output format, and so on). In response, the map server will return an XML file detailing the mandatory parameters for the spatial data that it returns to the browser in the client computer.

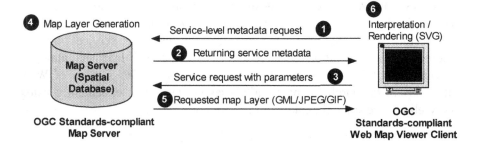

Figure 10-7. Interaction between a map server and a client viewer using the OGC WMS implementation specification

If the data are deemed to be useful, a request with the mandatory parameters specified in the service metadata is sent to the server. A WMS interface enables the map server to generate a map layer in a raster format (for example, JPEG, PNG, GIF etc.), re-format and re-project it according to the request's parameters, and return it to the client in the form of an XML document. A Web map viewer client can request different map layers from distributed Web servers. If the returned map layers cover the same geographical area and have the same physical dimension, then they can be overlaid in a single window to produce a composite map.

A WMS returns a map image file, not the original spatial data, to the user. In order to access spatial data in remote databases in object form (that is, points, lines and areas), interfaces that implement the *Web Feature Server* (WFS) *Specification* are required (OGC, 2002b). WFS-enabled map servers are capable of providing queries and transactional operations on individual spatial features. The working principle of a WFS is the same as a WMS. The user sends a request to a WFS-compliant server on the Internet. The server then executes the request and returns the results in GML to the client computer. For the GML data to be displayed, queried and edited in a regular Web browser, they first have to be converted into SVG, as explained in the previous section.

In order to enhance the implementation of a WMS and a WFS, several supporting specifications have been created or under active development. These include, for example, *Style Layer Description* (SLD), which extends a WMS to allow user-defined symbolisation of feature data (OGC, 2002b), and a *Map Coverage Service* (MCS), which extends a WMS to allow access to spatial data layers that represent values or properties of geographical locations rather than WMS-generated maps (OGC, 2003c). There are also several standards under active development including coordinate transformation, geocoding, information searching and mining, and style description.

WMS and WFS are both designed to work with catalogue services defined in the *OGC Catalog Services Specification* that provide a common architecture for automated clearinghouse or directory services on the Internet. This association between Web mapping and Internet catalogue services, which is commonly referred to as the *OGC Web Services* (OWS), allows users to search for potentially useable data more expediently using the Web services model explained in Section 4.4.

4.   WORKING PRINCIPLES OF WEB-ENABLED SPATIAL DATABASE SYSTEMS

The architecture and functionality of Web-enabled spatial database systems have constantly evolved in response to advances in Web-based concepts and technologies. The following sections explain how Web-enabled spatial database systems worked in different architectures over time relative to what they are today and where they are likely to be heading into the future.

4.1   Interactive Web Architecture

In an interactive Web architecture, the Web server is the hub through which all accesses to the database are handled (Figure 10.8). The Web server contains a suite of application programs that add significant functionality to its central data processing role and power. Such additional functions include handling requests for information contained in graphics, video and audio files, connecting to external databases, interactive database query, and communicating with external application programs that may reside on the server or even on other compters that are networked to the server.

The server program that connects a Web server to a database is commonly called *Web-to-database middleware*. Earlier Web-to-database middleware was created using the method of a *Common Gateway Interface* (CGI). A CGI is a script file that performs a specific function in the Web server. It contains commands written in a particular programming language such as PERL or VB.NET. Each of the script files on a particular Web server is uniquely identified by a URL. Whenever a request to a matching URL is received, the corresponding CGI script file is called. The concept of using CGI is relatively simple, but the need to execute CGI scripts individually for each request impedes system performance. For example, if there are 100 requests for the same service the script must be loaded 100 times, thus consuming a lot of CPU resources on the Web server.

The performance limitation of the CGI approach led to the introduction of *Server Application Programming Interfaces* (Server API). These interfaces are implemented as *dynamic-link libraries* (DLL) residing in the main memory of the server and are treated as part of the Web server application. As a result, they are generally faster than CGI because there is no need to load them for each individual request. However, the use of a server API is also subject to some disadvantages. Because a server API shares the same memory space with Web server software, any API error can potentially bring down the entire server. Besides, API are server-specific and operating system-dependent, which means that a different API must be developed for different Web servers running on the Microsoft Windows, Linux, UNIX and other operating systems.

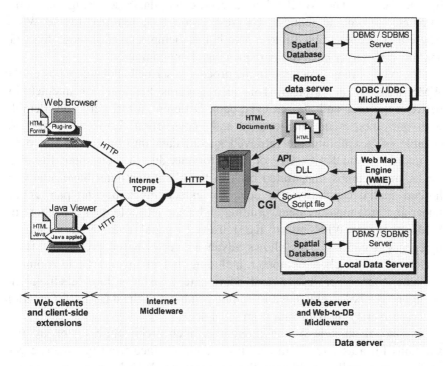

Figure 10-8. The interactive web architecture using CGI and server API

For Web-enabled spatial database applications, CGI and server API interact with the database indirectly through back-end spatial data processing functions that are known by various product names. These include commercial products such as ArcGIS Server, Arc Internet Map Server (IMS), MapInfo MapXtreme and GeoMedia WebMap. There are also open source products available free of charge, such as ALOV Map from the University of Sydney, Australia, MapServer from the University of

Minnesota and in late 2005 Autodesk announced that their MapGuide product would become an open source project. These back-end spatial data processing functions may be generically described as *Web Mapping Engines* (WME), as noted in Section 2.1 and Figure 10.1.

A WME is a constantly running program that "listens" for requests from client computers, for example, selecting specific map features within a user-defined window, identifying attributes of specific objects, overlaying map layers, creating buffers on features, and performing spatial joins on data tables. When the Web server receives a request from a client computer, for example, to retrieve all land parcels within a user-defined polygon on the browser window, it will call a CGI script or a server API function which in turn parses the request and translates it into commands that are understood by the WME. The WME then accesses the database using the input parameters from the request. The results are then returned to the WME where they are assembled into a map before being relayed through the CGI and the Web server back to the requesting client computer.

The type of map generated in the process described above depends on the WMS used by the system. This also governs the level of interactivity between the user and the retrieved data. Different WME vary considerably with one another in functionality and construction. Low-end WME are capable of assembling the retrieved spatial data into an image file only, commonly in the JPEG, GIF or PNG formats due to their small file size. Raster image maps transmitted back to the client computer have relatively limited spatial capability that does not extend much beyond pan, zoom, linear and area measurement, printing and basic attribute query. Higher-end WME using a WFS return the retrieved spatial data as a stream of independent entities. This allows advanced vector-based processing to be performed on the client computer, including dynamic panning and zooming, attribute queries, projection and coordinate transformation, labelling and changing symbols, selecting and editing a sub-set of the data, buffer creation and map overlaying with data from other sources.

On the client side of the Interactive Web architecture, there are two solutions for a user to interact with the data obtained from the Web server. One is functionally to enhance the conventional Web browser by software extensions. A variety of software tools are now available that enable Web browsers to function beyond their native capability of viewing raster images only. Autodesk, for example, provides a free plug-in for conventional Web browsers to be used as clients for MapGuide servers. A plug-in is an external program designed to handle data that are not originally supported by Web browsers, for example spatial data in vector format and video and audio files. A plug-in is application-specific, and is commonly supplied by the application developer to end users free of charge. It must be installed and is automatically invoked by the browser when needed.

The alternative is to build standalone spatial data viewers using an application development language such as Java, Javascript, ActiveX and VBScript (see Section 5.3). Examples of such viewers include ArcExplorer from ESRI, GeoMedia Viewer from Intergraph, and LandViewer and TIGER Map viewer from the United States Census Bureau. These viewers vary considerably in their capabilities to manipulate and query spatial databases. They are generally provided free of charge for the convenience of end users.

The spatial database in the Interactive Web architecture, which is commonly called the *data server*, can be hosted by the same computer as the Web server or externally in one or more remote computers. If the data server is a remote computer, the Web server has to access the data server through one of the *Open Database Connectivity* (ODBC) protocols discussed in Chapter 6. The data server includes both the spatial data repository and spatial data management software, such as Arc Spatial Database Engine (ArcSDE) or Oracle Spatial, that process spatial commands from the WME. It also typically includes a regular DBMS to handle queries on non-spatial data and to provide the necessary tools for database administration tasks such as user account management, database backup, and replication.

4.2   The Object Web Architecture

The Object Web architecture, also known as *Object/Web computing architecture,* was conceptualised in 1995 by the OMG in a joint project with the W3C to investigate the integration of CORBA with HTTP and Java. This project was prompted by three perceived benefits (Orfali et al, 1999):

- To overcome the limitations of using CGI and server API on the server, as explained above.
- To provide a scaleable and robust server-to-server Web infrastructure.
- To extend Java with a distributed object infrastructure.

Simply put, the Object Web architecture is the integration of Web protocols with distributed object technology. The advent of the Object Web represented a new direction of the development of the Web and provided the conceptual framework for building enterprise-class spatial database systems in modern organisations.

There are two current competing Object Web standards. One is of these is Microsoft's DCOM/ActiveX. This standard, which is built on the basis of OLE (see Chapter 6), enables direct communication between client computers and servers and, as a result, bypasses the constraint of the HTTP/CGI, as noted above. It is realised by creating objects using Visual C++ or Visual Basic programming, registering them on the server, and then using VBScript to enable them in the form of ActiveX controls on Web pages.

The other standard is CORBA/Java as presented jointly by the OMG and Javasoft. As shown in Figure 10.9, CORBA/Java is a multi-tier client/server application model consisting of (a) Java clients, (b) CORBA business objects and (c) data servers. This standard promotes the use of Java client applications to provide platform-independent Web-enabled solutions.

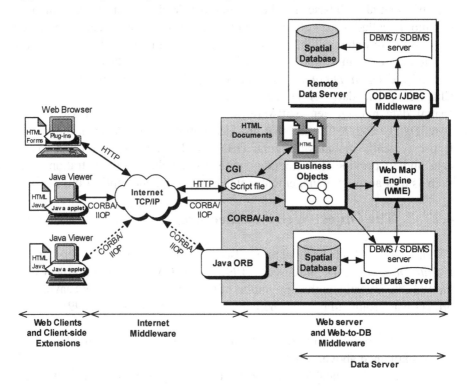

Figure 10-9. The object web architecture based on the CORBA/Java standard

The CORBA business objects in the Web server provide the application logic and encapsulate existing databases. These business objects were conventionally written in COBOL, C++ or SmallTalk, but are now increasingly developed using Java *servlet* technology. Servlets can be thought of as applets that run on the server side. They have access to the entire family of Java API and communicate with Java clients directly using the Java ORB or JDBC without going through Web server software, thus avoiding the delays inherent in processing through a CGI script for each user access as noted above. In addition, Web protocols such as HTTP and the OMG's Internet Inter-ORB Protocol (IIOP) provide alternative means of network communication between client browsers and servers. Since both HTTP and IIOP use TCP/IP to provide lower level packet-oriented transport and address protocols, they can be run on the same network for different functions. Specifically, HTTP is used to download Web pages, applets and images, whereas CORBA/IIOP is used to provide the Java client-to-server communications.

The advent of the Object Web architecture does not mean the automatic exclusion or demise of solutions based on CGI and server API approaches. The Object Web architecture represents an advance in database concept and technology rather than a replacement of existing architecture using CGI and server API. For Web-enabled spatial database systems, most if not all of the current WME are designed using a hybrid Interactive and Object Web architecture. Such a hybrid approach combines the simplicity of using CGI and server API one the one hand, and the performance and interoperability of object-oriented technology on the other hand. As the Web continues to grow, new concepts and methods such as Web services, semantic Webs and synchronised multimedia, have been developed to create the next generation Web. The vision of this new Web architecture, as described in the following section, will bring Web-enabled spatial database technology to a new height.

4.3   The Next Generation Web Architecture

Pragmatically, the Web architecture of the future will be more a conceptual framework for development of the Web, rather than a physical blueprint for constructing operational distributed computing systems like the Web architectures explained above. Different people have different ideas of what the next generation Web will be like. As the theories and technologies of the Web evolve, the concepts and visions of the Web will evolve correspondingly. However, there is a common collective goal to turn the Web from a medium for delivering information content to essentially passive clients, into a medium for large-scale, cross-platform and interactive distributed computing using the emerging theories and technologies shown in Figure 10.3.

Many software products and methods are now available that support the implementation of Web-enabled spatial database systems in the context of the Web of the future. One of the most broadly embraced concepts to implement the new breed of spatial database solutions pertains to *Web services*. The concept of Web services were introduced briefly in Chapter 6 from the perspective of spatial database interoperability. The following discussion explains how this particular concept is used in practice to implement Web-enabled spatial database systems.

The term "Web services" is defined in different ways and the technology can be used for numerous applications ranging from relatively simple request-response types of functionality to very sophisticated transactional processing in business-to-business collaboration and process management (Newcomer, 2002). The term "services" refers to software components that can be used to build larger, more comprehensive applications. These components are typically composed of XML text, defined according to a specification such as the *Simple Object Access Protocol* (SOAP) and described using a specification such as the *Web Service Description Language* (WSDL). Individual services are "invoked" through an API or remote procedural call (RPC) running on a remote desktop computer or a mobile computing device (for example hand-held computer, cellular phones or field data loggers). Thus, a Web Service can be generally defined as a vendor-independent, standards-based means of activating remote cross-platform procedures over the Internet.

As explained in Chapter 6, the Web services model is a tripartite relationship involving service users, service providers, and brokers. The service provider develops and deploys a service on a Web server and makes it discoverable through an online registry or catalogue of a service broker. For each service, the registry provides a *contract* that defines the name of the service, the service's input and output parameters, and how to communicate with the service. The contract is an XML document that follows a rigidly defined XML schema defining its structure. The most common type of contract is the WSDL document. Users of applications on the Web can discover deployed services by browsing and querying service registries on the Web and once one is found, it is invoked remotely according to the service description. The user program that deploys a service can be built using any technology that can understand the Web Service protocols including another service. In Web services terminology, the processes of finding and invoking a service is collectively called *binding*.

Figure 10.10 illustrates the concept of using Web services to move Web-enabled spatial database systems beyond the conventional role of information provider to become a service provider. The services are built on an architecture that is made up of a stack of software layers. Components of these software layers are loosely coupled, and interact with one another using standard protocols. At the foundation of the stack are the standards on which the Web is built today, that is TCP/IP, HTTP and HTML. The next layer of software components is made up of software tools and interfaces that allow client computers to communicate with servers, and to discover what they serve and how to obtain their services.

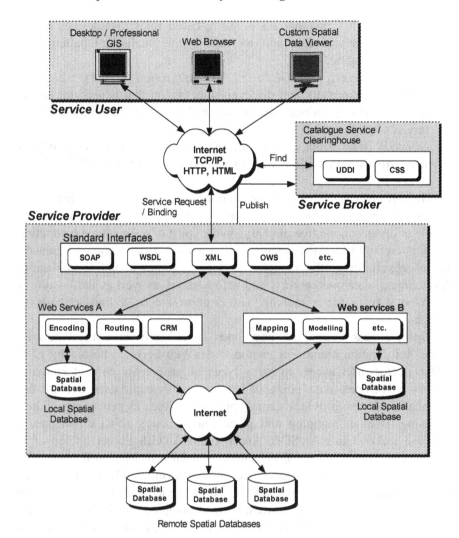

Figure 10-10. Delivering spatial database applications using the Web services model

The software tools and interfaces referred to in Figure 10.10 are implemented using standards that include:

- *XML,* which is the basic foundation on which Web services are built and which provides a language for defining data and the methods of processing them.
- *SOAP,* which is a collection of XML-based rules that defines the format of communication between a Web Service and its clients.

- *WSDL*, which is another set of XML-based rules that defines a Web services interface, data and message types, interaction patterns and protocol mapping.
- *Universal Description, Discovery and Integration* (UDDI), which is a Web services registry and discovery mechanism that is used for storing and categorising business information and for retrieving pointers to Web services interfaces.
- *OpenGIS Web Services* (OWS), that comprise the OGC family of standards whose functions were explained above in Section 3.5.

The top layer of the stack is the service layer consisting of spatial data processing programs. These programs can be developed and offered by people or organisations free of charge in the spirit of open source computing, or commercially by ASP companies in the information technology industry. These organisations offer complete service packages including application development, database creation and maintenance, as well as the hosting of the Web servers for a monthly subscription fee or a transaction-based charge.

This business model has now been successfully introduced into the spatial information arena. For example, ArcWeb Services hosted by ESRI (2004a) offers user access to many types of geographic content including base maps, business data, points of interest, and dynamic data such weather information. It also provides spatial applications such as geocoding, address matching, routing, mapping and place finding using ESRI's own spatial database technologies ArcSDE, ArcIMS and ArcGIS Server on top of its commercial data holdings (ESRI 2004b and 2004c). Microsoft also offers MapPoint Web Services based on the company's MapPoint business mapping technology. This is an XML-based service that allows developers to integrate location-based services, such as mapping, driving direction and proximate search, into their applications, using an extensive set of mapping related contents such as business listing and points of interest. This particular Web service was reported by Microsoft to serve 15 million transactions per day in early 2004.

Web services represent a great leap forward in Internet concepts, technology and functionality. The emergence of a service orientation in Web activity means that the Web is no longer used only as a platform for delivering information, but also as a springboard for delivering services. The Web services approach offers several advantages that substantiate the general benefits of Web-enabling spatial information, as noted in Section 2.2. The Web services approach has especially enhanced interoperability among spatial database systems, the integration of spatial and non-spatial database systems, the rapid deployment of applications at considerably lower cost, and the popularisation of spatial information technology. Further, the Web services approach has nurtured a collaborative paradigm of using spatial information and the development of national and international spatial data infrastructures (see Section 5.2).

It would be wrong to imply that the recent emergence of Web services heralds open-ended possibilities since inherited legacy aspects of the Web still constrain what is possible at the present time. From the perspective of spatial information, the most important limitation is the inability of Web applications to handle and understand semantics. The present HTML and XML markup languages specify how information is depicted and stored. However it is not possible for computers to interpret and capture the meaning and significance of the data represented by these markup languages. Given this reality, much of the decision making in deploying Web services cannot be automated, and human intervention and involvement are indispensable. As a result, even with the introduction of Web services concepts and technologies, the Web is still largely an information delivery medium rather than a content or knowledge management platform to support decision making.

The concept of a *Semantic Web* was proposed by Tim Berners-Lee and his colleagues at the W3C to address the limitations of the current Web by augmenting HTML/XML documents with an additional layer of metadata and logic rules (Berners-Lee et al., 2001). This new Web is made possible by a set of standards being coordinated by the W3C, such as the Resource Definition Framework (RDF) and Web Ontology Language (OWL), that provide a framework for enterprise integration and the sharing and reuse of data on the Web.

There is considerable interest in extending the idea of the Semantic Web to Web services generally (McIlraith et al., 2001). A Semantic Web Service represents yet another substantial enhancement to Web technology. In the Web services environment, a programmer can build applications with the ability to search registries such as a UDDI server for a list of available Web services. While the application may be able to find a Web service without human intervention, there is no way for it to determine the usefulness of the data that it retrieves. A Semantic Web service uses a semantic markup language based on RDF, OWL and related standards to enable a wide variety of agent or mediator technologies (see Chapter 6) for automated Web service discovery, invocation, composition (that is, chaining services to perform complex tasks) and interpretation of the results.

The concept of a *Semantic Geospatial Web* put forth by Egenhofer (2002) signifies some initial interest of the spatial database research community in the semantic Web idea. By explicitly representing semantics using multiple spatial and terminological ontologies (people, interfaces, search systems, and information resources), the Semantic Geospatial Web will enable users to retrieve more precisely the data that they need. However, as explained in Chapter 6, given the complexity of developing and using spatial ontologies, it is arguably very likely that the idea of semantically enhancing the Web and Web services for spatial applications will remain a goal rather than becoming a reality in the foreseeable future.

5.   IMPLEMENTING A WEB-ENABLED SPATIAL DATABASE SYSTEM

From an implementation perspective, the greatest challenge for Web-based database systems is to build software applications that are platform-independent, and hence can run on any computer connected to the Internet using a Web browser. This Web-based approach to spatial database processing and analysis is drastically different from the conventional methods that are implemented on local area networks (LAN) or Intranets using proprietary software products and a finite set of known hardware and peripherals. The following discussion highlights the factors that must be considered in designing and implementing a typical Web-based spatial database system.

5.1   Design Considerations

When designing a Web-enabled spatial database system, there are a number of factors that govern the direction and ultimate success of the project. It is important to understand the absolute and relative importance of these factors, which are grouped broadly into eight categories. These are qualified by a larger number of secondary factors that provide further definition and meaning for each of the primary factors, as shown in Table 10.2.

It is important to note that these factors can seldom be considered separately in practice. Rather, given the high level of interrelations, they must be considered in tandem, sometimes encompassing all eight primary factors and most of the secondary factors. For example, the purpose of setting up a Web-enabled spatial database system involves establishing who the target audience is, what technology can be most suitably used, how the enabling system must be configured, what data formats are acceptable, what costs will be incurred in implementing the system, and which legislation or regulations must be observed along the way.

Table 10-2. Design and implementation requirements

Primary Factors	Secondary Factors
Purpose of implementing the spatial database system	o Mission and business goals of the organisation. o Providing the general public information versus supporting transactional processing in business operations. o Delivering maps versus providing attribute spatial data. o Providing maps versus interactive querying. o Browsing versus downloading. o Data viewing versus data analysis. o Internet versus intranet versus extranet.
Audience or users of the system	o Internal (same or different locations) versus external users. o Level of computer literacy and skills. o Exposure to spatial database technology. o Accessibility to the Internet.
Technology supporting the system	o Server requirements (UNIX-based versus high-end PCs). o Client computer requirements (thick versus thin clients). o Bandwidths of local and wide area networks. o Plug-n versus Java applets versus Active X controls. o Web browser compatibility. o Graphical user interface. o Integrations with other technologies (e.g. GPS).
Architecture of the system	o Operating systems. o Security and firewall requirements. o System Architectures (see Sections 4.1 to 4.4). o DBMS software, Web server and Web Mapping Engine. o Off-the-shelf applications versus custom applications.
Spatial data to be served by the system	o Raster versus vector data. o Possible applications. o Standards and quality. o Sources and volume of framework / application data. o Intellectual property and copyright. o Updating frequency.
Application requirements and constraints	o Business logic and processes. o Work flow. o On-line Analytical versus On-line Transaction Processing.
Cost of capital investment in the system	o Cost of hardware / software. o Cost for systems operation and maintenance. o Cost to clients. o Free distribution of data versus commercial selling of data.
Legislation and regulations for collection, access to and application of spatial data	o Domain-specific federal and provincial/state legislation. o Domain-specific federal and provincial/state regulations. o International, national and provincial/state data standards. o Multilevel hardware/software standards. o Communication and telecommunication infrastructure.

It is not difficult to establish how the influences of technology, systems, data, application and cost affect one another. When a decision pertaining to one of these factors is changed, it is necessary to modify decisions on the related factors accordingly.

In order to address the complexity of the interplay of different factors, an iterative approach is typically used during the design process. This process starts with a definition of the purpose of implementing a Web-enabled spatial database system. To be successful, the purpose must be well defined with respect to the mission or business goals of the organisation on the one hand, as well as taking into account the technical, fiscal and human issues on the other hand. Other categories of factors are then considered in turn with respect to the purpose of the system. Upon completion of the first iteration of considerations, the connections and interrelationships between individual factors are reviewed and iteratively refined until a consensus is reached that the results represent the best balance and compromise among the requirements of the various factors that need to be considered.

Defining these factors and their interrelationships is a highly conceptual and time-consuming activity. These tasks are normally carried out in formal focus group meetings that involve the project manager, database administrators, systems analysts, application programmers, Web designers, and subject matter specialists. It is also advisable to include the end users in this process as products that are created without considering the perspective of end users may quickly fall into abeyance. Since, the manner of conducting these meetings is very similar to the data modelling processes as described in Chapter 3 it is not repeated here.

5.2   Approaches to Implementation

Web-enabled spatial database systems can be implemented in different ways according to their respective purposes. The four general alternative approaches to the implementation of spatial database systems are described in the following sections. In practice, these approaches are not mutually exclusive but are often used in conjunction with one another for different implementation scenarios.

5.2.1   Information-centric versus Application-centric Approach

The primary objective of an information-centric approach to Web-based spatial database implementation is to provide users with information that is necessary in their daily work or life. Use of spatial database systems to access information about weather, traffic and road conditions are often constructed using this approach, as rapid access to information is the primary consideration from an end-user perspective. Web-based spatial information providers, such as MapQuest, Google Maps and Microsoft's MapPoint and

Virtual Earth also fall under this approach (see Chapter 12). The application-centric approach, on the other hand, aims to provide domain-specific data primarily for professional users in engineering, collaborative research, and multi-agency decision making, however recent advances in Web-based public participation using spatial information technology has broadened the traditional pool of users. Land and resource management systems are typical examples of application-centric spatial databases.

5.2.2 Organisational versus Infrastructural Approach

An organisational approach to Web-enabled spatial databases is used for implementing systems solely for the purpose of serving the business needs of a particular organisation in the public and business sectors. Such systems are characterised by a relatively well-defined user community within or outside the organisation, hence the focus is much more specific than in other cases. In contrast, the infrastructural approach is often adopted by public sector agencies to implement spatial database systems designed to provide basic topographic and socio-economic information to the general public. In this approach, the Web is used as the backbone of global, national, and regional spatial data infrastructures. The Canadian GeoConnections website (http://www.geoconnections.org), which is the home of the Canadian Geospatial Data Infrastructure (CGDI) as well as the GeoBase and GeoGratis spatial data portals, and the Geospatial One-Stop (http://gos2.geodata.gov) of the United States government are national examples of this approach. In these cases the focus is as much on providing access to data as publicising and encouraging the use of national and international standards in their use.

5.2.3 In-house Development versus Outsourcing

The earlier generation of spatial database systems (i.e., GIS-driven systems) were often developed in-house from conception through design to implementation by the implementing organisation. Advances in computing and information technologies gave rise to the now very large spatial database industry that has turned out tens or even hundreds of off-the-shelf software options that organisations could purchase to meet their application needs. Most of these software products allow the user to customise their systems by macro programming by in-house systems staff or outside contractors.

As spatial database systems and spatial applications became more complex and sophisticated, it was subsequently much more difficult for organisations to develop their systems in-house except for relatively simple *ad hoc* applications. As a result of this, the process of outsourcing has become a norm rather than an exception in many spatial database implementation projects (see Chapter 9). The debut of Web-based spatial database systems pushed forward this trend toward outsourcing much faster than previously as Internet technologies are built largely around the idea of a service provider.

The concept of a *spatial application service provider* grew out of the generic *application service provider* and Web services models in the computer and information technology industry (see Chapter 12). In essence, a spatial application service provider is a consulting firm that provides total spatial information services, including data, hardware and software, on an annual or monthly subscription bases. Over the years, there have been fluctuations in the use of this model in the spatial database industry (Waters, 2001). However, many organisations have adopted it and outsourced their spatial database system to external service providers. In Chicago, Illinois for example, the police department's crime reported system is now hosted by a service provider through the chicagocrime.org Web site (Figure 10.11). In Ontario, Canada a private company named Teranet (http://www.teranet.ca) has provided Web-based land and property registration services used by government agencies, real estate companies and law firms since the early 1990s.

While the use of a service provider has the advantage of reducing in-house workloads and transferring costs of maintenance to the service provider, issues that relate to control over directions and costs must be carefully considered before this option is adopted by an organisation. The primary argument against this approach concerns information security (including access, confidentiality and protection of privacy), which is particularly important in the context of the types of personal information that is conventionally managed by government agencies. A further factor relates to fear of legal and political liability of the outsourcing organisation if and when the service provider is unable to sustain its business and goes bankrupt. The unwillingness of organisations to leave their information services totally in the hands of an external provider has led to the rise of what has come to be known as the portal model of active service provision (Figure 10.12).

The portal model approach provides the base data (for example the street fabric of an urban area and associated block-face socio-economic data that are in the public domain) and the facilities for the dissemination of this information over the Internet. The outsourcing organisation, in the meantime, feeds and updates the system at the provider's site with information that it wants to release to the public by extracting or replicating it from the corporate database.

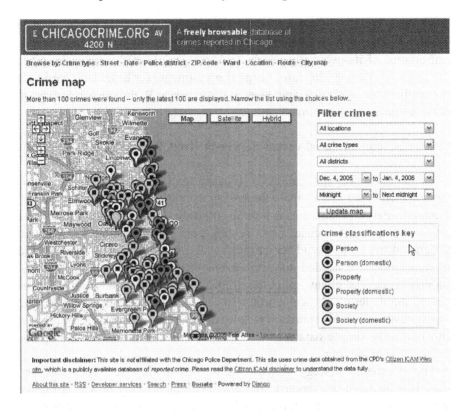

Figure 10-11. Example of a web-based spatial database application (using Google Maps) hosted by an application service provider

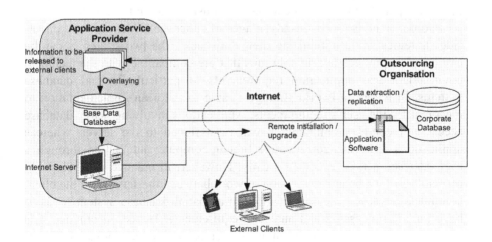

Figure 10-12. The portal model of an application service provider (ASP)

All data transmitted from the outsourcing organisation to the service provider can be done over the Internet. In this way, the organisation retains total control of its own information resources, while at the same time it is able to avoid the need for initial capital investment (for base data, hardware and application software) and to operationalise a Web-based information within a relatively short time frame.

5.2.4   Client-oriented versus Server-centred Approach

As a client/server application, Web-based spatial database systems can be implemented using one of three strategies, namely server-side development, client-side development, and a hybrid approach. As the name implies, the server-side approach focuses on providing both spatial data and application on-demand from a powerful server that has access to the spatial database and software products needed to perform the required data processing and analysis tasks. The client computers in this strategy serve no more than the role of dumb terminals to submit requests for services, and receive and display responses. This approach is best suited for mass access to a database, with relatively simple data retrieval applications and little need for spatial analysis and modelling.

The client-side strategy seeks to move part of the data processing and analysis tasks from the server to the client computer(s) by downloading the necessary data from the server and by installing software applications required for local processing, onto the client machine(s). This strategy is most suited to scenarios where users at different locations share a common database but have relatively unique, sophisticated and computationally-intensive spatial analysis applications that must be run on the database.

In order to utilise fully the advantages of either the server-side and client-side strategies while minimising their drawbacks, the two strategies can be combined to produce hybrid solutions that are compatible with the technical environment and application requirements of particular spatial database systems. In theory, a hybrid strategy is able to tailor an implementation to meet exactly the user's requirements. Applications involving heavy database access and complex computation can be assigned to the server, whereas applications requiring considerable human-computer interaction through a user interface can be executed locally. However, in practice it is relatively difficult to find a balanced division of work between the server and the client computers because every application on a spatial database system tends to have unique processing and analysis requirements. Hence, in practice it is more common to encounter either server-centred applications or client-oriented approaches implemented rather than a hybrid of the two. Table 10.3 summarises the relative advantages and disadvantages of the three strategies.

Table 10-3. Comparison between server-side, client-side and hybrid strategies to web-based spatial database implementation

Strategies	Advantages	Disadvantages
Server-side	o Allows users to access large and complex data sets that are otherwise difficult to transmit across the Internet. o Uses a high performance server capable of handling sophisticated computations in complex applications. o Centralised control over data and processing to ensure applications are correctly deployed. o Possible lower overall implementation cost since there is no need to install costly hardware and software at every user site.	o Slow processing since each data access and processing step must be returned to the server for execution. o Performance is subject to available bandwidth and network traffic on the Internet between the server and the client. o Inability of applications to utilise fully the processing power of modern desktop and laptop computers.
Client-side	o Inability of applications to take full advantage of the processing power of the client computers. o Greater user control of data processing and analysis in applications, thus allowing interactive computing where intermediate results must be interpreted before the next processing step is taken. o Better data processing due to lesser needs for client-server communication and data transmission.	o The requirement to transmit large amounts of data from the server to the client(s) may cause delays. o Large and complex data sets may be hard to store and process on less powerful client computers. o Sophisticated analytical and modelling applications may run more slowly on less powerful client computers. o Sophisticated analytical and modelling applications often require expensive training in order to use them properly.
Hybrid	o Can make full use of the relative advantages of server-side strategies while minimising their relative disadvantages. o Can make full use of the relative advantages of server-side strategies while minimising their relative disadvantages.	o Difficult to determine a balanced division of work between server and client computers because of variability in the nature of applications, the processing power of the computers used, the technical background and training of the users, and the bandwidth that is available to access the server.

No matter which of the above strategies, or a combination of them, is adopted in practice, the process of setting up a Web-based spatial database system is similar to the process for implementing a conventional spatial database. This process usually comprises the same phases of a typical systems development life cycle (SDLC) and database development life cycle, as explained in Chapter 3. The concepts and methods of project management described in Chapter 9 can also be similarly applied to Web-based spatial database implementation projects.

However, because Web-based data processing has a number of characteristics and considerations that make it somewhat different from conventional database processes, special attention must be paid to multiple factors that are particular to this case. These can be summarised as follows:

- Emphasis on systems security testing to provide full protection against hacking and unauthorised access to the database.
- The use of functional design specifications to create a test database, while ensuring that the test data are representative of the complexity of the actual data in the production database.
- Importance of user interface design to enhance user-friendliness and maximize the likelihood of actual use while minimizing the need for support when the system is in operation.
- The need and common practice of using a replica database to ensure security of the production database and to replace transparently and expediently the production database in case of a system failure.

6.   SUMMARY

This chapter explained how Web-enabled spatial database solutions are not simply an extension of existing database designs and technical implementations. On the contrary, they have their own distinct architectures that integrate spatial database software tools, server-to-database middleware, distinct application programming languages, and Internet markup languages for end-user presentation of database content. The new breed of Web-enabled spatial databases are directed toward a host of equally new business functions such as e-commerce and e-government that embrace new user communities, many of which may have limited skill and experience in database use. The implications of this for user interface design as well as end-user training and support were discussed.

The chapter also reviewed technical aspects of the rapid growth in enabling technologies and standards of Web enabled spatial databases, including communications protocols, such as TCP/IP and its associated IP addresses, that are required to maintain uniqueness and order within a rapidly growing global network that uses the Internet as its communications medium. Growth of the Web and its enabling markup languages, such as the basic HTML standard, were briefly documented prior to focusing discussion and providing more detailed examples on the use of XML and the domain-specific and important spatial database XML dialect of GML.

Attention was given in the latter parts of the chapter to the emergence of Open Web mapping standards that have been driven by activities of the OGC and are now widely deployed as Web-centred spatial database front- and back-ends by the developer community. The discussion also considered interactive and object Web architectures that facilitate the deployment of most of the current applications that are found in the workplace. The chapter concluded with a discussion of the design and implementation considerations that are required for production of a Web-enabled spatial database system.

7. REFERENCES

Abiteboul, S., Suciu, D. and Buneman, P. (1999) *Data on the Web: From Relations to Semi-structured Data and XML*, San Francisco, CA: Morgan Kaufman Publishing.

Anderson, G. and Moreno-Sanchez, R. (2003) "Building Web-enabled Spatial Information Solutions around Open Specifications and Open Source Software", *Transactions in GIS*, Vol. 7, Issue 4, pp. 447-466.

Berners-Lee, T., Hendler, J. and Lassila, O. (2001) "The Semantic Web", *Scientific American*, Vol. 284, No. 5, pp. 34-43.

Castro-Leon, E. (2004) "The Web within the Web", *IEEE Spectrum* (North American Edition), Vol. 41, No. 2, pp. 42-46.

Chaudhri, A.B., Rashid, A. and Zicani, R. (2003) *XML Data Management: Native XML and XML-enabled Databases*, Boston, MA: Addison-Wesley Publishing Co.

Egenhofer, M. (2002) "Toward the Semantic Geospatial Web", *Proceedings of the 10th ACM International Symposium on Advances in GIS*, McLean, VA.

ESRI (2001) *An Overview of the Geography Network*, an ESRI White Paper, Redlands, CA: Environmental Systems Research Institute, Inc.

ESRI (2004a) *Understanding ArcWeb Services: An Overview for Developers*, an ESRI White Paper, Redlands, CA: Environmental Systems Research Institute, Inc.

ESRI (2004b) *ArcGIS Server: ESRI's Enterprise GIS Application Server*, an ESRI White Paper, Redlands, CA: Environmental Systems Research Institute, Inc.

ESRI (2004c) *From the Desktop to the World: An Introduction to Developing and Hosting a High-availability ArcIMS Application*, an ESRI White Paper, Redlands, CA: Environmental Systems Research Institute, Inc.

Graves, M.R. (2003) *Using Terminal Services to Serve Geospatial Software and Data Resources to Corps Project Offices* (Technical Report ERDC/EL TR-03-13), Vicksburgh, MS: U.S. Army Engineer Research and Development Centre.

Kraak, M.-J. and Brown, A. (2000) *Web Cartography: Development and Prospects*, London, UK: Taylor & Francis.

Kreger, H. (2001) *Web Services Conceptual Architecture* (WSCA 1.0), Somers, NY: Software Communications Department, IBM Corp.

Lake, R. (2001) "The Hitchhiker's Guide to New Web Mapping Technologies", *GEOWorld*, Vol. 13, No. 12, pp. 38-41.

Lowe, J.W. (2002) "Spatial on a Shoestring: Leveraging Free Open-source Software", *Geospatial Solutions*, Vol. 12, No. 6.

Manola, F. (1999) "Technologies for a Web Object Model", *IEEE Internet Computing*, Vol. 3, Issue 1, pp. 38-41.

Newcomer, E. (2002) *Understanding Web Services: XML, WSDL, SOAP, and UDDI*, Indianapolis, IN: Addison Wesley Professional.

Nieh, J., Yang, J. and Novik, N. (2000) *A Comparison of Thin-client Computing Architecture* (Technical Report CUCS-022-00), Department of Computer Science, Columbia University, New York, NY.

OGC (2001) *Web Map Service* by Beaujardiere, J. de la, (Ed.), Wayland, MA: Open GIS Consortium Inc.

OGC (2002a) *Web Feature Service* by Vretanos, P. (Ed.), Wayland, MA: Open GIS Consortium Inc.

OGC (2002b) *Style Layer Description* by Lelonde, B. (Ed.), Wayland, MA: Open GIS Consortium Inc.

OGC (2003a) *Geography Markup Language* by Cox, S., Daisey, P., Lake, R., Patele, C and Whiteside, A. (Eds.), Wayland, MA: Open GIS Consortium Inc.

OGC (2003b) *Open GIS Web Mapping Server Cookbook* by Kolodziej, K. (Ed.), Wayland, MA: Open GIS Consortium Inc.

OGC (2003c) *Web Coverage Service* by Evans, J. (Ed.), Wayland, MA: Open GIS Consortium Inc.

OCG (2003d) *Web Map Context Documents*, by Humblet, J.-P. (Ed.), Wayland, MA: Open GIS Consortium Inc.

Orfali, R. and Harkey, D. and Edwards, J. (1997) *Client/Server Programming with Java sand CORBA*, New York, NY: John Wiley & Sons.

Orfali, R., Harkey, D. and Edwards, J. (1999) *The Essential Client/Server Survival Guide*, 3rd ed., New York, NY: John Wiley & Sons.

Oracle (2003) *Oracle Spatial and Oracle Locator*, an Oracle Technical White paper, Redwood Shores, CA: Oracle Inc.

Peng, Z.-R. and Tsou, M.-H. (2003) *Internet GIS: Distributed Geographic Information Services for the Internet and Wireless Networks*, New York, NY: John Wiley & Sons.

Plewe, B. (1997) *GIS On-line: Information Retrieval, Mapping and the Internet*, Santa Fe, NM: OnWord Press.

Rana, S. (2002) *Delivering Light-weight Online Geographic Analysis Using ArcIMS*, Working Paper Series #57, Centre for Advanced Spatial Analysis, University College London, London, UK.

Reichardt, M. (2004) "Building the Sensor Web", *Geospatial Solutions*, Vol. 14, No. 3, pp. 36-38.

Spohrer, J.C. (1999) "Information in Places", *IBM Systems Journal*, Vol. 38, No. 4.

Tang, W. and Selwood, J. (2004) *Connecting Our World: GIS Web Services*, Redlands, CA: ESRI Press.

W3C (2001a) *Scalable Vector Graphics (SVG) Specification 1.0, W3C Recommendation*, by Perraiolo, J. (Eds.) World Wide Web Consortium.

W3C (2001b) *Web Service Description Language (WSDL) 1.0, 3rd Ed., W3C Recommendation*, by Christensen, E., Curbera, F., Meredith, G. and Weerawarana, S. (Eds.) World Wide Web Consortium.

W3C (2003) *Simple Object Access Protocol (SOAP) Specifications 1.2, Part 0: Primer,* by Mitra, N. (Eds.) World Wide Web Consortium.

W3C (2004) *Extensible Markup Language (XML) Specifications 1.0, 3rd Ed., W3C Recommendation*, by Yergeau, F., Bray, T., Paoli, J., Sperberg-McQueen, C.M., and Maler, E. (Eds.) World Wide Web Consortium.

Wagner, M.J. (2001) "Will Geoportals Lead to Digital Data Fulfilment?", *GEOWorld*, Vol. 13, No. 12, pp. 32-36.

Waters, N. (2001) "Internet GIS: Watch for ASP", *GeoWorld*, Vol. 14, No. 6, pp. 26-28.

Chapter 11

SPATIAL DATA MINING AND DECISION SUPPORT SYSTEMS

1. INTRODUCTION

Spatial database systems were initially implemented primarily for data management purposes. As the focus of development was on data structure and transaction processing, these systems were in general weak in data analysis functions and, as a result, they could seldom be used to support executive and managerial decision making.

The increasing integration of spatial database systems with mainstream database technology has led to a growing recognition of the value of spatial data in organisational decision making processes. This emerging interest in using spatial data for decision support has benefited tremendously from parallel advances in knowledge discovery in large databases in the business world, using concepts and techniques now generally known as *data mining* or *knowledge discovery in databases*.

This chapter presents an overview of the principles and methods of data mining as they pertain to spatial decision support. The first part of the chapter examines the nature of spatial data mining and the current state of the technology. The second part explains the concepts of decision support and discusses how spatial knowledge obtained from data mining can be applied to support decision making in which location and spatial relationships are central. Throughout the chapter, the discussion focuses on concepts, techniques and implementation issues of spatial data mining and decision support rather than the technical details of underlying algorithms or approaches.

2.   DATA MINING DEFINITION, CONCEPTS AND TECHNIQUES

Spatial data mining is a special application domain of data mining. It has its theoretical foundations in conventional data mining and relies heavily on general data mining techniques to handle the attribute component of spatial data. A thorough understanding of the principles and methods of data mining is essential for users of spatial data mining. Hence, the following discussion provides an overview of the definition, concepts and techniques of data mining to pave the way for consideration of spatial data mining in Section 3.

2.1   The Origin and Nature of Data Mining

Data mining is the process of extracting interesting and hidden (that is, previously unknown) information from a huge amount of complex data typically stored in corporate databases and data warehouses (see Chapter 6). The term data mining is relatively new, but its origins can be traced to scientific research in the 1950s when computers were first used to analyse massive experimental databases by means of statistical and machine learning (also called artificial intelligence) methods. Since the late 1980s the deployment and growing importance of data mining as a database tool was driven by several interrelated factors. These include:

- The proliferation of database technology and the unprecedented volumes of data that are collected routinely at a rate never experienced before in human history by organisations in the public, business, academic, and research sectors. This activity has created a demand for automated tools sophisticated enough to sift rapidly through large databases to detect new information unknown to users and to present this information in a relevant and accessible format.
- The growing realisation that databases can be used as a basis for knowledge discovery, for decision support and to meet the needs of executives and senior managers to access information in order to make informed and logical decisions quickly in a competitive business environment.
- The inability of conventional methods of statistical analysis, SQL and OLAP techniques to detect and extract knowledge effectively and efficiently from databases characterised by large data volumes (a very large number of observations) and high dimensionality (a large number of attributes or variables).
- The surge in the data processing power of computers and parallel advances in the principles and techniques of database management,

machine learning, information theory, decision science, and many other related disciplines in science and engineering that have contributed to the emerging philosophy and methodology of knowledge management in business, industry, government, and scientific research.

Data mining is an integral part of modern database technology. It plays a critical role in the evolution of database systems from straightforward data management tools through to decision support, as shown in Figure 11.1.

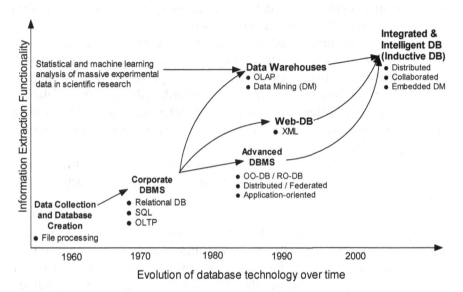

Figure 11-1. The evolution of database technology from data management to decision support

Before the mid-1960s, database systems were largely rudimentary stand-alone file processing systems. The advent of SQL in the 1970s and early 1980s accounted for the wide acceptance of relational database technology as a major paradigm for efficient storage, retrieval and management of large amounts of data in corporate database systems (see Chapter 2). Database development since the mid-1980s was characterised by new object-oriented and object-relational database models and the rapid growth of application-oriented spatial, temporal, and multimedia databases. Also during this time there was a spectacular growth of data warehouses and Web databases in a distributed computing environment (see Chapters 6 and 10 respectively). Further, OLAP was developed by using a combination of straightforward analytical functions such as summation, aggregation and consolidation to extract information from data warehouses for presentation of different perspectives to support organisational decision making. The debut of OLAP signified a major milestone in the transition of database systems from simple and complex knowledge discovery toward decision support functionality.

Data mining is a higher form of information extraction than OLAP by using machine learning to extract hidden and potentially useful information for use by decision makers and others. Data mining techniques were originally developed for use with data warehouses but are now embedded in several commercial database software products, such as Oracle, IBM's DB2 and Microsoft's SQL Server among others. This has given rise to a new generation of integrated and intelligent database systems that combine data management and decision support functionality in a single systems environment. This new breed of databases is sometimes referred to as *inductive databases*. These systems also allow collaborative decision support in a distributed network environment by deploying the data mining capabilities of database systems to work on distributed (networked) database nodes belonging to different organisations at different locations (or different databases at the same location belonging to different departments within the same organisation).

As a new method of information extraction, data mining differs from conventional SQL and OLAP methods in several important ways. These include:

- It is designed specifically for use with very large databases or data warehouses that typically contain millions of records (tuples) with hundreds and possibly thousands of attributes (also called *items* or *dimensions*) and which are, therefore, far too complex to be analysed effectively and efficiently by conventional methods.
- It is concerned with secondary analysis of large data sets that seeks to discover previously unknown or "hidden" knowledge, including unforeseen patterns, relations and trends that are otherwise impossible to detect due to the sheer size of the database and the complexity of the data content.
- It follows an inductive strategy of data analysis where users apply machine learning algorithms to gain knowledge progressively from the data without any *a priori* assumptions or hypotheses of what knowledge will be obtained at the end.
- It focuses on the detection of the characteristics of and correlations among a large number of attributes in a data set, but not the definitive identification of cause and effect between individual attributes (in other words, data mining is exploratory and non-deterministic rather than explanatory and deterministic in nature).

Data mining, therefore, goes well beyond the objectives of using legacy data to understand what happened in the past. It makes use of machine learning algorithms to sift automatically through each record and variable in a dataset to uncover information that may be hidden in or obscured by the complexity of the data. The outcomes of data mining include depiction of the past (models of details of the past) as well as insights into the future (models of details of the future). The ability to link past scenarios to what will possibly happen in the future makes data mining a crucial tool for supporting strategic decision making in organisations in the public and business sectors.

2.2   Data Mining and Knowledge Discovery in Databases

Data mining is commonly used as a synonym for knowledge discovery in databases (KDD) in the literature. The view adopted here considers KDD to be a more embracing process in which data mining is only one of the steps of a longer process. As shown in Figure 11.2, the process of KDD typically consists of the following sequence of steps:

- *Data integration and cleansing*, which combine multiple and possibly heterogeneous data sources into a single data warehouse or data store where the issues of erroneous, missing, and inconsistent data are rectified.
- *Data selection and transformation*, during which data relevant to the objective of a particular data mining task are retrieved from the database and transformed (for example, de-normalised, re-classified and aggregated) into a form appropriate for data mining techniques to be used. The selected data are sometime called an *itemset* to distinguish them from the source data set. However, the term "data set" is used more generally in this discussion to mean both full data sets and itemsets.
- *Data mining*, which involves the actual process of applying machine learning, visualisation, and statistical methods to extract and reveal information of interest that may be hidden in the database.
- *Knowledge discovery and construction*, which include the evaluation and interpretation of the extracted information, often through visualisation, and the subsequent incorporation of the information into a computerised knowledge base, or through documentation and reporting to end users such as subject experts and decision makers.
- *Deployment*, which is the practical use of data mining results in support of decision making in different application domains.

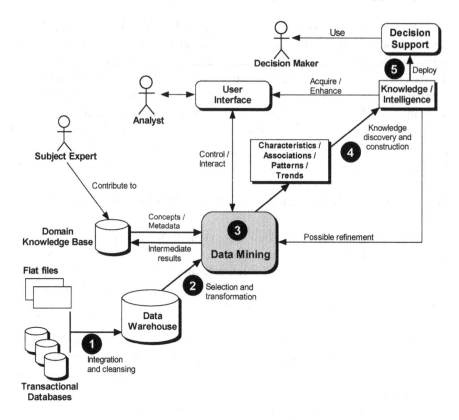

Figure 11-2. The steps of knowledge discovery in databases (KDD)

Knowledge discovery is an interactive and iterative process. With the aid of a suitable GUI, the analyst is able to control the data mining process by changing appropriate input data parameters to obtain different scenarios, and to cross-reference knowledge acquired using different mining techniques to ensure the consistency and integrity of results. The interactive and iterative nature of the approach also allows the analyst to improve continuously or to optimise the acquired knowledge by combining it with knowledge drawn from a knowledge base for higher levels of intelligence extraction in successive iterations of the process. The KDD process stops when the analyst is satisfied that no more insights of interest can be obtained by continuing the process (see Section 2.4).

Because of the interactive and iterative nature of KDD, it is not always easy to identify where data mining begins and where it ends. Han (1993) suggested that the steps of KDD could in fact be grouped into two relatively distinct functional stages. The first of these includes data integration, cleaning, selection and transformation, which can be performed by constructing a data warehouse and performing OLAP operations on the resulting warehouse content. In the second stage, data mining and evaluation techniques can be integrated into one, possibly iterative, process that is collectively called data mining.

From the broader perspective of spatial database design and implementation, Han's pragmatic definition of and approach to data mining is used in this chapter. In this conceptualisation, data mining is treated as a crucial component of database systems rather than as an optional extension to database technology. The use of data mining is also seen as only one component in a systematic approach to KDD rather than an *ad hoc* application in and of itself that uses a variety of statistical, SQL and OLAP methods as analytical tools. An ideal data mining strategy includes the use of theoretically sound algorithms capable of extracting information relevant to the user's needs, as well as visualisation methods that will enable decision makers to understand the outcomes of information extraction and apply them in an effective manner. Further, data mining is not an entirely automated process, but rather a confluence of human intelligence and machine learning or artificial intelligence that resides behind a battery of techniques designed to meet the challenges of discovering previously unknown knowledge from large databases and data warehouses.

2.3 Human Intelligence in Data Mining

Contrary to the popular belief that data mining is a machine-automated process from beginning to end, it is important to understand that people play a significant and indispensable role in the activity of data mining. There are several points during the data mining process where the injection of human intelligence is an essential rather than optional ingredient. These points of human intervention include:

- *Data preprocessing.* During data preparation, human intelligence is used to determine the usability of data from heterogeneous sources, to clean and transform the data into a form suitable for data mining, free of any adverse impact caused by errors, incompleteness and inconsistencies.
- *Data mining.* Choice of the mining model to be used with respect to the generic type(s) of knowledge to be discovered and required, and selection of the mining techniques and underlying algorithms both require human intervention. As data mining algorithms are often heuristic in nature, there are typically different techniques that focus on different aspects of the data objects available for a specific model. The choice of techniques and the sequence of applying them is a human decision that cannot easily be automated by relegating it to a machine.

- *Knowledge discovery and construction.* Data mining is an interactive and iterative process in which the analyst serves as the critical interface between syntactic knowledge generated by the computer and the semantic knowledge required by people to understand and reason about the real world that the data represent. Knowledge discovery and construction require a fine balance in the division of work between the computer, which performs the ground work of data processing and computation to distill data into information or knowledge implied by the data, and the human analyst whose intelligence decides where to begin and where to stop the process.

- *Presenting and visualising discovered knowledge.* The organisation and forms in which the discovered knowledge is displayed and communicated are still largely a mental process that sometimes includes considerable trial and error for optimal effectiveness and expressiveness. Despite substantial efforts in research and development, fully computer-generated information products are seldom considered as aesthetically pleasing as those produced with human intervention.

Han and Kamber (2001) used the term *data mining primitives* to denote the various aspects of using human intelligence noted above that help define and control a data mining inquiry. To be effective, a data mining task must be properly understood, its objectives precisely set, its process vigilantly monitored and corrected, and more importantly its results intelligently interpreted and precisely reported by people.

Data mining does not seek to discover all types of knowledge that may be hidden in a target data set. Instead, it seeks to detect only "interesting" and potentially useful knowledge. The term "interesting knowledge" was used earlier without defining precisely what "interesting" meant. In the context of data mining, the "interest" of knowledge is a quantifiable measure of its validity (its ability to stand the test against new data with a quantifiable degree of certainty), novelty (being unexpected and unknown), relevance (usefulness or perceived value for one or more applications), and ease of understanding (ability to be interpreted and distilled into knowledge) (Fayyad, et al., 1996).

The concept of novelty is particularly important in knowledge discovery. From the perspective of data mining, any knowledge that is expected or hypothesised *a priori* is not novel and, therefore, not interesting, as explained in Section 2.1. On the other hand, information that is not anticipated or that reveals previously unknown relationships between indicators is novel and interesting. This knowledge allows the data miner to be able to establish testable relationships between indicators within the various dimensions of a multidimensional data model.

The ability to identify and measure the extent to which discovered knowledge is interesting is critically important in data mining. Measurement of "interest" is typically determined by a certain threshold set by the analyst using information from a computerized knowledge base, or his or her own intelligence. This implies that the important principle and method of a *concept hierarchy* also plays an important role in discovering interesting knowledge. Han and Kamber (2001) define a concept hierarchy as the background knowledge that provides a sequence of "mappings" from a set of low-level concepts to high-level, more generalised concepts in a particular knowledge domain.

Figure 11.3 shows an example of the possible concept hierarchies within the knowledge domain of Canada's census of population and dwellings. These concept hierarchies can be represented graphically as illustrated in (a), or alternatively as various forms of *rules* as exemplified in (b), (c) and (d) in the figure. Rules are somewhat similar to relational views (see Chapter 2) in a database. They specify virtual relations that are not actually stored in the database but can be formed from the rule specifications. In the structure of a concept hierarchy, the nodes at different levels represent progressive degrees of information abstraction from the highest (at the root) to the lowest (at the leaves). With the aid of one or more concept hierarchies, the analyst can logically and systematically "roll up" or generalise (for example, demographic characteristics of Canada) and "drill down" or specify (for example, demographic characteristics of individual census tracts in the Toronto Census Metropolitan Area, Ontario) during the data mining process. It is also possible to "drill across" a concept within a given level of a concept hierarchy to examine temporal variation on a given concept between, for example, nodes such as census subdivisions at various points in time.

Concept hierarchies provide the background knowledge to control the exploration of the data set at different semantic levels and at different stages of the data mining process, and according to the specific needs of the end users of the discovered knowledge. In this example, it is impossible to extract knowledge not collected and stored explicitly in the population census database without the help of the concept hierarchy levels shown in Figure 11.3. Such knowledge includes, among many other possibilities, the distribution of families living below the poverty line across Canada, the differences between provinces on fundamental indicators of well-being such as the cost of living, the regional disparity between northern and southern Ontario, and the relationship between the level of education of the heads of households and the number of families living below poverty line at a single point in time or over different points in time. Note that data can conventionally only be extracted for and analysed at the smallest level at which data are aggregated spatially. In the case of the concept hierarchy shown in Figure 11.3 this refers to census blocks.

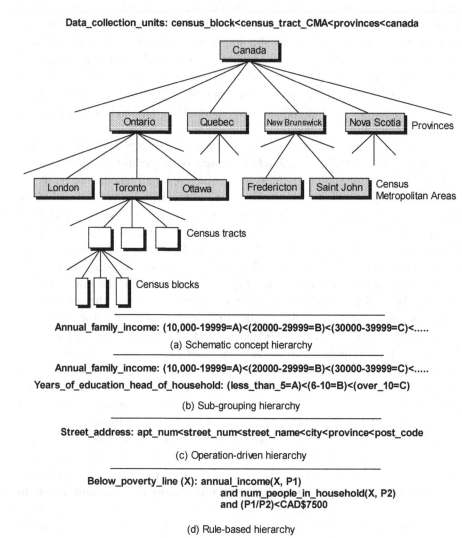

Data_collection_units: census_block<census_tract_CMA<provinces<canada

Annual_family_income: (10,000-19999=A)<(20000-29999=B)<(30000-39999=C)<.....

(a) Schematic concept hierarchy

Annual_family_income: (10,000-19999=A)<(20000-29999=B)<(30000-39999=C)<.....
Years_of_education_head_of_household: (less_than_5=A)<(6-10=B)<(over_10=C)

(b) Sub-grouping hierarchy

Street_address: apt_num<street_num<street_name<city<province<post_code

(c) Operation-driven hierarchy

Below_poverty_line (X): annual_income(X, P1)
          and num_people_in_household(X, P2)
          and (P1/P2)<CAD$7500

(d) Rule-based hierarchy

Figure 11-3. An example of concept hierarchies in the context of Canada's census population statistics

2.4   Data Mining Concepts and Techniques

In the last several years, the growing importance of data mining as a database tool has resulted in the development of numerous data mining techniques ranging from visual interpretation and understanding to algorithmic logic and probability rules expressed in mathematical form. Since the theoretical foundations of data mining have drawn on concepts and techniques of various subjects in science and technology, it is not uncommon for the same term to mean different things, or different terms to mean a same thing in the data mining literature. The confusion is compounded by technology vendors positioning their products as data mining tools without rigorously defining what they mean by data mining.

Figure 11.4 reiterates the concepts of data mining as defined in Section 2.2 with special reference to the relationships between source data, analytical techniques and the resulting models of discovered knowledge. Beneath the use of methods of primarily multivariate statistical analyses, which have in themselves played a relatively important role in the development and application of data mining techniques, is the process and methods of visualisation.

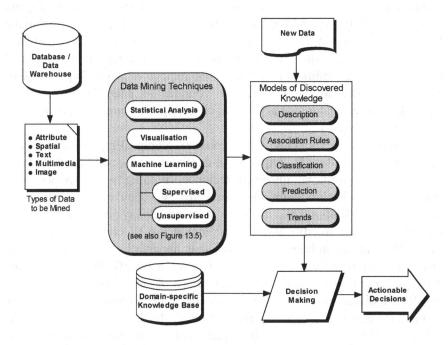

Figure 11-4. The concepts and techniques of data mining

This group of techniques includes a variety of two dimensional (2-D), three dimensional (3-D), and animated graphics-based techniques for representing the complex patterns and relationships in large volumes of data that are impossible to describe by mathematical formulae and equations. Visualisation has grown out of a graphical approach of data analysis in scientific research commonly called *visualisation in scientific computing*

(ViSC) or *scientific visualisation* (McCormick et al., 1987). In the context of data mining, visualisation is much more than the idea of "making something visible". It is made up of three elements, namely:

- *Computation*, which includes the hardware, software and algorithms that turn alpha-numeric data into graphical images, together with the mechanisms that facilitate human-computer interaction.
- *Cognition*, which is the human ability to develop mental representation, identify patterns and create order.
- *Graphic design*, which includes the conceptualisation and construction of pictorial displays using accepted principles of graphics communication.

In theory, visualisation is an ideal data mining tool because graphical displays are one of the most intuitive ways for people to learn, explore, and reason effectively (Ribarsky et al., 1999). Visualisation is a natural technique for extracting information from image-based databases. It is also particularly useful in situations where the discovered knowledge cannot be presented effectively using traditional methods that require *a priori* hypotheses or cannot be scaled to handle massive volumes of data. Moreover, it is particularly relevant for spatial data which are tightly bound with map-based visualisation methods.

In practice, however, interpreting complex graphical representations can be a daunting task. It requires in-depth domain-specific skills and knowledge, as well as advanced technical training and experience in scientific visualisation. Further, there are no standard semantics for describing visualisations and there are no standard ways of formalising and quantifying visual interpretations for common use by end users. In Figure 11.5, for example, although visualisation is apparently able to provide an effective synoptic view of a very large data set, it is difficult to describe the displays verbally and to relate the observations of the information visualised in different display windows. This problem of formalisation and linkage compounds exponentially as the number of graphical displays increases. As a result, although it is generally agreed that visualisation is an essential component and effective tool of data mining, its limitations must be recognised. The ways visualisation is used in support of spatial data mining is discussed further in Section 3.4.

Contemporary data mining is dominated by machine learning methods. These methods derive their name from the fact that the machine (that is, the computer) uses a learning algorithm first to find out the characteristics of a training data set and then produce a model to which new data are mapped in order to generate classification, patterns, predictions, and trends. Machine learning can be conveniently classified into *supervised machine learning* and *unsupervised machine learning* according to the level of human intervention required in the learning process. Supervised machine learning, also called *predictive data mining*, is directed toward *problem solving*. It is called "supervised" because it requires the data analyst to identify a target field or dependent attributes in the data set that is being mined. A chosen algorithm sifts through the data trying to detect patterns and relationships between the independent and dependent variables. It then applies the detected patterns and relationships to build a model of discovered knowledge that can be used to predict the behaviours or characteristics of new data objects or data sets.

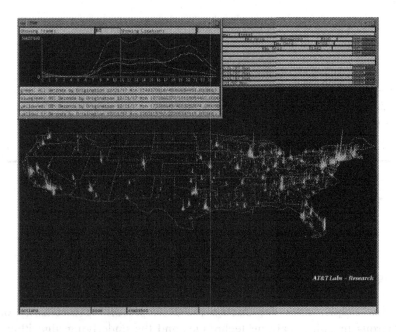

Figure 11-5. A landscape visualisation of telephone networks in the United States (Photo courtesy of AT&T Shannon Laboratory, Florham, NJ)

On the other hand, unsupervised machine learning, also referred to as *descriptive data mining*, is a form of *exploration-oriented* data mining that seeks to detect aspects of the properties of a data set in a concise manner. Algorithms in this category of data mining make no assumptions or hypotheses about the target data set and attempt to find associations, clusters and trends in the data independent of any pre-defined objective.

Numerous techniques have been proposed for data mining by machine learning. Figure 11.6 depicts a typology of commonly used supervised and unsupervised machine learning techniques. Based on the relative diversity of the forms of supervised and unsupervised machine learning in the diagram, it is clear that many strategies have been implemented to address this issue. However, on the basis of the kinds of discovered knowledge, the variety of approaches can be reduced to two major categories of supervised techniques, namely *classification* and *prediction*, and five categories of unsupervised techniques namely, *class/concept description, association, clustering, outlier analysis* and *time-series analysis*. Each of these categories of techniques can be classified further, as shown, according to the nature of the algorithms used. It is important to note that some algorithms, notably decision tree induction and neural networks, can be used for both supervised and unsupervised learning techniques.

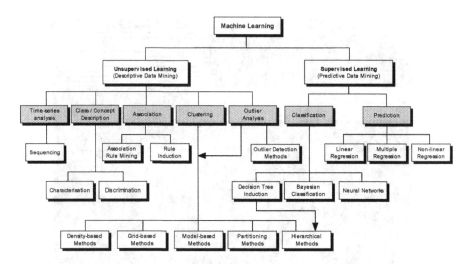

Figure 11-6. Classification of machine learning data mining techniques

This chapter does not discuss the concepts or the mechanics of the numerous machine learning techniques and the underlying algorithms that have been proposed or are currently in use. There are many recent texts in which comprehensive accounts of machine learning techniques in data mining can be found, including Bremer (2003), Fayyad et al., (1996), Han and Kamber (2001), Hand et al., (2001), Miller and Han (2001), and Whitten and Frank (2000). The following discussion provides a brief and necessarily general introduction to the seven major categories of techniques noted in Figure 11.6 to provide the required background for the study of spatial data mining as discussed in Section 3.

2.4.1 Classification

Classification is the grouping of unlabelled data objects into predefined classes or categories that typically, but not always, have mutually exclusive characteristics. It is a predictive data mining technique used for data objects with discrete values. Classification algorithms normally use a training data set where all data objects are already associated with predefined classes. The algorithms learn from the training data set and build a classification model to predict the classification of new data objects. Many techniques have been proposed for classification tasks in data mining, including *decision trees*, *neural networks* and *Bayesian classification*. These are described as follows:

- *Decision trees* derive their name from the resulting model that is represented in a tree-like structure. A decision tree is generated from a training data set in a top-down, general-to-specific direction by recursively partitioning the data set until a completely discriminating tree is obtained. The discriminating tree is then "pruned" to generalise branches that are too specific with respect to a predefined threshold.
- *Neural networks* refer to analytical techniques modelled after the hypothesised processes of learning in the cognitive system and neurological functions of the human brain, and are capable of predicting new observations from known observations after executing a learning process from existing data.
- *Bayesian classification* includes a variety of statistical techniques that predict class membership probabilities computationally or graphically using Bayes' Theorem of conditional probability.

2.4.2 Prediction

Prediction is the data mining function that is used to determine possible values of missing data or to forecast the values and distribution of attributes in a set of objects. Prediction can be carried out using a variety of techniques, including *classification* (for discrete categorical data) and *regression analysis* (for continuously distributed numerical data). Prediction by classification makes use of the classification methods noted in Section 2.4.1. There are several methods of regression analysis in data mining, including:

- *Ordinary least squares simple linear regression*, models the relationship between a random *response* (or *dependent*) *variable* Y as a linear function of another random *predictor* (or *independent*) *variable* X in the form of $Y = \alpha + \beta X + \varepsilon$, where α and β are respectively the value of the intercept on the Y axis and a coefficient for the slope of the "best fit" least squares regression line for the relationship between the observed and predicted values of Y based on the observed values of X. The difference between the observed and predicted values is captured in ε, which is a randomly distributed error term.

- *Ordinary least squares multiple linear regression,* is an extension of the simple linear regression model to include more than one predictor (X) variable for the unknown values of the response variable, Y.
- *Non-linear least squares simple and multiple regression,* which models the relationship between the response variable (Y) and the predictor variable(s) $(X_1,....,_n)$ by adding nonlinear polynomial or other terms to the linear model, that is, $Y = \alpha + \beta_1X_1 + \beta_2X_2^2 + \beta_3X_3^3 + \varepsilon$, where X, Y, α, β and ε are defined as in the simple linear regression model. This form of regression model is nonlinear in the variables X_2 and X_3 although the intrinsic form of the model remains linear.

2.4.3   Class/Concept Description

Class/concept description provides a summary of the general properties of individual classes or concepts in a data set. It includes statistics such as count, sum and average, as well as indicators of data dispersion, such as the variance or discrimination by comparing the general properties of a particular class of data objects with those of one or more contrasting classes. Class description can also be presented as a cross-tabulation, a chart, a graph and a map.

2.4.4   Association Rule Mining

Association rule mining, which is also called *dependency analysis* and *linkage analysis*, seeks to detect the *correlations* among attributes or items of interest in a data set. Correlations are expressed in the form of an *association rule*, $X \Rightarrow Y$ (c%, s%), where X is called the *antecedent* and Y the *consequence*, c% is the *confidence* (also called *predictability* or *conditional probability*) that, for a given data set, the attributes that satisfy X are likely to satisfy Y, and s% is the *support* (also called *prevalence* or *probability*) that measures how often variables X and Y occur together as a percentage of the total transactions or records in the data set. Association can also be expressed as an *induction rule* in the form of "IF X THEN Y", meaning that if event X occurs, then event Y will likely follow.

2.4.5 Clustering

Clustering techniques, also known as methods of *database segregation*, identify clusters or scenarios embedded in a data set where a cluster is defined as a collection of data objects that are "similar" to one another and "dissimilar" to objects in other identified clusters. Cluster analysis is applied in multiple fields in both scientific and social scientific research. It is widely used in data mining as a form of unsupervised learning because it does not rely on predefined classes and class-labelled training techniques. This property distinguishes clustering from classification as data mining techniques.

Clustering is a very computationally-intensive technique and the results can be easily affected by the type of data being mined (for example, nominal, cardinal and ratio-scaled) and "noise" caused by errors or missing data. Many clustering techniques have been developed and can be generally classified into the following four categories:

- *Partitioning methods*, which comprise "standard" clustering techniques that develop a partition of the data set under examination such that data objects in a cluster are more similar to each other than they are to objects in other clusters.
- *Hierarchical methods*, which perform a sequence of partitioning operations that can be done bottom-up (that is, repeating amalgamation of group of data objects until some pre-defined threshold is reached) or top-down (that is, recursively dividing the data set until some predefined threshold is reached). Grid- or pixel-based clustering, which is commonly used in remote sensing image analysis, is essentially a hierarchical method.
- *Locality-based methods*, which group data objects based on local relationships that use density or random distribution statistics as the clustering principle.
- *Neural networks*, which perform clustering using one of the above methods after executing a learning process based on existing data.

2.4.6 Outlier or Deviation Analysis

Outliers, also called *exceptions*, are observations whose data values appear inconsistent with or substantially different from the values of other observations in a data set. Deviations from the norm within a data set can be caused by data collection discrepancies or computation errors (often called "noise" within the data), but it can also indicate irregularity in the properties or behaviours of the feature a data value represents. In a sense, outlier analysis can be regarded as a special case of clustering. While clustering focuses on similarity, outlier analysis seeks to identify cases of dissimilarity. Hence, outlier analysis can be carried out using clustering techniques or by using techniques such as regression analysis or specialised algorithms that are optimised for detecting incidences of anomalies in a data set.

2.4.7   Time Series Analysis

Time-series analysis is also called *trend detection*. It is concerned with the detection of temporal characteristics, such as sequences and subsequences, sequential patterns, periodicities, trends and temporal deviations, that are hidden in a data set. Many techniques have been developed in statistics and econometrics for time-series analysis. From a data mining perspective, the most useful one is arguably *sequencing*. This method extends the association rule noted earlier in Section 2.4.4 by adding time comparisons that are related by unique identifiers between transactions or events. By taking pair-wise combinations of all transactions or events that have the same identifier and computing the time difference between each pair, it is possible to identify all before- and after-the-fact transactions or events.

3.   SPATIAL DATA MINING CONCEPTS AND TECHNIQUES

In order to use data mining concepts and techniques in the spatial domain, it is necessary to enhance them theoretically and technically to accommodate the characteristics of spatial data and to meet the specific requirements of end users in spatial decision making. The following sections explain the conceptual and technical enhancements that are required to discover previously unknown and potentially useful knowledge in large spatial databases, and how these enhancements are implemented in practice.

3.1   Characteristics of Spatial Data Mining

The ability of data mining to extract knowledge from large databases makes it a tool that is well suited to spatial databases which have over time become increasingly large and, in some cases, difficult for conventional spatial analysis methods to handle effectively. The challenge of spatial data mining is to detect spatial knowledge from the patterns and relationships that exist within spatial data that may not have had spatial dimensions explicitly defined, computed, and stored in the database.

Spatial data mining is far more complex than attribute-oriented data mining because of several factors. Some of these are related to the inherent nature of spatial data, for example:

- *Spatial data structures*. Spatial data typically carry locational and topological information, often organised by sophisticated indexing structures and accessed by spatial access methods. These data are organised by themes into different tables and are often optimised for forms of transaction processing (for example, normalised) that are not necessarily efficient for analytical processing in conventional data mining.
- *Spatial data volume*. Enterprise-level spatial databases nearly always contain substantial amounts of data that are often heterogeneous in format and quality and that require considerable amounts of computing power to clean and select them for use in spatial data mining. The large data volumes not only impose considerable computing overhead for data mining, but they may also exaggerate broad patterns (that is, global variance) contained in the data set and, as a result, make small but critical patterns (that is, local variance such as outliers) harder to discover.
- *Spatial data collection*. Much of the spatial data in use today have been collected by sampling and are provided in spatially and temporally aggregated form. These characteristics mean that salient information can be lost due to sample design and interpretation in the data collection, computation and compilation processes.
- *Spatial dependencies*. Spatial features are often intrinsically interrelated or interconnected such that it is often difficult or impossible to discover knowledge hidden in the data without some prior knowledge of the characteristics of the data sets under investigation. Such knowledge is not always easy to find and even if it ever exists it is difficult to validate or verify with confidence.
- *Temporality of spatial data*. Spatial features are also often intrinsically interrelated or interconnected in time. Because spatial data records the state of the features only at a particular point in time, the information of previous states is seldom captured and stored in spatial databases. This represents a crucial void in the information that is available and necessary for spatial knowledge discovery. Further, many geographical phenomena are cyclical in nature (that is, seasonal, annual and circulatory) with strong temporal patterns that may totally overshadow salient local variances over space.

Other factors are related to spatial mining techniques and the concepts of spatial knowledge. These include, for example:

- *Spatial data mining techniques.* Spatial data mining requires geometric computation and spatial operations that are only available in spatial database systems, which implies that spatial data mining demands a tight integration with and heavy reliance on relatively sophisticated spatial database technologies.
- *Spatial data conceptual models.* The apparent lack of a universally accepted model of geography, as suggested by Goodchild (1992) and reiterated by Sui and Goodchild (2001), has caused difficulties in formalising the spatial domain. This has, in turn, resulted in the implementation of spatial database systems that use different philosophies and techniques from mainstream databases. The difficulty or inability to integrate data represented by different models impedes both the application of formal geographical knowledge to the process of spatial knowledge discovery and putting discovered knowledge formally into the proper geographical perspective for decision making.
- *Different concepts of spatial space and spatial knowledge.* Spatial knowledge is most often represented in *Euclidean space* in terms of Cartesian coordinates and geometry, as well as non-Euclidean spaces when the time dimension is included (for example, travel time between different locations) and human cognition is taken into account (that is, personal values and perceptions). While spatial knowledge in Euclidean space can be represented and structured relatively easily, it is much harder to define and formalise spatial knowledge in non-Euclidean spaces. The interaction between spatial knowledge in Euclidean and non-Euclidean space is even harder to define and formalise.

Spatial data mining, therefore, is characterised by both the inherent complexity of spatial data and the sophistication of the implementing technologies. It differs from conventional attributed-oriented data mining in several ways, for instance:

- Spatial data mining is concerned with spatial knowledge in a continuous 2- or 3-D geographical space, whereas attribute-oriented data mining is concerned with spatial knowledge in a discrete object space.
- Unlike conventional data mining, spatial data mining deals not only with numerical and categorical data, but also with extended data objects such as points, lines, areas and surfaces.

- Spatial data mining focuses mostly on the discovery of local knowledge, but attribute-oriented data mining is directed more toward discovering global knowledge.
- Spatial data mining algorithms rely heavily on the concept of a neighbourhood on the premise that the characteristics of spatial data objects are affected more by their immediate neighbours than those objects that are farther away from them in geographic space.
- Spatial data mining predicates (for example, overlay, close_to, intersect, beside, and so on) are implicit and large in number, whereas conventional data mining predicates (for example, equal_to, more_than, less_than, and so on) are explicit and more limited in number.

Conventional attribute-oriented data mining algorithms often perform poorly, or do not work at all in some instances when they are applied to spatial data sets. In order to handle effectively the characteristics of spatial data and the specific requirements of spatial data mining, new concepts and techniques have been proposed as explained in Section 3.2 through Section 3.4 below.

3.2   Spatial Concept Hierarchies

Section 2.3 introduced the principle of concept hierarchies as a means of controlling data mining using existing human intelligence. In the context of spatial data mining, a spatial concept hierarchy provides the knowledge base for the analyst to drill down and roll up the data set flexibly and systematically at different levels of geographical abstraction. In essence, a spatial concept hierarchy is an extended spatial data cube that models a spatial data warehouse and facilitates OLAP operations on it (Han and Kamber, 2001). It functions in very much the same way as concept hierarchies for attribute-oriented data mining except that it has additional dimensions. These include:

- An attribute dimension, which is the exact equivalent of the attribute-oriented concept hierarchy as described in Section 2.3, and is used in controlling the discovery of interesting knowledge from attributes associated with locations and geometries.
- A spatial-to-attribute dimension, whose primitive-level data are spatial but whose generalisation, starting at a high level, becomes non-spatial. For example, in a criminal investigation database maintained by a city's police department each individual crime incident is spatial because its location is recorded in a given geographical space (either by GPS co-ordinates or by street address) and can be displayed and analysed

spatially. However, when crime incidents are aggregated in spatial data mining to generate an annual report of crime statistics, the resulting data become increasingly generalized and non-location-specific, even although they still relate to incidents reported within the city's boundaries. This is because in the annual report, the number of incidents in each category of crime is important, but the specific locations of the crime incidents are no longer important.

- A spatial-to-temporal dimension, whose primitive-level data are spatial but whose generalisation over time is not. For example, in the same criminal investigation database example noted above, the daily break-and-enter incidents can be analysed to detect the variation of occurrence at different time periods in a day, on different days of a week, and in different months of a year. The original crime incident data collected are spatial, but the discovered knowledge is temporal because the focus of interest is the changing number of crime incidents over time rather than over space.

- A spatial-to-spatial dimension, whose primitive-level and all of its high-level generalised data are spatial. Most tasks in spatial data mining use this dimension of spatial hierarchies. For example, the daily precipitation data recorded at all weather stations in an environmental database are recorded individually over space and their generalisation in the form of isohyets of rainfall on a weather map is also spatial since isohyets, like the precipitation data at the individual stations, convey the idea of pattern and distribution in geographical space.

Section 2.3 also noted the importance of the measurement of the degree of interest in data mining activity. In spatial data mining three types of measurement of spatial interest can be identified as:

- *Numerical measures*, which apply to numerical data, for example, daily absenteeism data for individual schools can be rolled up to compute weekly, monthly, quarterly and annual statistics for specific school districts or for a whole municipality as numerical spatial measures.
- *Classification measures*, which apply to categorical data, for example, rolling up high-, medium- and low-value residential properties into a generalised residential property value category which can in turn be rolled up with other generalised categorical value classes for industrial, commercial and institutional land uses into a higher level of generalisation of urban values in a land value classification scheme.
- *Spatial measures*, which apply to spatial objects when they are generalised or rolled up to form larger units of spatial knowledge

representation in spatial data mining, for example, townships → counties → provinces/states → countries.

Spatial data mining requires not only a good knowledge of the spatial data being explored and working skills in data mining techniques, but also an intimate knowledge of the geography of the area and the subject matter that are represented by the data. It also requires a good understanding of the objectives of specific spatial data mining tasks (for example, strategic business planning versus local business improvement) with respect to different levels of the concept hierarchies. Failure of the analyst to possess the prerequisite knowledge and skills undermines the purpose of spatial data mining, often with serious consequences because decisions may be made on the basis of incomplete or erroneous knowledge discovered in spatial databases.

3.3 Machine Learning Techniques of Spatial Data Mining

The growing interest in spatial data mining has resulted in the emergence of a large number of techniques and algorithms that are beyond the scope of this book to explain (for a good coverage of the principal techniques see Miller and Han, 2001). New techniques and algorithms continue to emerge while existing approaches are being refined and improved. This section provides only a general overview of the status of machine learning techniques of spatial data mining as they stand at the time of writing. In order to help understand the close link between spatial and attribute-oriented data mining, the discussion classifies spatial data mining techniques in the same way conventional data mining techniques were classified in Section 2.4. Spatial data mining techniques are regarded simply as functional extensions of conventional data mining techniques, constructed on the same first principles but using algorithms designed specifically to handle the characteristics and requirements of spatial data and spatial data mining, as explained below.

3.3.1 Spatial Classification

In essence, the objective of spatial classification is no different from attribute-oriented classification in conventional data mining, that is, to group unlabelled data objects into predefined classes or categories that possess common characteristics, as explained in Section 2.4.1. Conventionally, spatial classification is achieved by the methods of attribute-oriented classification first to categorise individual data objects, and then to merge data objects into the same categories by means of spatial clustering as described in Section 3.3.5.

Chelghoum et al. (2002) propose an integrated spatial classification method that extends the concept of a decision tree to take into account the specific characteristics of spatial data and spatial databases, namely the structuring of spatial data in thematic layers and the use of spatial relationships. The underlying algorithm is able to consider several thematic layers and at the same time extend discriminating criteria to address any neighbourhood effects that may be present. In this way, their method uses a combination of attribute values and spatial relationships of neighbouring data objects to determine the best criteria for classification, rather than merely considering attribute values alone.

The increasing use of real-time spatial data collection methods, for example discrete and continuous feed GPS-based field data loggers and earth observation remote sensing, have resulted in substantial amounts of spatial data being collected continuously over time. These data collections are, in general, called *spatial data streams*. Because of their dynamic nature and extraordinarily large volume, spatial data streams typically cannot be classified effectively in a reasonable amount of time using existing methods. To overcome this problem, Ding et al. (2002) developed a new method for decision tree classification on spatial data streams using a data structure called a *Peano Count Tree* (P-tree). This is a spatial data structure that provides a lossless compressed representation of a spatial data set and facilitates efficient classification as well as the use of other data mining techniques. Experimental results show that the P-tree method is significantly faster than existing classification methods, making it a preferred method for the mining of spatial data streams.

It is also important to note the potential the method of *spatial aggregation* (SA) (see Yip and Zhao, 1996) has for spatial classification. Briefly, SA is a generic framework for organizing computations around image-like, analogue representations of physical processes in data interpretation and control tasks. It transforms a numerical input field (such as temperatures recorded across some spatial extent) to successively higher level descriptions by applying a small set of operators to each layer, given appropriate metrics, neighbourhood relations, and equivalence relations. SA is implemented using *Spatial Aggregation Language* (SAL), which provides abstract data types (ADTs) and interface operators for constructing field-based transformations. The language consists of a C++ library of component implementations from which a user can mix-and-match and customise components for particular applications, including data mining.

3.3.2   Spatial Prediction

In general, spatial prediction is any prediction method that incorporates spatial dependence. The regression methods for prediction in attribute-oriented data mining noted in Section 2.4.2 can be logically extended to include spatial data mining. A commonly used extension of classical ordinary least squares regression is the *generalized linear model* (GLM). This extension differs from classical regression methods by allowing the modelling of responses (dependent variables) that are non-normally distributed and, therefore, should not be modelled within the classical Gaussian analysis framework (Lehman, et al., 2002). With GLM, predictors (independent variables) are introduced in a model essentially in a linear or curvilinear form as in classical regressions. The *generalized additive model* (GAM) is a non-parametric extension of GLM that allows the introduction of non-linear responses to predictors. The shape of the response form is driven by the data and is not predefined, thus allowing the study of response shape and potentially improving predictions by allowing modelling to be closer to and more representative of the actual data (Lehman, et al., 2002).

Recent extensions of the GLM, which like all regression analyses is a *global data analysis technique* that operates on variances calculated by averaging data and assuming that estimated parameters are constant across the study area, include the development of a new form of regression analysis, namely *geographically weighted regression* (GWR) that explicitly includes the spatial dimension in the model (Fotheringham et al., 2002). GWR removes the need to assume that any geographic variation is confined only to the error term ε by explicitly including additional terms (u,v) for the location of data points in space such that the basic model can be rewritten as $y(u,v) = \beta_0(u,v) + \beta_1(u,v)x_1 + \varepsilon(u,v)$. This approach uses a weighting scheme that allows data points that are closer together in space to be weighted more heavily that those that are further away. Moreover, parameter estimates can be mapped in 2- or 3D space and other diagnostic measures applied, such as local standard errors, local measures of influence and local goodness of fit.

GWR has not been widely used to date as an exploratory or predictive approach to spatial data mining, however its substantial potential for discovering important local effects that reveal significant spatial variation in large spatial databases suggests that this technique will grow in importance over time.

The method of classical *trend surface analysis* (TSA) is an advanced method of spatial analysis that is useful for spatial prediction. TSA is the term originally used for the general family of mathematical and graphical techniques widely used in a number of fields such as geophysics, geology and geography as a data exploration tool (Chorley and Haggett, 1965). The concept of TSA is based on the assumption that the spatial distribution of a particular phenomenon can be represented by some form of continuous surface, usually a defined geometric function. It is assumed that an observed spatial pattern can be regarded as the summation of such a surface and a 'random', or local, term. The surface is a function of the two orthogonal coordinate axes. Mathematically, this can be represented by $Z = f(x, y) + \varepsilon$, where the variate Z at the point (x, y) is a function of a position relative to the two coordinate axes, plus an error term ε. This expression is in fact the generalised form of GLM noted above in which the function $f(x, y)$ can be expanded with various terms to generate polynomial equations that define the trend surface. This technique is widely used in various spatial data applications as its construction coincides with an explicit underlying spatial framework.

3.3.3   Spatial Class/Concept Description

Spatial class/concept description can be expressed using spatial characteristics and spatial discriminant rules in the same manner as their attribute-oriented counterparts explained in Section 2.4.3. A spatial characteristic rule is a general description of a set of spatially referenced data objects, for example, the characteristic land use structures of North American cities. A spatial discriminant rule, on the other hand, is a general comparison of the contrasting or discriminating features of a class of spatially referenced data objects from other classes, for example, a comparison between tundra, boreal and Carolinean forests in Ontario.

In essence, spatial class/concept description is a generalisation process. Generalisation inevitably leads to loss of information, but it makes spatial knowledge simpler and easier to comprehend. Spatial data objects are always collected and stored with detailed information at primitive concept levels. In order to discover potentially interesting spatial knowledge hidden in the detailed information it is often necessary first to summarise a large spatial data set at a high concept level and then apply spatial data mining techniques on the generalised data. This process of turning detailed information at a primitive concept level into knowledge at a high concept level is called *generalisation-based mining*.

There are two approaches to generalisation-based mining (Koperski and Han, 1995):

- *Spatial-data-dominant generalisation.* In this approach, spatial data are first generalised according to a specific spatial concept hierarchy (for example, the administrative structure of a country) defined by the analyst or subject expert. The generalised spatial data objects (that is, the merged regions) are then used to cluster the non-spatial data together. After generalisation of non-spatial data, every region can be described or compared at a high concept level using one or more spatial predicates.

- *Non-spatial-data-dominant generalisation.* As its name implies, this approach starts with the generalisation of task-related non-spatial data objects based upon their format. In the case of numerical attribute data, this involves the conversion of discrete values into range values (for example, from 6°C, 4°C and 8°C to a class of 0-to-9°C) or descriptive concepts (for example, from Cold to Somewhat Cold). In the case of categorical attribute data, this involves the conversion of symbolic values to high level concepts (for example, low-value, medium-value and high-value residential land values into a single residential land value class). The next step is to merge spatial data objects with identical high-level values into larger spatial data units. The end product of the generalisation is a spatial data set containing a smaller number of spatial data objects with high-level descriptions.

Spatial class/concept description, therefore, is typically a 2-pronged process that involves a generalisation and a characterisation or discrimination phase. In some cases, the spatial concept hierarchy used for spatial-dominant generalisation may not be available *a priori*. If this is so, it is necessary to subject the data set to spatial clustering analysis as described in Section 3.3.4, and as a result the spatial class/concept description becomes a 3-pronged process.

3.3.4 Spatial Association

Korperski and Han (1995) defined the rule of spatial association as $P_1 \wedge P_2...... \wedge P_m \rightarrow Q_1 \wedge Q_2... \wedge Q_n$ (c%, s%), where at least one of the predicates $P_1,.......P_m$; $Q_1,.....Q_n$ is a *spatial predicate*, and c% and s% are *confidence* and *support* as defined in Section 2.3.4. There are various kinds of spatial predicates that can be used to constitute a spatial association rule, including topological relations (for example, intersect, overlap, disjoint), spatial orientation (for example, east_of, left_of) and distance expressions (for example, close_to, far_away_from). In a typical spatial data base, a large number of associations may exist among objects but only a small number of them are of real significance or interest to users. Thus, Korperski and Han (1995) suggested that it is necessary to use two thresholds, namely *minimum confidence* and *minimum support*, to filter out associations describing a small percentage of objects and rules with low confidence. They proposed a top-down tree search approach that made use of a spatial concept hierarchy to control the spatial association rule mining process by progressively eliminating uninteresting associations.

Co-location is a special type of spatial association. It is defined as the occurrence of two or more spatial objects at the same location or at significantly close proximity to one another. Co-location differs from ordinary spatial associations in that there is no natural notion of a transaction between the antecedent and consequence spatial objects, and user-defined neighbourhood information is an important factor in constructing co-location rules (Shekhar and Huang, 2001).

Spatial autocorrelation is the method used in exploratory spatial data analysis to measure the correlation of a variable with itself (that is, the relationship among multiple occurrences of values on the same variable over space). Spatial autocorrelation is present when occurrences of similar values cluster together spatially. There are two common measures of spatial autocorrelation. One of these is Moran's I, which has a value between -1 and $+1$, where a positive value implies similarity among nearby or neighbouring data objects. A negative value implies dissimilarity, and a zero value indicates independent and random distribution of data values. The other commonly used measure is Geary's C, which is suitable for use in the analysis of aggregated spatial data, such as population statistics reported by census tracts and land use figures reported by planning zones. Spatial autocorrelation is used to measure the strength of the relationships among spatial objects of the same type. It helps to uncover the extent to which the occurrence of an event or feature at a certain point in space will constraint, or make more probable, the occurrence of another event or feature in its neighbourhood. Thus, spatial autocorrelation analysis is particularly useful for knowledge discovery about spatial association (Griffiths, 2003).

Up until recently, all spatial association mining techniques assumed that the data to be explored are represented in a single table of a relational database. Such an assumption obviously imposes serious limitations on spatial mining because spatial data for the most part are organised using multiple, related tables. Lee (2004) proposed a multivariate or multi-relational association approach that looks for patterns involving multiple tables within a relational spatial database environment. Extending a single-table data mining algorithm to a multiple-table one is by no means trivial, and the process is much more computationally demanding with spatial data mining than with attribute-oriented data mining.

3.3.5 Spatial Clustering

The objective of spatial clustering is to find the optimal number of clusters, defined in the same way as in Section 2.4.5, within a given data set and where the clusters are, thus providing knowledge about the overall spatial distribution patterns of objects in the data set. There are two approaches to spatial clustering (Koperski and Han, 1995):

- *Spatial-data-dominant clustering.* In this approach, task-relevant spatial data objects such as points and polygons are first grouped into clusters using an efficient clustering algorithm. An attribute-oriented induction is then performed for each cluster to extract rules describing its general properties.
- *Non-spatial-data-dominant clustering.* In this approach, attributes of task-relevant data objects are first generalised to a high concept level, which are then used as the basis for clustering the spatial data objects.

Han et al. (2001) suggested that many of the algorithms developed for conventional attribute-oriented data mining can be generally applied or adapted for spatial clustering. For example, traditional partitioning methods such as k-means are able to capture simple distance relationships and are therefore useful for spatial data mining. Similarly, density-based methods, which define clusters as regions of homogeneous characteristics, can be used to detect clusters of arbitrary shapes.

There are also clustering techniques that are especially useful for spatial data mining applications. These include grid-based methods for raster spatial data and constraint-based methods that allow the inclusion of spatial restrictions on the clustering process. Such methods are useful for establishing physiographic and ethnocentric constraints such as physical obstacles like rivers and mountains as well as human barriers in the form of international borders.

3.3.6   Spatial Outlier Analysis

Section 2.4.6 defined an outlier as an observation whose value appears significantly different from those of the rest of the observations in a data set. If the observation is geographically referenced, it is a spatial outlier. Practically all contemporary techniques of spatial outlier analysis are constructed using anomalies in attributes as the primary means of identifying spatial outliers. More generally, however, a spatial outlier can also be identified by its exceptional size and shape, as well as its specific rate and/or mode of change over time when compared with its spatial neighbours.

Methods of spatial outlier analysis can be generally grouped into two classes:

- *Graphical outlier detection*, which is based on visualisation of spatial data to highlight anomalies.
- *Quantitative outlier detection*, which provides statistical tests to identify data objects with values deviating significantly from those of their neighbours or the remainder of the data set.

Arguing that these approaches might lead to the identification of some true spatial outliers being ignored and some false ones being identified, Lu et al. (2003) proposed a multiple iteration approach whereby each iteration identifies only one outlier and modifies the value of this outlier so that it will not impact the subsequent iterations negatively. Lu et al. (2003) also proposed an alternative non-iterative algorithm that uses the median as the neighbourhood function, thus avoiding the negative impact caused by the presence of neighbouring points with relatively very high or very low values in the dataset.

3.3.7   Spatial Time-series Analysis

Of all the spatial data mining techniques, spatial time-series or spatio-temporal analysis is probably the most complex, logically and technically, to grasp. Spatiality and temporality are two unique dimensions of geography in terms of concept and representation in database systems. However, although these two dimensions are intrinsically and inherently interrelated, conventional database systems and data mining techniques have so far failed to treat them in an integrated way. Data in spatial database systems are typically modelled, collected and structured with little or no consideration of time. Such spatial data can at best be used to discover spatial knowledge that is a snapshot at particular point in time that can hardly be regarded as spatial time-series analysis in any sense of the term.

The *Event-based Spatial-temporal Data Model* (ESTDM) proposed by Peuquet and Duan (1995) signified one of the pioneering attempts to provide a time-based, conceptual-level model for analysing spatial data. Claramunt and Thériault (1995) proposed a similar model with associated query operators. Hornsby and Egenhofer (2000) have also described a visual language, called *change description language*, that describes temporal changes relating to spatial objects. These and several other efforts have provided the fundamental concepts for spatio-temporal data analysis and data mining. However, despite the apparent interest and accomplishment in temporal data mining generally (Roddick and Spiliopoulou, 2002), more work is required to develop a practical approach to spatial time-series analysis. This need is more urgent today than ever before in view of the growing use of spatial database systems in support of location-based services and socio-economic activities in space over time, including migration patterns, urban dynamics, sustainable resource development and global change studies.

3.4   Visualisation Techniques of Spatial Data Mining

Section 2.4 defined the concept of visualisation and explained the merits and limitations of visualisation as a method of conventional data mining. In the context of spatial data mining, the applicability of visualisation must be re-visited because it provides the most intuitive way of interpreting spatial data and presenting the results. Visualisation does not require the high level of understanding required to interpret complex mathematical algorithms or parameters. It is particularly effective for dealing with heterogeneous and noisy spatial data.

Adrienko and Adrienko (1999) noted that visualisation can be used in different phases of spatial data mining. It can be used to pre-process a data set with a view to exposing extreme or strange attribute values that may be errors and require validation or verification. During data mining, visualisation can be used to display intermediate results, thus enabling the analyst to interact with the system and adjust the course of action if necessary. Because of the nature of spatial information, it is always easier to detect spatial characteristics in patterns and trends visually than interpreting them in the form of statistical tables. Visualisation therefore also plays a significant role in the evaluation and interpretation of results.

On the basis of these observations, Adrienko and Adrienko (1999) suggested that visualisation should be an integral part of spatial data mining. There are two visualisation-based approaches to spatial data mining in such an integrated environment:

- A *visualisation-dominant* or *"geography-to-mathematics" approach*, in which the analyst first evaluates the data by visualisation (for example, using dynamic maps) to obtain some geographically interpretable information, and then validates these results using spatial data mining methods prior to their use for knowledge construction.
- A *data mining-dominant* or *"mathematics-to-geography" approach*, in which the analyst starts with spatial mining methods to explore the data and then uses visualisation for an in-depth analysis of the discovered spatial knowledge deemed interesting to the user.

In order to support the integration, it is necessary to have a dynamic link that connects the data mining and visualisation components and allows the analyst to interact with these components simultaneously.

Ribarsky et al. (1999) used the term *discovery visualisation* (DV) to refer to a human-centred visual data mining approach for clustering. DV promotes the concept of continuous interactions between the analyst and the machine. The process of DV starts by automatically generating a visual overview of the data using a fast clustering method. The resulting clusters are then followed as time-dependent features by the analyst who interactively selects the cluster regions for in-depth analysis and improvement.

More recently, Guo (2003) proposed a similar human-centred spatial data mining environment incorporating a coordinated suite of computational and visualisation methods to explore high-dimensional data for uncovering patterns in multivariate spaces. This environment is characterised by:

- An interactive feature selection method for identifying potentially interesting, multidimensional subspaces from a high-dimensional data space.
- An interactive, hierarchical clustering method for searching multivariate clusters of arbitrary shape.
- A suite of coordinated visualisation and computational components centered around the above two methods to facilitate a human-led exploration.

There are many visualisation software packages that can be used for spatial data exploration (Kiem, 2001). Some of these have been developed commercially and others as prototypes resulting from academic research. Experiments by Adrienko and Adrienko (1999), Ribarsky et al. (1999) and Guo (2003) have demonstrated that an integrated approach is viable, effective and efficient. Such an approach leverages human visual cognition on immediate spatial knowledge during the spatial data mining process. It combines the computational power of the computer and the knowledge of the analyst to create spatial data mining solutions that cannot be accomplished as effectively and efficiently by using machine learning techniques alone. However, merging two relatively complicated technologies is not a trivial technical task. Both spatial data mining and visualisation are computationally intensive processes that have not yet been implemented cost-effectively on the desktop environment for popular use. As a result, total integration in spatial data mining has up to now remained more an ideal than a reality.

3.5 Implementation Issues of Spatial Data Mining

Figure 11.7 illustrates the reference model of implementing spatial data mining proposed by the *Cross-Industry Standard Process for Data Mining*

(CRISP-DM, 2000), a consortium formed by the world's leading data mining companies and sponsored by the European Commission to develop an industry-neutral and tool-neutral process model. The CRISP-DM reference model provides a framework for carrying out data mining projects, the success of which is dependent as much on the use of technology as it is on human expertise, as explained in the earlier discussion.

From a project management perspective, the CRISP-DM reference model serves as a common conceptual framework for project sponsors, end users and systems staff to discuss spatial data mining and to increase for all participants the understanding of important issues. It enables systems staff to develop an implementation strategy using industry standard and contemporary software engineering practices. At the same time, it allows project sponsors and end users to have more reasonable expectations as to how the project will proceed and what will be delivered at the end.

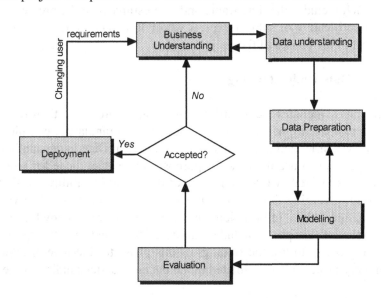

Figure 11-7. The CRISP-DM reference model (Source: After CRISP_DM, 2000, with modifications)

The life cycle of a spatial data mining project as prescribed by the reference model contains six phases, which generally follow the steps of KDD that was discussed in Section 2.2. The sequence of the phases is not strict, as the arrows in Figure 11.7 indicate merely the most important interdependencies of the phases and the flow of information between them. Spatial data mining projects, like all information system projects, is cyclic in nature (see Chapter 9). This implies that the data mining process does not stop when the results are deployed as the experience learned during the mining process and from the deployed solutions can trigger new, often more focused, data mining tasks. The objectives and nature of each of the six phases of the life cycle of a typical spatial data mining project are discussed in the following sub-sections.

3.5.1   Business Understanding

The focus of this initial phase of the project life cycle is on identifying the objectives of the implementation in the context of organisational business requirements. The project team attempts to find out what executive and managerial decisions have to be made, what knowledge already exists to support decision making, and what knowledge is lacking. The project team then converts this understanding of the business into a spatial data mining problem definition. The deliverable of this phase is a preliminary project plan, which includes the time frame and cost estimates of the project, as well as the identification of subject experts who will be able to help one or more of the rest of the phases of the project.

3.5.2   Data Understanding

This phase includes the identification of data sources and their respective quality and usability in terms of accuracy, scale, granularity, completeness, currency, classification and format (see Chapter 7). Data understanding also entails the right to use the data for purposes other than those for which the data were collected, while being mindful of the possibility of violating legislation governing freedom of information and protection of privacy for exposing personal information in the discovered knowledge. The deliverables of this phase include data description and quality reports. It is common practice to use the findings of this phase to fine-tune or adjust the preliminary project plan completed in the business understanding phase.

3.5.3   Data Preparation

Data preparation includes all activities required to obtain one or more spatial data sets that are ready for use by data mining. Since the quality of the spatial knowledge discovered is directly related to the quality of the spatial data source, most of the data preparation activities are concerned with data cleaning that aims to assure the quality of the data with respect to the requirements of the spatial data mining tasks identified in the previous phase. Typical data cleaning activities include eliminating or correcting erroneous data objects, removing redundant data items, resolving geometric or attribute conflicts, and combining data from different tables or sources into a single data set in a unified scale or classification scheme.

An important issue in data preparation is concerned with imprecise (also referred to as uncertain) and incomplete data, which are often the norm rather than the exception in spatial databases. Methods that employ the use of *fuzzy set theory* and *rough set theory* have been used to deal with imprecise and incomplete data sets respectively (Han and Kamber, 2001). Fuzzy set theory models imprecise geographic and class value boundaries by allowing the concept of partial class membership rather than strict Boolean class membership. This makes it possible for an observation or spatial object to have partial membership in multiple classes rather than only one exclusively (Kruse et al., 1994). Rough set theory, on the other hand, provides the mathematical foundation for eliminating incomplete data under rigorous computational control such that the loss of information is kept to a minimum (Hu et al., 1997). Where necessary, imprecise and incomplete data sets can be cleaned by using these methods to make them useful for discovering spatial knowledge with a known level of certainty or confidence.

3.5.4   Modelling

Modelling is the spatial data mining phase when various techniques are applied to the data set being mined. Since spatial data mining is more often than not interactive in process, modelling requires expertise and skills in spatial data mining techniques as well as a good working knowledge in particular application domains. There are several implementation issues that need to be considered when applying a particular modelling technique, for example:

- The suitability of the particular data mining techniques with respect to data format and the objectives of deployment, since individual techniques are designed for different types of data and for discovering spatial knowledge suitable for different decision making purposes.
- The choice of algorithm, since the same spatial data mining technique, for example spatial prediction, may be constructed using different algorithms, each with its own merits and limitations.
- The sequence of applying spatial and attribute-oriented techniques when both are required for a particular task (for example, generalisation-based mining).
- The sequence of applying different techniques when more than one technique is required to obtain different aspects of the inherent spatial knowledge that is required to support decision making.

- The interface requirements that will best facilitate interaction between the analyst and the computer during the data mining process.
- The points at which the analyst can intervene to change input parameters or to change from one strategy to another.

3.5.5   Evaluation

Before the results of modelling are returned to the end user, it is important for the analyst to evaluate thoroughly the discovered knowledge with respect to the objectives of the spatial data mining task as identified in the business understanding phase. There are two key considerations in this context. The first is to identify the knowledge that is of real interest to the end user by filtering it out from hundreds or even thousands of rules and/or visualisation displays. The second is to examine critically the results against the objectives to ascertain that all important aspects of the business have been adequately considered.

3.5.6   Deployment

The discovered knowledge is seldom delivered to the end user directly in the form that it was originally generated. With the exception of IF-THEN induction rules and decision trees, spatial data mining results are not always easy to comprehend and understand. Hence, the discovered spatial knowledge needs to be organised and presented in a way that the end user can apply it to support decision making, as explained in the following discussion.

4.   SPATIAL DECISION SUPPORT CONCEPTS, SYSTEM COMPONENTS AND APPLICATION

The concept of a decision support system (DSS) is defined by Gorry and Scott Morton (1971, p. 57) as "a type of interactive computer-based system that helps decision makers to utilise data and models to solve unstructured and semi-structured problems". This definition is sufficiently general to remain valid to describe the newer generation of DSS that have appeared over the intervening thirty five years. However, the types of systems that fall into this class of software have broadened considerably in their scope, especially with the advent of the Internet. Independent of the type of system, the process of decision making and thereby decision support involves a relatively small number of phases.

4.1 The Phases of Decision Making

The use of computers to assist with decision making was influenced by the work of Simon (1977), who identified two basic types of decisions, namely highly structured or programmed and highly unstructured or non-programmed. Structured decisions involve routine and repetitive processes that are relatively easy to solve using one or more of a series of standard techniques. On the other hand, unstructured decisions are multi-faceted and have no clear cut solution. Simons placed both types of decisions into a general framework that comprised three sequential phases, namely intelligence, design and choice.

Two additional phases, namely implementation and monitoring, can be added to those identified by Simon (Figure 11.8). The decision process begins with intelligence gathering to identify and define a decision problem. A problem only exists if some individual, group or division within an organisation takes on the responsibility of identifying, addressing and resolving it. During this phase the decision maker seeks to determine whether a problem exists and, if so, what its characteristics and magnitude are before it is defined. To establish this may involve assembling initial data that relate to the problem, establishing whether required data are available and, if not, how these data can be obtained. Ownership of the decision problem is also established during the intelligence phase.

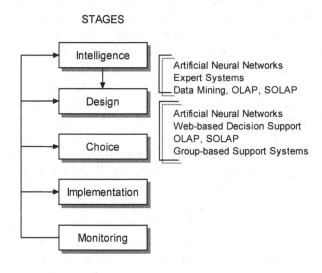

Figure 11-8. Phases of decision making

During the design phase, participants create a model of the decision problem by refining and constructing relationships between decision components identified during the intelligence phase. The decision model takes form during this phase and criteria are set for evaluating alternative courses of action. Further, potential alternative solutions may be suggested by the form of the model, and the model itself may be recast by identifying alternative solutions. As shown in Figure 11.8, various techniques, among others, can be used to assist in designing the decision model. Also, techniques discussed earlier in this chapter, including data mining, OLAP and ROLAP and can be used to structure decision alternatives.

Independent of the actual decision model, the design process requires conceptualising the problem and abstracting it to measurable quantitative and/or qualitative forms. The attributes of interest are identified and relationships among them are established. Inevitably during this process simplification is required by making assumptions about the data since it is not possible to measure or know everything with precision.

The extent of simplification is important as simpler decision models suggest lower development costs, easier use, and a faster solution. However, a simple model may be less representative of the decision problem than desired and may result in poor decision outcomes. Hence, it is important to balance simplicity and complexity during the design phase. A strategy often used for this is scenario building, where multiple decision models can be used with varying assumptions to capture a range of possibilities that reveal the implications of changing a parameter or relaxing a decision rule.

After defining one or a series of models to support the decision problem, the choice phase allows selection of a solution to the model, rather than the problem for which the model was constructed. The problem is only considered solved if a solution to the decision problem is successfully implemented. Preliminary model solutions are evaluated to determine their validity or, more specifically, whether they provide a set of reasonable or feasible decision alternatives. Choice is the critical act of making a decision, however the boundary between the design and choice phases is often unclear as it is common for decision makers to iterate between modifying design parameters and making a final choice. For example, it is possible to identify new alternatives that were not initially apparent when performing an evaluation of existing options.

Once proposed outcomes seem reasonable, the decision model can be implemented. A successful implementation can only be evaluated after a decision is taken and acted upon. Often an evaluation may not be feasible until some time has passed, and some decisions may not warrant post-implementation evaluation. However, it is usually prudent to engage in a process of evaluation so lessons can be learned for similar decision processes in the future. If the most feasible solution is implemented and subsequently proves to be a failure then the process can iterate through the feedback loop from monitoring, and additional information that may have been neglected can be used to reassess the decision process at any of the earlier phases. Also, it is worth noting that the process can iterate at any of the earlier phases if the decision participants are unhappy with their progress.

Clearly, various approaches to decision support can be employed at different phases of the decision process. The next section discusses the characteristics of DSS.

4.2   Characteristics of Decision Support Systems

Despite the considerable diversity in DSS applications, virtually all operational systems include a core set of characteristics that distinguish them from other types of computer-based systems. Specifically a DSS is distinguished by the following attributes:

- It is a methodology.
- It is computer-based.
- It uses data, typically stored in a database, that relate to a particular problem domain.
- It often includes multiple models and techniques.
- It has an easy-to-use graphical user interface.
- It must be capable of expressing the decision makers' own ideas.
- It is typically iterative and highly interactive.
- It supports all phases of the decision making process.
- It can be used by a single user in a stand-alone environment or networked to run on an LAN or Web-based to run across the Internet.

These attributes accommodate multiple decision support styles, from highly consultative approaches that provide decision participants with decision options which may be discussed, refined and re-evaluated, to highly autocratic approaches where a single optimal or near optimal decision outcome is produced by the system. Decisions are rarely made only once, but rather the same sort of decision is made at various points in time, especially for strategic decision making. Hence, approaches to decision support should be adaptive to changing circumstances and changing needs. This means that a user should be able to add, delete, combine, rearrange or change basic elements considered in the decision process and to be able to move individuals and groups into and out of the decision process as required.

Like any computer-based system that requires high levels of user interaction, a DSS must be easy-to-use and intuitive, or else it will quickly fall into disuse and fail to satisfy its objectives. Hence, a well designed user interface is centrally important, especially for systems that are used by non-experts who need information quickly in easily understandable and visual formats, such as graphs and charts. With an emphasis on enhancing the effectiveness (accuracy, timeliness and quality of decisions) rather the efficiency (the cost of making decisions) of decision making, the decision maker must have complete control over all steps in the process. Hence, DSS should always be regarded as a complement to rather than a replacement for a human decision maker.

Ideally, end users of a well designed DSS should be able to construct and modify aspects of the decision process without the need for expert intervention. That is, the system components should function like a set of building blocks that can be assembled visually, depending on the decision task, and used by non-experts as desired. However, not all decision problems can be addressed in this manner and DSS often require some form of expert intervention, especially during the design phase. Even in cases where expert intervention is involved, modular systems that allow end user experimentation with inputs and outputs are likely to be more successful in achieving their purpose than complex and less easily manipulated systems (Uran and Jansen, 2003).

4.3   Decision Support System Components

Although a DSS can take on many possible designs, several core components are required (Figure 11.9). These include the user interface,

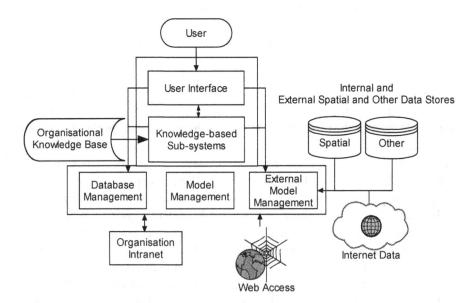

Figure 11-9. High-level decision support system components

knowledge-based sub-systems including an organisational knowledge base, a model management sub-system, external models that may be used during the decision support process, and a database management sub-system.

The user interface includes not only use of the hardware, software and databases that comprise the DSS but also important user factors that relate to ease-of-use and general human-machine interaction. In fact, Whitten and Bentley (1997) suggest that the user interface is so fundamental to a DSS that, from the user's perspective, it is 'the system' in much that the same way an Internet browser is 'the Internet' for many users. In addition to providing flexibility in interaction with the system the user interface should provide users with access to help by example, where the software guides users through the input and modelling process. In addition, some systems actually track, store and allow DSS developers to analyse user interactions with a system in order to make improvements to the system's usability.

Users of DSS are typically either senior managers, who carry decision making responsibilities, or a group leader who has responsibility to generate feasible decision alternatives. In many organisations it is also common for an IT specialist or data analyst to act as an intermediary or decision support agent in DSS use. As noted above, use of an expert technical intervention in the decision support process can allow senior managers to benefit from the use of a DSS without actually ever having to learn how to use the system themselves.

There is a growing trend, especially with the use of spatially enabled DSS, to achieve broad-based participation in decision making by facilitating access for the general public or specific stakeholder groups in decision processes. This is increasingly implemented across the Internet, where the public can become involved as decision contributors using an on-line tool operating through a standard Web browser.

Both the user interface and knowledge-based sub-systems shown in Figure 11.9 interact with the data management and model management components, as well as with models that are external to the DSS. The data management sub-system contains the DSS database. This may be wholly or partially contained within a corporate data warehouse. Alternatively, it may be a conventional relational DBMS, an object-oriented DBMS, or it may be some form of hybrid database system that draws upon all or some of the above, allowing access to networked data stores that are internal to the organisation, as well as external data on the Internet. In decision problems that involve spatial decision support, either all aspects or some aspects of the DSS database will be spatially enabled with digital maps and other data that are locationally referenced.

In general, the database management component of a DSS performs a number of functions, including the following:

- Facilitates entry or extraction of data for inclusion in the DSS database.
- Allows update (addition, deletion, edit, and change) of data records and files.
- Manages data dictionary and metadata entries.
- Integrates data from different sources and in different formats.
- Retrieves data from the database primarily using SQL for queries and reports.
- Facilitates spatial and other analyses through interaction with a model base and/or analysis toolbox.
- Provides full security against unauthorised access to protect user-specific scenario results and personal data (such as user profiles).
- Provides backup and recovery against corruption or system crash.
- Tracks user interaction with the data via the DSS user interface.

The model base in a DSS typically contains standard as well as custom-programmed (either through scripting or macro programming) statistical, forecasting, simulation, and evaluative functions. Here, the ability for a user to select, run, integrate, evaluate (numerically and visually) and change the parameters of models and functions is critical to the success of the system. Generally, four types of models exist with a DSS model base, namely:

- *Strategic models*, which are used to support mid- to long-range decision making on issues such as organisational restructuring, locational decision making, and policy directions.
- *Tactical decision support models*, which relate more to shorter term decisions such as the allocation of staff or resources within an organisation and the deployment of resources such as community policing and traffic management.
- *Operational decision models*, which are typically used to support short-term or day-to-day issues that relate primarily to the internal functioning of an organisation, such as the allocation of time and money to staff training and other activities. Hence, the focus is more on daily management that is neither strategic nor tactical in nature.
- *Analytic models*, which cut across aspects of strategic, tactical and operational decision analysis.

The model management sub-system should be easily accessible from the user interface to allow rapid assembly of model components. While many of the DSS that are in use are stand-alone or networked, there is a growing trend toward use of the Internet and Web as the primary enabling technologies for delivering decision support functions to organisations

(Bhargava and Power, 2001). This has required developing new approaches to DSS design and implementation, as discussed in the following section.

4.4 Web-based and Web-enabled Decision Support Systems

Networked deployment of DSS is now more common in the workplace than stand-alone systems (Power and Kaparthi, 2002). Modern enterprise-level database management systems provide out-of-the-box functionality for decision support tasks, running either across a controlled Intranet, a controlled Extranet, across the Internet using secure logon and authentication, or via an open interface that is accessible to any potential user. Three primary forms of DSS architecture characterise Web environments, namely:

- A thin client Web browser interface and TCP/IP to link to a back-end server which hosts the information infrastructure that drives the decision support capabilities. Server-side scripting and technologies including CGI, Java, *active server pages* (ASP) or *Java server pages* (JSP) are used to enable the connection, process and route requests to allow client workstations to interact with the decision support components that reside on the server. Emphasis in this context is put on server functionality and the client browser simply provides universal access to the information infrastructure.
- A fat client that allows DSS applications to transfer small parts of the processing load to the client computer through the use of *Java applets*, *Active X controls* or *browser plug-ins* that must be installed on the client computer for the application to work. This approach seeks to move some of the processing requirements from the server to the client computer and thereby improve network performance by reducing the information transfer load. While speed of network response is not so much of a constraint on a dedicated Intranet, it may be a significant issue for the analysis of decision problems that are data-centred and computationally intense, especially across the Internet.
- A distributed approach that manages aspects of the DSS components installed across multiple Web, application and database servers using one or some combination of CORBA, DCOM, *Java remote method invocation* (RMI), or *Enterprise Java Beans* that form part of the Java 2 Enterprise Edition platform for developing and running distributed multi-tier applications. This is a far more complex architecture that allows building block components to be pieced together by the decision

analyst(s) as required, independent of where the components may reside on a network.

Although the Web is a common and central element of the above architectures, it is important to distinguish between Web-based and Web-enabled DSS. The former refers to systems that are implemented entirely using Web technologies. In contrast, the latter uses network-based legacy components, especially associated with the previously discussed database or model management sub-system, of the computing infrastructure to host parts of the DSS. The user accesses these legacy components and interacts with them via a Web browser. Hence, the Web is co-incidental to the DSS architecture rather than central to it. Power and Kaparthi (2002) view Web-accessible or enabled DSS architectures as 'quick and short-term fixes' to utilise the possibilities offered by the Internet which can lead to piecemeal and unsatisfactory results.

Designing and implementing a fully fledged Web-based DSS is preferable, especially in the context of an enterprise information infrastructure, to a basic two-tier client-server Web-enabled architecture running in a networked environment. This is because while a Web-enabled approach may be expedient for a customised and specialised DSS that runs somewhat independently of an overall information technology infrastructure, it is likely to lack the robust and integrated architecture that characterises a Web-based enterprise-level DSS.

Most commercial Web-based DSS are designed around a three or higher tiered architecture. In the three tier case, the simpler two tier model is supplemented by the insertion of a middle dynamic processing tier in which decision support functions are processed by application scripts or packages running on a dedicated Web server. This server interacts with one or more separate but networked database servers in the third tier of the system (Figure 11.10). The three tier architecture offers much more flexibility in the way that client requests are processed and allows communication for processing queries and analysis with the back-end database(s).

The three tier model is extensible and can be expanded in the second tier by creating multiple layers which may each perform specialised decision support functions. Further, multiple application servers can also be introduced to perform additional specialised decision support functions. This added complexity may require the introduction of one or more *transaction processor (TP) monitors* or a *proxy server* that forwards connection requests to the database server or to other proxy servers to balance the processing load. Second tier tools such as Oracle's Connection Manager also allow *multiplexing* of database access in order to channel multiple client sessions through a single network connection, thereby increasing the number of concurrent users without compromising system performance. More complex architectures using the *Internet Inter-Object Request Broker Protocol* (IIOP) can also be invoked in the second tier so that object-oriented and distributed programs written in different programming languages can communicate within the overall DSS environment.

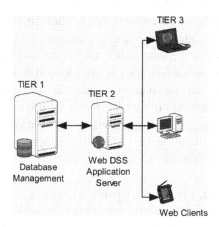

Figure 11-10. Three tier web-based DSS architecture

Current developments in the Java family of tools, open source server-side scripting and object broker technologies have opened many new possibilities for Web-based computing. These developments are rapidly finding their way into DSS technologies, however the core functions and basic purpose of a DSS remain essentially true to Gorry and Scott Morton's (1970) definition and the first systems developed during the 1970s. The enabling environment has changed considerably, but a DSS must be usable and must be proven to be successful in supporting operational decisions for it to be deemed successful.

In addition to the many operational DSS that have been developed in widely ranging fields, such as clinical use in medical diagnosis (Coiera, 2003), air traffic control and airline planning (Yau, 1993) and supporting the information needs of government departments (Wild and Giggs, 2004), there has been considerable and sustained effort invested in developing systems that use spatial data to assist with decision processes involving issues associated with space and place. The inclusion of spatial data and the assessment of spatial decision problems is discussed in the following section.

5.   SPATIAL DECISION SUPPORT SYSTEM APPLICATIONS

Decision support concepts were introduced into the spatial domain in the early 1990s. Complex and poorly defined decision problems that characterise aspects of urban and regional planning (van Leeuwen and Timmermans, 2004) as well as numerous other application areas are well suited to the spatial data processing functions of GIS. Between the first discussions of SDSS in the late 1980s (Densham and Goodchild, 1989) and the current time, the use of GIS for spatial decision analysis has evolved into an important driver of innovation. Since it is not possible to review comprehensively all aspects of spatial decision support, the main aspects of the previous discussion of DSS are reviewed within the spatial domain, and two types of operational spatial decision support systems (SDSS) are described in the final section.

5.1   Spatial Decision Support Systems

Numerous SDSS applications have been developed in the past fifteen years, including Web-based SDSS, collaborative SDSS, spatial knowledge-based SDSS, environmental DSS, and group SDSS among others. As noted above, the objectives of SDSS are fundamentally the same as in the non-spatial domain, however the nature of the decision problems that SDSS address are somewhat different, given the fundamental nature of space and spatial problem solving (Densham, 1991).

Several factors distinguish the nature of spatial decision problems from other types. Specifically, they are characterised by:

- A large and open-ended number of possible decision alternatives exist relative to non-spatial problems that typically involve a small and fixed number of alternatives.
- Each spatial alternative is likely to be characterised by a large number of criteria, none of which may be clearly defined at the outset of intelligence gathering.
- Some of the evaluative criteria may be inherently qualitative and difficult or impossible to measure, while others may be easily measurable with quantitative data.
- In most cases decision problems need to be assessed by multiple participants, each of whom may have varying capability of understanding and using spatial data, as well as a different personal knowledge base or perspective that will govern their point of view on decision alternatives and their evaluation.

- It is likely that spatial decision problems will involve the participation of groups as well as individuals in collaboration with each other.
- Most spatial decision problems have a high degree of uncertainty in terms of the likely consequences of decision outcomes.
- The outcomes of spatial decision problems, given their nature, are spatially variable, hence solutions that are globally applicable are often hard to determine.

These factors each influence the phases of decision support described in Section 4. In the intelligence gathering phase GIS and remote sensing software play important roles for problem scoping as they allow decision makers to scan various data sets that relate to a geographic location (or multiple locations) for which a decision needs to be made. This could be a decision such as where to locate instruments to gather data on water quality in a river's watershed, or where and how much land of given uses should be released for development at the periphery of an urban area. Presenting this information in map form so that it is easily visualised offers considerable advantages over other forms of data visualisation during the preliminary phase of the decision support process.

The design phase involves structuring and assembling, with the assistance of various decision support tools and models, feasible alternatives or strategies that can be evaluated prior to choosing the best alternative. Current spatial data processing software has both strengths and weaknesses in this regard. Numerous researchers have noted that most current proprietary GIS packages, while strong in standard data manipulation and integration tasks, lack the sort of modelling and evaluation tools that decision support software requires (Densham, 1991; Feick and Hall, 1999). Hence, the use of commercial GIS software in spatial decision support has often been limited to arriving at a set of decision alternatives.

An alternative strategy is to develop a SDSS that uses open data standards and is either stand-alone, networked, or Web-based, depending on its intended use and users. This strategy involves a significant amount of programming and development work over a potentially lengthy time period to produce a tool that may be used only periodically. Hence, there are relatively few examples of independently programmed and generically applicable SDSS in routine use.

In each of the cases described above, most spatial decision problems and their solutions are difficult for users lacking moderate to high levels of technical skill and knowledge to address. Hence, the principles of accessibility and user friendliness remain obstacles for SDSS, although as applications continue to be ported to and developed in Web environments these constraints will become less restrictive.

The choice phase of spatial decision support uses the outputs of the two previous phases to focus on the selection of a preferred alternative. Current spatial decision support software falls somewhat short of integrating techniques for the identification of decision alternatives with their evaluation. Although substantial inroads have been made in the past half decade to achieving inter-operability in spatial data access and processing, many of the core GIS software packages do not currently incorporate evaluative techniques such as spatial *multi-criteria analysis* (MCA) for assessing decision alternatives (Jankowski, 1995; Malczewski, 1999; Ascough II et al., 2002; Feick and Hall, 2004). Although there are numerous examples of the use of these techniques in the spatial decision support literature (see for example, Malczewski, 2006), they are predominantly case study-oriented and are not normally packaged to run within an enterprise-level infrastructure as part of a SDSS.

The lack of MCA and related evaluation tools within mainstream GIS products is not surprising overall, since other DSS techniques such as data mining have only relatively recently attracted the attention of spatial data researchers (Miller and Han, 2001; Fisher, 2005). Moreover, spatial extensions of OLAP and relational on-line analytic processing (ROLAP) are still comparatively new tools in the spatial data domain (Rivest et al., 2001). However, the increasing inroads of leading enterprise database companies such as IBM and Oracle into the spatial database industry will likely see closer integration and packaging in future years of a variety of spatially enabled decision support functions.

Two important and related developments in SDSS have invigorated interest in their use and refocused the need to include spatial decision support functions within an organisation's database environment. The first concerns the fact that SDSS must be flexible enough to incorporate the inputs of multiple decision participants including individuals as well as group. This requirement has given rise to considerable interest in collaborative GIS and group SDSS (Jankowski and Nyerges, 2001a, 2001b; Balram and Dragicevic, 2006) and is, in part, associated with the substantial current interest in the concept of public participation GIS (PPGIS) (see Chapter 12). The second development is a function of the migration of spatial data applications in general and GIS in particular to the Web. Web-based and Web-enabled architectures for spatial data management are discussed extensively elsewhere in this book and are the focus of dedicated textbooks that are rich in technical detail (see for example, Green and Bossomaier, 2002; Peng and Tsou, 2003).

Rinner and Jankowski (2002) describe the technical foundations and applications of SDSS using Web technologies. However, the recency of the use of the Web for this purpose means that there is only a modest number of comprehensive Web-enabled or Web-based SDSS relative to the large number of classical DSS that exist in the non-spatial data domain. Despite this, examples can be found in the spatial domain for each of the Web DSS architectures identified in Section 4.4, with distributed SDSS being somewhat under-represented relative to the others.

The next section takes two different examples of SDSS architectures and approaches and briefly describes their implementation and function.

5.2 Spatial Decision Support System Applications

To complete the discussion of SDSS, two applications are described. It is not possible to discuss these in detail, hence issues relating to the design and choice stages of decision making outlined in Sections 4.1 and 5.1 are the focus of interest.

5.2.1 Stand-alone Spatial Decision Support for Multiple Participants

Most of the applications of SDSS reported in the literature have been implemented within commercial GIS software as customised extensions, and are written in a variety of languages such as Arc Macro Language (AML), Avenue and more latterly Python, Visual Basic and VB.NET. A relatively smaller number of applications have been designed and programmed as stand-alone systems that utilise industry standard spatial data formats, such as ESRI *shape files*, and a commercial relational database system such as Microsoft Access. One such tool, named TourPlan, was developed in the 1990s by researchers at the University of Waterloo, Canada to assess and balance multi-participant views on the suitability of cadastral land parcels for development purposes (Hall et al., 1977; Feick and Hall, 1999, 2004). The tool was intended for use in small island developing nations and was prototyped using data from the Cayman Islands. It was used initially to assess consensus among a purposively selected sample of participants from various sectors of Caymanian society on the suitability of land in the District of North Bay for future tourism development.

TourPlan was designed to run on a modestly powered, stand-alone or networked personal computer. The software, written in Visual C++ for map geometry management and manipulation and Visual Basic for all interface components as well as database interaction and numerical data analysis, was designed to fulfil two primary tasks. First, functions had to be written to manage map display and geometry relative to basic and more complex spatial object manipulation and analysis. Combined with these functions the software was designed to allow multiple users to select, using a wide variety of interactive techniques, land parcels with tourism development potential under various user-defined development scenarios (Figure 11.11). This stage of use of the tool coincides with both the intelligence and design phases of decision support outlined earlier.

Functions were programmed to retain user site selections within a user-defined scenario file. In addition users could specify through an intuitive interface which parcels were tagged for which development purposes in their selected set over pre-set time periods. For example, a user could specify parcels that were immediately developable for tourism, others that were developable within 2 to 5 years, other parcels that should be developed for commercial land use, and other parcels for which any tourism development was vetoed (such as environmentally sensitive or culturally important land) indefinitely.

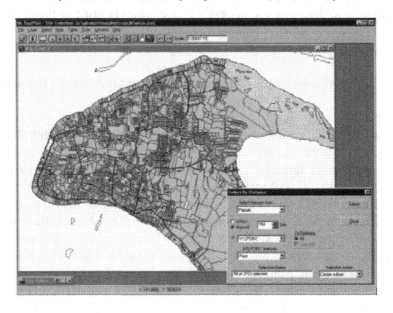

Figure 11-11. Selection of cadastral parcels by distance criteria (Source: Feick and Hall, 1999)

Once a participant had created a scenario-based selection set this was saved and merged by the decision analyst into a combined database table containing the selections and associated selection criteria for all participants. Next, for the design and choice phases of decision support, participants first used a series of quantitative evaluation criteria calculated by the software, such as the distance of candidate parcels from sandy beaches or from environmentally protected parcels and defined these as either a cost or a benefit to the decision objective in order to derive attribute criterion weights. One of three available multi-criteria evaluation methods, namely linear weighted summation, net concordance-discordance, and subtractive summation, were programmed to allow evaluation of the suitability of candidate parcels. The evaluation process was programmed to run through a series of intuitive interfaces designed not to confuse users with the complexities of the specific evaluative techniques being used.

Ultimately, the system output from the participant evaluations a new database table that could be filtered to identify the top n or n% parcels ranked by a specific participant, by everyone, or by subsets of participants. These selections were also viewable in map form using intuitive wizards or assistants to help the user present outputs in a format understandable to them. The derived information allowed the decision analyst, planners and planning boards members to develop short-, medium- and long-term land use development strategies that met with the consensual approval on the part of the decision participants.

TourPlan represents a generic approach to a specific problem. However, the decision support strategy and the techniques programmed into the tool can be used in any context for any type of spatial development problem that involves seeking consensus on development options for cadastral parcels or for defined land use areas. However, the tool lacks Web-readiness, interoperability with other tools (although OLE-based connections with spreadsheet and other software were programmed) and, although this was considered in its design and implementation, the relative complexity of using the GIS-based and MCA functions proved to be difficult for non-expert users.

5.2.2   Decision Support with Spatial On-line Analytic Processing in a Web Environment

In contrast to the approach used by TourPlan, much more recent emphasis has been placed on using the Web as a means of facilitating SDSS and incorporating mainstream DSS concepts, ideas and techniques into SDSS design and function. Work by Bedard et al., (2003) and Rivest et al., (2005) from Laval University seeks to incorporate tools derived from data warehousing, data mining and OLAP into the spatial domain. Their research merges BI components of mainstream information technology with spatial information technology using Web map services (WMS) to address a variety of issues in the province of Quebec, Canada.

The decision support objective in the example discussed is to allow epidemiologists, health planners and analysts to achieve rapid exploration of complex spatial, temporal and associated attribute data in order to reduce environmental health risks through early intervention. A commercial Web map server, namely JMap was used to couple tightly the spatio-temporal exploration of a multi-dimensional data cube with an intuitive Web browser interface. Exploration is facilitated through two modules, the first of which allows configuration of the spatial database and the second is a visualisation module that facilitates interactive display and data exploration for the analyst or decision maker (Rivest et al., 2005).

In general, OLAP applications typically run within a three tier SDSS architecture (a multi-dimensional spatial database running in a dedicated DBMS, an OLAP server than can be either Web-based or Web-enabled, and clients that accesses and mine the database via the OLAP server). The fundamental operationalisation of the JMap Spatial OLAP or SOLAP module by Bedard and his colleagues is currently achieved using a ROLAP architecture without an OLAP server, using instead functions programmed as a pseudo-middle tier between the JMap Web map server software and a relational DBMS. The client computers access the SOLAP functions through custom programmed tools accessed in the browser interface of JMap. This approach allows the WMS provided by JMap to be enhanced substantially with highly accessible and easy-to-use tools for knowledge discovery and decision support, including the data mining techniques of *drill down, roll up, drill across, slicing and dicing,* and *swapping* and to produce results in map, tabular, and visual graphic form.

An example of this approach from Rivest et al. (2005) deals with respiratory disease from individual hospitalisation data in the Canadian province of Quebec over a fifteen year period for three spatial hierarchy levels, namely community health centres (local), regional health authorities (regional), and the provincial level. This example serves to illustrate the capabilities of this approach to spatial decision support. The multi-dimensional model used comprises the following dimensions:

- Disease.
- Type of case (incidence, death or hospitalization).
- Age group.
- Sex.
- Location and time reporting.

as well as the following measures:

- Number of cases.
- Standardised rate.
- Comparative figure.
- Numerous statistical indicators.

Using these database components and additional derived measures it is possible for example to drill down into the spatial hierarchy of the database as shown in Figure 11.12 by a mouse click from the regional level (top) to the local level (bottom). Figure 11.13 shows an example of a temporal drill across function on a complete database level (per years of data maintained) resulting in a new table containing the other elements at the same level (each year for which data are maintained in the dataset) in the initial dataset.

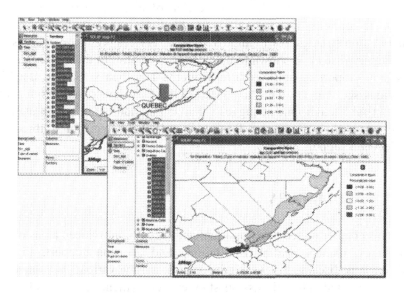

Figure 11-12. Example of a spatial drill down operation from regional to local level (Source: Rivest et al., 2005)

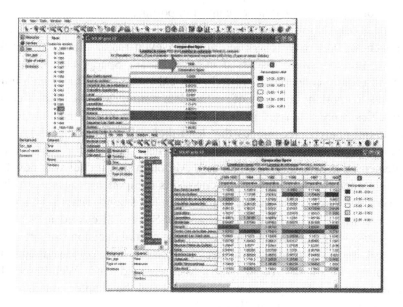

Figure 11-13. Temporal drill across operation with resulting table of all other elements at the same level of detail (Source: Rivest et al., 2005)

The investigation of complex data using SOLAP and JMap primarily focuses on the intelligence and design phases of decision support. Little or no capability exists for the identification of alternative courses of action, however implicitly it is possible to use the functionality to identify potential intervention strategies or to improve health services provision in areas where respiratory disease is shown to be most problematic. In this regard the actual decision support capabilities differ from the TourPlan SDSS discussed in the previous section. The fact that JMap SOLAP is developed within a highly customisable environment suggests that adding additional design and choice functions to the software is a feasible in order to account for all three phases of decision support.

6.   SUMMARY

This chapter united two important and recently introduced areas of spatial databases, namely spatial data mining and decision support. First, data mining was discussed including its origins and the various ways in which it is implemented. This discussion stressed deployment of the results of data mining techniques to assist in decision making. The view of data mining as knowledge discovery was treated as synonymous with the intelligence gathering phase of decision support, where techniques such as OLAP and the newer concept of SOLAP are of central importance.

Since data mining techniques are dominated by machine learning approaches, these were classified and the seven major classes of machine learning methods were discussed in some detail. Spatial data mining was noted to require conceptual and technical enhancement of basic data mining concepts. Important concepts of spatial data mining, such as spatial concept hierarchies, were discussed in Section 3. Also, a variety of machine learning and spatial analytic techniques that lend themselves to spatial data mining were discussed. The CRISP-DM reference model for data mining implementation was also summarised. Although this model was developed as a reference for data mining in general, its component stages were shown to lend themselves transparently to the mining of spatial data.

The concept of decision support and DSS were discussed in Sections 4 and 5. DSS were defined and their use in the spatial decision support domain was explained. The phases of the decision process were reviewed along with the core components of a DSS. Web-enabled and Web-based DSS development and deployment was discussed and the chapter concluded with a review of two examples of SDSS.

7. REFERENCES

Adrienko, G. and Adrienko, N. (1999) "Knowledge-based Visualization to Support Spatial Data Mining", *Advances in Intelligent Data Mining (Lecture Notes in Computer Science # 1642)* In Hand, D.J., Kok, J.N. and Berthold, M.R. (Eds.), Berlin, Germany: Springer-Verlag.

Ascough II, James C., Rector, H. D., Hoag, D. L., McMaster, G. S., Vendenberg, B. C., Shaffer, M. J., Weltz, M. A. and Ahjua, L. R. (2002) "Multicriteria spatial decision support systems: overview, applications, and future research directions", In Rizzoli, A. E. and Jakeman, A. J. (Eds) *Integrated Assessment and Decision Support, Proceeding of the First Biennial Meeting of the International Modelling and Software Society,* Manno, Switzerland, pp. 175-180.

Balram S. and Dragicevic, S. (Eds.) (2006) *"Collaborative Geographic Information Systems",* Hershey, PA: Idea Group Publishing.

Bedard, Y., Gosselin, P., Rivest, S., M. J. Proulx, Nadeau, M., Lebel, G. and Gagnon, M. F. (2003) "Integrating GIS Components with Knowledge Discovery Technology for Environmental Health Decision Support", *International Journal of Medical Informatics,* pp. 79-94.

Bhargava H. K. and Power D. J. (2001) "Decision Support Systems and Web Technologies: a Status Report", Proceedings of the Seventh Americas Conference on Information Systems, Boston, Mass, pp. 229-235.

Bremer, M.A. (Ed.) (2003) *Knowledge Discovery and Data Mining: Theory and Practice,* Institution of Electrical Engineers (IEE), London, UK.

Chelghoum, N., Zeitouni, K. and Boulmakoul, A. (2002) "A Decision Tree for Multi-layered Spatial Data", *Proceedings Symposium on Spatial Theory, Processing and Applications,* Ottawa, ON.

Chorley, R.J. and Haggett, P. (1965) "Trend Surface Mapping in Geographic Research", *Transactions of the Institute of British Geographers,* 7, 47-67.

Claramunt, C. and Thériault, M. (1995) "Managing Time in GIS: An Event-oriented Approach", In *Recent Advances in Temporal Databases* by Clifford, J. and Tuzhilin, A. (Eds.), Berlin, Germany: Springer-Verlag.

Coiera E. (2003) *The Guide to Health Informatics,* 2nd Edition, London: Edward Arnold.

CRISP-DM (2000) *CRISP 1.0 Process and User Guide,* Cross-Industry Standard Process for Data Mining (CRISP-DM).

Densham, P. (1991) "Spatial Decision Support Systems", In Maguire, D.J., Goodchild, M. F., Rhind, D. W., (Eds) *"Geographical Information Systems: principles and applications",* London: Longman, pp. 403-412.

Ding, Q., Ding, Q. and Perrizo, W. (2002) "Decision Tree Classification of Spatial Data Streams Using Peano Count Trees", *Proceedings ACM Symposium on Applied Computing,* Madrid, Spain.

Ester, M., Frommelt, A., Kreigel, H.-P. and Sander, J. (1999) "Spatial Data Mining: Database Primitives, Algorithms and Efficient Database Support", *Data Mining and Knowledge Discovery,* Vol. 4, No. 2/3, pp. 193-216.

Fayyad, U.M., Piatetsky-Shapiro, G., Smyth, P. and Ulthurusamy, R. (Eds.) (1996) *Advances in Knowledge Discovery and Data Mining,* Menlo Park, CA: AAAI/MIT Press.

Feick R. D. and Hall, G. B. (1999) "Consensus Building in a Multi-Participant Decision Support System", Journal of the Urban and Regional Information Systems Association, No. 11, pp. 17-24.

Feick, Robert D. and. Hall, G. B (2004) "A method for examining the spatial dimension of multi-criteria weight sensitivity in a Spatial Decision Support System", *International Journal of Geographic Information Science*, Vol. 18, pp. 1-26.

Fisher, P. (Ed.) (2005) *"Developments in Spatial Data Handling"*, Berlin: Springer.

Fotheringham, A.S., Brunsdon, C., and Charlton, M.E., (2002), *Geographically Weighted Regression: The Analysis of Spatially Varying Relationships*, Chichester, UK: Wiley.

Gahegan, M. and Brodaric, B. (2002) "Computational and Visual Support for Geographical Knowledge Construction: Filling in the Gaps between Exploration and Explanation", *Proceedings Symposium on Geospatial Theory, Processing and Applications*, Ottawa, ON.

Gahegan, M., Wachowicz, M., Harrrower, M. and Rhyne, T. M. (2001) "The Integration of Geographic Visualization with Knowledge Discovery in Databases and Geocomputation", *Cartography and Geographic Information Systems*, Vol. 28, No. 1, pp. 29-44.

Goodchild, M.F. (1992) "Geographical Data Modeling", *Computers and Geosciences*, Vol. 18, No. 4, pp. 410-408.

Gorry, G. A. and Scott Morton, M. S. (1971) "A Framework for Management Information Systems", *Sloan Management Review*, Vol. 13, No. 1, 56-79.

Green, D. R. and Bossomaier, T. (2002) *Online GIS and Spatial Metadata*, London: Taylor and Francis.

Griffith, D.A. (2003) *Spatial Autocorrelation and Spatial Filtering: Gaining Understanding through Theory and Scientific Visualization*, Berlin: Springer-Verlag.

Guo, D. (2003) "Coordinating Computational and Visualization Approaches for Interactive Feature Selection and Multivariate Clustering", *Information Visualization*, Vol. 2, No. 4, pp. 232-246.

Hall, G. B., Feick, R. D., and Bowerman, R. L. (1997) "GIS-based Decision Support Architecture and Applications for Developing Countries", *South African Journal of Geoinformation*, Vol. 17, No. 3, pp. 73-80.

Han, J. (1999) "Data Mining" in *Encyclopedia of Distributed Computing* by Urban, J. and Dasgupta, P. (Eds.), Boston, MA: Kluwer Academic Publisher.

Han, J. and Kamber, M. (2001) *Data Mining Concepts and Techniques*, San Francisco, CA: Morgan Kaufmann Publishers.

Han, J., Kamber, M. and Tung, A.K.H. (2001) "Spatial Clustering Methods in Data Mining: A Survey" in *Geographic Data Mining and Knowledge Discovery* by Miller, H.G. and Han, J. (Eds.), London and New York: Taylor & Francis.

Hand, D.J. (1998) "Data Mining: Statistics and More?", *The American Statistician*, Vol. 52, No. 2, pp. 112-118.

Hand, D., Mannila, H. and Smyth, P. (2001) *Principles of Data Mining*, Cambridge, MA: MIT Press.

Hornsby, K. and Egenhofer, M.J. (2000) "Identity-based Change: A Foundation for Spatio-temporal Knowledge Representation", *International Journal of Geographical Information Systems*, Vol. 14, No. 3, pp. 207-224.

Hu, X., Cercone, N and Ziarko, W. (1997) "Generation of Multiple Knowledge from Databases Based on Rough Sets Theory" In *Rough Sets and Data Mining* by Lin, T.Y. and Cercone N. (eds.), Boston, MA: Kluwer Academic Publishing.

Jankowski P. (1995) "Integrating Geographical Information Systems and Multiple Criteria Decision Making Methods", *International Journal of Geographic Information Systems*, Vol. 9, No. 3, 251-273.

Jankowski P. and Nyerges T. (2001a) *GIS for Group Decision Making*, London: Taylor and Francis.

Jankowski P. and Nyerges T. (2001b) "GIS-supported Collaborative Decision Making: results of an experiment", *Annals of the Association of American Geographers*, No. 91, pp. 48-70.

Keim, D.A. (2001) "Information Visualization and Visual Data Mining", *IEEE Transactions on Visualization and Computer Graphics*, Vol. 7, No. 1, pp. 100-107.

Koperski, K. and Han, J. (1995) "Discovery of Spatial Association Rules in Geographic Information Databases", In *Advances in Spatial Databases (Lecture Notes in Computer Science 951)*, *by* Egenhofer, M. J. and Herring, J. R. (Eds.), Berlin: Springer-Verlag.

Kruse, R., Gebhardt, J. and Klawonn, F. (1994) *Foundation of Fuzzy Systems*, Chichester, UK: J. Wiley & Sons.

Lee, I. (2004) "Mining Multivariate Associations with GIS Environment" In *Innovations in Applied Artificial Intelligence (Lecture Notes in Artificial Intelligence #3029)*, Orchard, C., Yang, C. and Ali, M. (Eds.), Heidelberg, Germany, Springer-Verlag.

Lehman, A., Overton, J. and Austin, M. P. (2002) "Regression Models for Spatial Prediction: Their Role for Biodiversity and Conservation", *Biodiversity and Conservation*, No. 11, pp. 2085-2092.

Lu, C.-T., Chen, D. and Kou, Y. (2003) "Algorithms for Spatial Outlier Detection", *Proceedings 3rd IEEE International Conference on Data Mining (ICDM'03)*, Melbourne, FL.

Malczewski J. (1999) *GIS and Multicriteria Decision Analysis*, New York: John Wiley and Sons.

Malczewski, J. (No Date) "*Internet Resources for Geo-Information-Based Decision Analysis*", http://publish.uwo.ca/~jmalczew/gimda/intres4.htm, last accessed Friday, 3rd February 2006.

Malczewski, J. (2006) "GIS-based Multicriteria Decision Analysis: a Survey of the Literature", Unpublished paper, Department of Geography, University of Western Ontario, London, Ontario, Canada.

McCormick, B.H., Definti, T.A. and Brown, M.D. (1987) "Visualization in Scientific Computing", *SIGGRAPH Computer Graphics Newsletter* (Special Edition), Vol. 21, No. 6.

Miller, H.G. and Han, J. (Eds.) (2001) *Geographic Data Mining and Knowledge Discovery*, London and New York: Taylor & Francis.

Oracle (2003a) "Spatial Analysis and Mining" Chapter 8 of *Oracle Spatial User's Guide and Reference*, 10g Rel. 1(10.1), Redwood City, CA., Oracle Corporation.

Oracle (2003b) *Oracle Data Mining Concepts*. 10g Rel. 1(10.1), Redwood City, CA., Oracle Corporation.

Peng Z.-R. and Tsou M.-H. (2003) *"Internet GIS: Distributed Geographic Information Services for the Internet and Wireless Networks"*, New York: John Wiley and Sons.

Peuquet, D.J. and Duan, N. (1995) "An Event-based Spatiotemporal Data Model (ESTDM) for Temporal Analysis of Geographical Data", *International Journal of Geographical Information Systems*, Vol. 9, No. 1, pp. 7-24.

Power, Daniel J. and Kaparthi, S. (2002) 'Building Web-based spatial decision support systems, *Studies in Informatics and Control*, "Vol. 11, No. 4, pp. 291-302.

Ribarsky, W., Katz, J., F. Jiang and Holland, A. (1999) *Discovery Visualization and Visual Data Mining*, Technical Report #99-14, Georgia Institute of Technology, Graphics, Visualization and Usability Center, Atlanta, GA.

Rinner C. and Jankowski, P. (2002) "Web-based Spatial Decision Support - Technical Foundations and Applications", In *"The Encycolpedia of Life Support Systems (EOLSS)*, Theme 1.9 *"Advanced Geographic Information Systems*, (Ed.) Claudia Bauzer Medeiros, Oxford, UK: UNESCO/EOLSS Publishers.

Rivest, S., Bedard, Y., and Marchand, P. (2001) "Towards Better Support for Spatial Decision Making: defining the Characteristics of Spatial On-Line Analytic Processing (SOLAP)", *Geomatica*, Vol. 55, No. 4, pp. 539-555.

Rivest, S., Bédard, Y., Proulx, M.J. Nadeau, M., Hubert F. and Pastor, J. (2005) "SOLAP: Merging Business Intelligence with Geospatial Technology for Interactive Spatio-Temporal Exploration and Analysis of Data, *Journal of International Society for Photogrammetry and Remote Sensing*, Vol. 60, No. 1, pp. 17-33.

Roddick, J.F. and Spiliopoulou, M. (2002) "A Survey of Temporal Knowledge Discovery Paradigms and Methods", *IEEE Transactions on Knowledge and Data Engineering*, Vol. 14, No. 4, pp. 750-767.

Shekhar, S. and Huang, Y. (2001) "Discovering Spatial Co-location Patterns: A Survey of Results", In *Advances in Spatial and Temporal Databases (Lecture Notes in Computer Science #2121)*, by Jensen, C.S., Schneider, M., Seeger, B. and Tsotras, V.J. (Eds.), Heidleberg, Germany: Springer-Verlag.

Simon, H. (1977) *The New Science of Management Decision*, Englewood Cliffs, NJ: Prentice Hall.

Sui, D.Z. and Goodchild, M. (2001) "GIS as media?" Guest editorial. *International Journal of Geographical Information Science*, Vol. 15, No. 5, pp. 387-390.

Uran, O. and Jansen, R. (2003) "Why are spatial decision support systems not used? Some experiences from the Netherlands", *Computers, Environment and Urban Systems*, No. 27, pp. 511-526.

van Leeuwen, Jos P. and Timmermans, Harry, J. P. (Eds.) (2004) *"Recent Advances in Design and Decision Support Systems in Architecture and Urban Planning"*, Dordecht, the Netherlands: Kluwer Academic Publishers.

Whitten, J. L. and Bentley, L. B. (1997) *Systems Analysis and Design Methods*, 4th Edition, Burr Ridge, IL: Irwin.

Whitten, I.H. and Frank, E. (2000) *"Data Mining: Practical Machine Learning Tools and Techniques with Java Implementation"*, San Francisco, CA: Morgan Kaufmann Publishers.

Wild R. H. and Giggs K. A. (2004) "A Web Portal/Decision Support System Architecture for Collaborative Intra-Government Planning", *Electronic Government*, Vol. 1, No. 1, pp. 61-76.

Yau, C. (1993) "Interactive decision support system for airline planning", *IEEE Transactions on Systems, Man and Cybernetics*, Vol. 23, No. 6, pp. 1617-1625.

Yip, K. and Zhao, F. (1996) "Spatial Aggregation: Theory and Applications", *Journal of Artificial Intelligence Research*, No. 5, pp. 1-26.

PART 4

THE FUTURE

Chapter 12

TRENDS OF SPATIAL DATABASE SYSTEMS

1.   INTRODUCTION

In Chapter 1 it was noted that change and growth are the only constants in the world of information technology. Increasingly fierce competition within the spatial information industry as well as the growing complexity of user needs have greatly accelerated the pace and drastically diversified the nature of change in recent years.

Conventionally, the spatial information industry sought to support basic needs in spatial data capture, provide easily understandable spatial data structures and models, and facilitate cartographic presentation of the results of spatial analysis and map-based manipulation of data. Propelled by a general trend toward systems integration in the IT industry, recent developments in spatial databases have concentrated mostly on building extensions of database systems, cross-platform interoperability, the development of user-friendly interfaces and decision support capabilities using spatial information. Indeed, spatial database systems as they exist today are quite different from the creations that existed in the past. As spatial database systems continue to evolve in response to advancing technologies and changing user requirements, they will also be different from those now in use in the years to come.

This chapter identifies the major current trends in spatial database design and implementation. Sections 2 through 5 review emerging concepts and techniques that are shaping the trends of spatial database systems from the perspectives of technology, data, application and people. Following this, Section 6 discusses the implications of these trends for the spatial database research community. Section 7 speculates on the likely paths of future developments.

2.   TRENDS OF SPATIAL INFORMATION TECHNOLOGY

The growing integration of spatial information technology with mainstream information technologies (IT) is probably the most important development of spatial database systems up to the present time. This has changed the ways that spatial databases are designed and implemented and the ways spatial data are collected and applied. However, different people tend to have different interpretations and expectations for spatial data relative to their increasing use in mainstream IT concepts and methods. Also, concepts and terminology associated with new technologies are often confusing and even contradictory due to the lack of universally accepted definitions. The following sections track the main trends in spatial information technology development and explain what is meant by spatial and mainstream IT integration.

2.1   Spatial and Mainstream Information Technology Integration

Since the debut of the Canada Geographic Information System (CGIS) in the early 1960s spatial database systems have been perceived more as a distinct and specialised field of information technology than as a part of mainstream IT. During most of the 1980s and 1990s, spatial data were structured and stored in GIS to optimise access and presentation with little or no interaction with other business processes. Moreover, all software products and applications were typically obtained from a relatively small number of GIS-specific software vendors. While the latter has not changed drastically up until relatively recently, the inclusion of spatial data handling in mainstream database software has grown consistently and the emergence of a healthy open source geospatial software community has meant that the mainstream and spatial database worlds have converged.

This convergence has meant that conventional barriers between spatial database systems and mainstream IT have been increasingly removed since the mid-to-late 1990s to the point now where all major database vendors have developed spatial data handling functions in their products (for example, Oracle Spatial, IBM Spatial Extender, IBM Informix Spatial DataBlade, and IBM Informix Geodetic Datablade). The increasing use of mainstream IT since the 1990s is the major factor underlying the substantial growth and advances of spatial information technology, as illustrated in Figure 12.1.

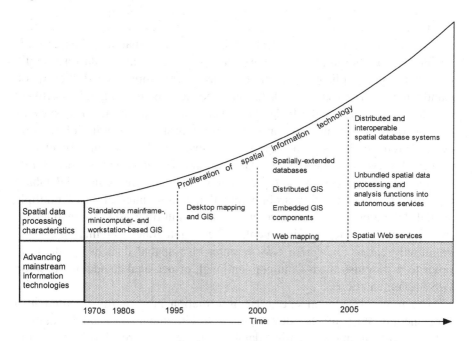

Figure 12-1. Evolution of mainstream IT as the driving force behind the growth and advances of spatial information technology

The trend of spatial and mainstream IT integration was driven by a combination of several subtle but interrelated factors. These include:

- Advances in computer hardware, software and standards, which have helped to overcome the long-standing problems of incompatibility between spatial and non-spatial data representation and processing.
- The advent of the Internet and the new paradigm of networked computing that stresses standardisation, interoperability and usability, which have effectively removed the boundaries and barriers between spatial and other branches of IT.
- The growing demand for novel and sophisticated spatial applications, which have forced spatial database software vendors to look for methods and tools outside the traditional realms of GIS technology.
- The growing recognition of the importance of spatial information as a commodity with value for modern society and the resulting business

opportunities that motivated mainstream IT companies to enter the spatial database marketplace.

The injection of mainstream IT concepts and techniques into spatial database systems has fundamentally changed the way spatial database systems are designed and implemented. In the past, spatial databases were designed by modelling the real world and collecting spatial data for specific objectives and application domains. They were implemented mainly on standalone high-powered workstations and servers running GIS software coupled with a relational DBMS. However, today's spatial database systems are mostly implemented as part of an organisation's overall information system architecture using a variety of mainstream IT components such as universal servers, Web servers, data warehouses, data marts and *mobile computing*. These systems are typically implemented using a shared database in a distributed computing architecture that allows integration of multiple spatial data sources and concurrent access by multiple users in an Internet, intranet or extranet environment. These integrated architectures are commonly called *enterprise GIS* or enterprise spatial database systems in order to distinguish them from conventional, or personal standalone desktop GIS implementations.

The move toward an enterprise information architecture resulted in a fundamental shift in the development environment of spatial database systems. Traditionally, spatial database systems were developed using proprietary data structures, macro or scripting languages, and data processing procedures. These systems are now implemented largely using mainstream systems development methodologies characterised by industry standards, commercial database products and generic application programming tools. As a result, spatial applications are able to access relatively seamlessly mainstream IT resources such as computer-assisted drafting (CAD) software, transaction databases and business intelligence (BI) databases. At the same time, mainstream IT applications are also able to embed spatial functionality without extensive macro programming and data conversion. Examples of embedded spatial functionality include *helper programs*, *plug-ins* and *Java applets* for displaying and querying spatial information using Web browsers, and application programming interfaces (API) that allow word processors, database systems and spreadsheet applications to use maps and spatial statistics generated directly from spatial databases.

It is now commonplace for vendors of spatial database software to engage in business or technical partnerships with their counterparts in the commercial database, Web services, BI and enterprise resource planning (ERP) marketplaces. This has resulted in the development of a number of spatially-enabled enterprise applications, collectively referred to as *location-based services* (LBS) (see Sections 2.3 and 4.2). The advent of the new breeds of business-oriented spatial applications, together with the presence of today's prominent IT companies including Yahoo and Google, have propelled spatial database systems into being an integral part of mainstream IT.

2.2   Mobile Computing and Technology Integration

Recent advances in wireless communication technologies are now adding a new dimension to technology integration that plays a pivotal role in shaping the trend of spatial and mainstream IT integration. Wireless communication devices such as mobile phones, pagers and computers have decreased in size, weight and cost, and increased in functionality, portability, security and reliability. Popular acceptance of these devices in the consumer market has rapidly changed the way spatial information is delivered and used. This in turn has changed, and will continue to change, the way spatial database systems are designed and implemented.

Mobile phones allow users to communicate not only by voice, but also by text messaging, images and video using the *Short Message Service* (SMS) and *Multimedia Messaging Service* (MMS) protocols respectively. Mobile phones can be connected to the Internet directly using *Wireless Application Protocol* (WAP) browsers. They can also be connected to a mobile computer, and used to connect it to the Internet through wireless 'dial-up'. WAP browsers use a special derivative of (HTML), called *Wireless Markup Language* (WML), which allows Web pages to be constructed so that they can be displayed on the smaller screens of mobile phones and other wireless communications devices.

Viewing Internet pages on a mobile computer is usually easier than on a mobile phone, due to the larger, higher resolution colour screens as well as more powerful and user-friendly navigation tools. In this context, a *mobile computer* is generally defined as a lightweight, compact, portable device with a flat colour screen, an interactive GUI accessed using a pen or stylus, compatibility with desktop computer applications, the ability to communicate over a wireless network with other devices, available memory expansion slots, and the ability to connect to the Internet. Mobile computers can be divided into four main categories, as described in Table 12.1, namely laptop computers, personal digital assistants (PDA), hand-held computers, and Tablet PCs.

From the perspective of mobile or field-based spatial database applications, the advent of Tablet PCs in 2002 is particularly noteworthy.

Tablet PCs differ from the other three classes of mobile computers in three important ways, as noted in Table 12.1 and explained below.

Table 12-1. Types and characteristics of mobile computers

Type	Characteristics	
Laptop Computers	O	"Toughened" for field use.
	O	Wireless communications capability.
	O	Built-in global positioning system (GPS) device.
	O	Screen designed for outdoor viewing and touch-screen navigation.
Personal Digital Assistants (PDAs)	O	Palm-held size.
	O	Word processing, spreadsheet and Web browsing capabilities.
	O	Connection to and communication with PCs.
	O	GUI for navigation and data input using a pen-like stylus.
Hand-held Computers	O	Similar to PDAs in size but with larger screen and a keyboard.
	O	Connection to and communication with mobile phones, GPS receivers, laser rangers and spatial database on a PC or workstation.
Tablet PCs	O	A scaled down PC using a full-featured Operating System (OS).
	O	Large and high-resolution screen, external keyboard, and hand writing and voice recognition capabilities.
	O	Built-in GPS receiver for position fixing.

- *Full-featured operating system.* Tablet PCs use an extension to the Windows XP Professional operating system, called *Windows XP Tablet PC Edition.* This is a full-featured operating system that allows Tablet PCs to run any existing Windows application including databases, graphics, hand-writing recognition and voice recognition, and to connect to both stationary and mobile computing devices over local and global communications networks.

- *Enhanced mobility.* Because of their compact size, light weight, high-capacity memory, pen-based input and, most importantly, long battery life, Tablet PCs are more mobile than other types of mobile computers. They support the IEEE 802.11 family of protocols (Cooklev, 2004) and other broadly supported wireless standards. As a result, they can be used for real-time data updates and collaboration in inter-group or inter-office scenarios in ways not practically possible with a PC before such as data entry while a user is walking or working in the field (when it is impossible to type).

- *Digital ink.* Tablet PCs use a pen device that replaces the traditional mouse action. Users tap and press on the screen with this device to interact with applications. The pen interface allows a user to write on the screen using *digital ink* (also known as mark-up and redline). Each ink stroke (or group of strokes) and its associated properties (for example, colour, width, and attributes) can be edited and stored just like traditional graphics and text. Spatial database software vendors, notably

Autodesk and ESRI, have already released software products that accept and store field data entered through digital ink as one or more data layers on top of existing map or spatial database layers.

Tablet PCs have overcome most of the limitations of display screen size, storage capacity and processing power imposed by laptops, PDAs and hand-held computers. Digital ink allows Tablet PCs to be used not only for the retrieval, processing and display of spatial information, but also for real-time capture and editing of field data through the use of GPS devices, electronic field survey equipment and hand-writing (including sketch graphics and text-based data). The processing power and data storage capacity of a Tablet PC are comparable to a desktop computer, hence it can be used as a "thick" client in a typical client/server computing environment. As such, it is able to minimise the amount of data traffic that would otherwise be required in a thin client architecture. At the same time, it is able to perform sophisticated spatial data processing and analysis in real time in the field under rapidly changing application and decision making scenarios, such as emergency response and management, search and rescue operations, and utility asset inspection and maintenance.

The use of mobile computers for spatial information capture and processing signifies a major breakthrough in the evolution of hardware platforms for spatial database systems. Prior to the advent of mobile computers, hardware platforms (including mainframes, minicomputers, workstations, and desktop PCs) could be used essentially only in the office. Rapid advances of mobile computers have provided the portable hardware platforms necessary for mobile or field-based computing. Practically all major spatial database software vendors have already developed software products for use specifically on mobile computers (see Table 12.2). These software products and related applications, together with the increasing availability and affordability of persistent broadband and wireless connections to the Internet, have enabled workers in the field to access an enterprise database for accurate and up-to-date spatial information whenever it is required.

Figure 12.2 shows the generalised architecture of a spatial database system for mobile or field-based applications. Such a system is typically set up using a combination of information technologies including multi-tier client/server computing, the Internet, wireless communications, telemetry and location positioning. There are two features that distinguish mobile spatial database systems from their conventional counterparts. One is the mobility of the client computer. The other is the real-time determination of its location. Various technologies can be used to determine or track the location of a mobile computer but these generally fall into two main approaches

(Figure 12.2), namely satellite-based GPS and ground-based communication base stations (Grejner-Brzezinska, 2004).

Table 12-2. Mobile software products of major spatial database software vendors

Vendor	Mobile Software Products	Brief Description
Autodesk	OnSite Enterprise	A client/server solution that includes (i) Autodesk GIS Design Server using a Java servlet to deliver interactive vector maps and design information to mobile devices, and (ii) OnSite Viewer, which is a user interface on the client PDA or Tablet PC used in the field.
ESRI	ArcPad	A customisable mobile application that uses and processes spatial data, in shapefile and several popular raster formats, that is served up over the Internet using ArcIMS.
	ArcPad Application Builder	A development tool for customised mobile applications within the desktop environment for deployment on ArcPad devices in the field.
	Mobile ArcGIS Desktop System	The ArcGIS Engine customised for field tasks that require GIS analysis and decision making, typically performed on high-end Tablet PCs leveraging features such as pen-based computing and digital ink technology.
MapInfo	MapX Mobile	A specialised version of the regular MapInfo MapX Active X Control used for creating map-based applications for mobile computing devices.
	MapXtend	A development environment for creating wireless spatial applications using J2EE (Java 2 Enterprise Edition) and J2ME (Java2 Micro Edition) technology.
Intergraph	IntelliWhere	IntelliWhere provides an open and scaleable environment that leverages Intergraph's GeoMedia technology, using the industry-standard WAP and Microsoft's COM architecture to serve map images and accompanying location-based information.

(Sources of information: Technical brochures of software products and Web sites of Autodesk (www.autodesk.com), ESRI (www.esri.com), MapInfo (www.mapinfo.com) and Intergraph (www.intergraph.com/gis)).

The ability for a mobile computer to determine its own position and make use of this locational information in spatial data processing has allowed a large number of spatially-enabled business applications to be developed. These include, for example, real-time supply of driving directions, emergency response locations, traveller information, advertising and marketing and real-time environmental data collection. Such LBS applications were generally recognised around the turn of the last century as being the next frontier of spatial information technology (Wilson, 2001). While the promise offered by LBS applications was not realised to the extent that was initially anticipated, the advent of the devices that made such services technically feasible represents one of the major innovations of spatial database development in the last decade. The characteristics and applications of LBS technologies are explained in more detail in Section 4.2.

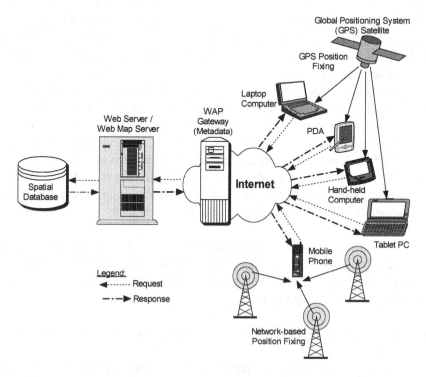

Figure 12-2. Architecture of a spatial database system for mobile or field-based applications (see also Figure 12.7)

2.3   The Meaning and Impacts of Technology Integration

The convergence of spatial and mainstream IT did not necessarily preclude, as some predicted, the demise of spatial database systems as a distinct and special type of database systems (Batty, 2004a). This is because the design and implementation of spatial database systems, whether using standalone or integrated technologies, requires specific skills, knowledge and ability that are relatively distinct from those required for conventional database systems. These characteristics include, among others, the distinct nature of spatial data representation, the use of map projections and coordinate systems, the distinct nature of spatial statistics, and the pervasive need to respect cartographic presentation. Acquisition of a working knowledge and mastery of these topics requires years of education and practical training as well as extensive workplace experience. Also, there are always professional spatial data users in government, business, and academic research whose application needs and objectives call for spatial analytical and visualisation functions that conventional commercial database systems cannot fulfill, regardless of the implementation technologies and system architectures. Thus, in spite of the growing integration of technologies and the rapid disappearance of the conventional boundaries between spatial and mainstream IT, spatial database systems will continue to thrive as a distinct class of database systems.

Conventional domain-specific GIS and LBS-oriented spatial database systems serve distinct purposes and user communities, but are basically constructed using similar technologies and spatial data resources. This has led to the emergence of a new framework of spatial database development as illustrated in Figure 12.3. In the top layer of this framework are the two classes of systems are generally used independent of one another. However, the use of common data standards and communication protocols allow them to interoperate, when and where necessary, functionally to compliment one another.

The middle layer of the framework in Figure 12.3 is the spatially-enabled information infrastructure, where most of the integration of spatial and mainstream IT occurs. This information infrastructure layer is made up of the following five components:

- Information policies and standards at global, national and organisational levels.
- Database access, communication and integration protocols (including middleware, markup languages, Web services, the Internet, and ISO/OGC/W3C specifications).
- Database and application development environments (for example Java and Microsoft Visual Studio.NET).
- Database management systems (including database engines, database administration, performance evaluation, spatial database extensions, and spatial data types).
- Data stores (including spatial databases, data warehouses, data marts, and federated database systems).

The foundation layer of the framework is made up the various spatial data collection methods and existing spatial data resources that are used to create and update the data stores that reside in the middle layer. These include *in-situ* data collection using *Sensor Web* technology (see Section 3.2), data resulting from business transactions, field data collection, digitising of existing maps and images, and conversion from external non-spatial database systems.

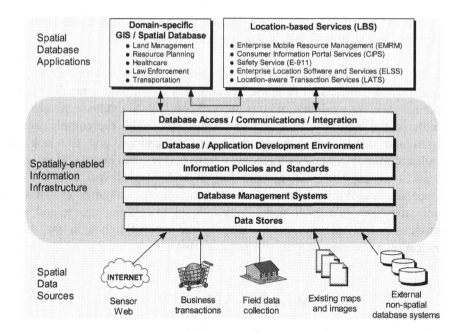

Figure 12-3. A three-layered framework of spatial database development

The integration of spatial and mainstream IT has brought numerous perceived benefits to spatial and commercial database vendors, systems developers and end users alike. These include:

- Better stewardship of spatial data as a corporate resource in an integrated enterprise database environment.
- Greater flexibility and scalability in database implementation using a combination of proprietary and non-proprietary technologies.
- More robust functionality of spatial database systems by incorporating conventional database concepts and techniques (such as SQL, OLAP, data warehousing, business intelligence and decision support) in spatial data processing and analysis.
- Shorter systems development cycle by deploying re-useable software components from common off-the-shelf systems, domain-specific object-oriented libraries, and free open source code.
- Providing a more cost-beneficial business model of implementing spatial information by using technical and human resources for spatial and non-spatial data within a single database environment.
- Improved communication among different departments of an organisation resulting from a higher level of interoperability of spatial and non-spatial database applications.

- Improved customer services by making it easier for external users to access and share an organisation's spatial data resources.

On the other hand, there are also obvious limitations in the use of mainstream IT for spatial database systems. These limitations result mainly from the inherent weakness of conventional commercial database systems to handle spatial data structures and operations efficiently, the complexity of spatial and non-spatial interoperability in applications, and the lack of wireless communication coverage away from major urban centres. However, advances in database and communication technologies will alleviate or overcome these limitations in time. Meanwhile, the commitment of the spatial database industry to technology integration remains strong, and this continues to play a key role in shaping the present trends in the data, application and people aspects of spatial database.

3.   SPATIAL DATA TRENDS

As the basic building block of spatial database systems, data have always played a central role in their design and implementation. In the past, the focus was largely on data structure, digital data collection by field surveys, map digitisation and database creation. As large volumes of spatial data have been accumulated, and as user requirements have become more sophisticated and stringent, attention has shifted gradually to focus on spatial data usability, interoperability, real-time processing, delivery and accessibility.

3.1   Spatial Data Usability and Interoperability

The concept of data usability is concerned with the "fitness for use" of a particular data set. Hunter et al. (2003) define data usability as an umbrella term encompassing some forty elements describing the quality, marketability, human perception and cognition, relationships with software and tools, as well as applications. Obviously, data usability is application-specific. This means that a data set usable for one application may not be necessarily usable for another due to, for example, different requirements in scale, projection, attribute classification, and update frequency. Further, there is also a temporal dimension to the concept of usability. The usability of a spatial data set tends to diminish when the content becomes gradually outdated over time. As a result, a data set that is fit for use at one point in time may not necessarily be fit for use at another point in time.

The practice of evaluating spatial data usability is not new. Numerous techniques and standards have been adopted by mapping agencies and professional bodies to help users identify the qualities of a given data set and determine its suitability for use. Four main reasons underlie the pervasiveness of the need to evaluate data usability as part of an ongoing process:

- *The separation of spatial data providers* including government agencies, academic and research institutions, and commercial value-added suppliers from spatial application developers and end users.
- *The increasing complexity of spatial problems* that often require data from different sources for solutions.
- *The critical decisions that spatial applications are designed to assist with* that call for a good knowledge of the qualities of the data sets involved in order to minimise uncertainties in the decision making process.
- *The legal liability of using spatial data* of unknown qualities that forces suppliers and users of these data to be vigilant about the appropriateness of the data they use.

Spatial data usability has focused primarily on structure (i.e., formats by which different data sets are stored) and geometry (i.e., coordinate systems, map projections, and geodetic datum used). The solutions for usability have often involved conversion or transformation of one data set to another through a *de facto* transfer standard. This approach to spatial data usability does not work well in the current distributed computing environment for two important reasons. First, the resulting data sets are duplicates of the original data at a certain point in time. It is difficult to synchronize change in these data sets with changes in the original data and, as a result, their usability may deteriorate quickly. Second, converting and transforming data from different sources into a single data set or store of uniform specifications is technically problematic, costly and time-consuming. These difficulties often deter users from taking advantage of existing spatial data resources in their spatial database implementation projects.

As noted earlier in the book, technologies are now either available, or under active development, that enable the redefinition of the concept and practice of spatial data usability. With the aid of the *Web Services Description Language* (WSDL), *Universal Description, Discovery and Integration* (UDDI), Simple Object Access Protocol (SOAP), and global and local communication networks, users are now able to obtain a variety of spatial data services on the Internet, as explained in Section 3.3 (see also Chapter 6). Once data are obtained, their usability with respect to the requirements of a particular application can be established, and their suitability with respect to intellectual property and copyright can also be established relative to existing spatial data resources.

At present, spatial data usability is perceived mainly from the perspective of interoperability and the use of standards, a service-oriented architecture, and on-demand access to distributed spatial databases. The ability of users to use a portion of a remote data set either for data integration or for data processing removes the conventional practice of mass conversion or transformation of data sets. It also makes possible the development of new models of using spatial data such as the geospatial data portals and application service providers discussed in Section 3.3. The aim of spatial data usability is no longer limited to the passive evaluation of a few potentially usable data sources, but it is tightly coupled with the planning, design, implementation and on-going operation of spatial database systems. This implies that the current scope of spatial data usability goes well beyond the conventional realm of data format conversion and geometric transformation. Semantic heterogeneity and solutions through the use of geographic ontology and intelligent agents remain a key area in spatial database research as discussed in Chapter 6.

3.2   Real-time Spatial Database Processing

Conventionally, spatial databases are concerned only with cross-sectional and static data. The content of the database represents a snapshot or a series of snapshots that are typically unlinked from the real world at one or more definitive points in time. As spatial database technology has become more integrated with mainstream IT, in particular wireless communication technology, the differences between spatial information users and general information users is fast disappearing. Hence, increasingly, spatial information users require access to or use of real-time dynamic databases for time-dependent decision making, such as traffic or weather conditions for trip planning and the locations and status of fleet vehicles throughout a normal working day. The ability to collect, process and analyse such real-time spatial data will become an increasingly essential function of spatial databases in the future and this will completely change the paradigm for spatial database design and implementation.

There are two important trends in real-time spatial database systems, as suggested above. One is location-tracking, the other is *in-situ* location-based data collection and processing.

3.2.1   Location-tracking

Location tracking involves the functionality of spatial database systems to determine, record and retrieve at a later time the movement of a computing device attached to or carried by people, field equipment, and vehicles. Location-tracking is based on GPS technology. It is now commonplace for GPS to be found, in addition to conventional hand held receivers, in automobiles (business and private), mobile telephones and even in wrist watches. By means of the relatively new *Bluetooth* communication protocol, these receivers are able to communicate using wireless with laptop computers or PDAs over relatively short distances, thus enabling them to access spatial databases without a physical connection.

Several ground-based positioning systems are now in use or under development. These include *Enhanced Observed Time Difference* (E-OTD) systems, which determine positions by triangulation from cellular telephone towers, and the *Radio Frequency Identification* (RFID) system, which enables readers at specific locations to detect the tags or *transducers* of objects that come within close proximity to them. The growing ability of utilities providers, telecommunications companies and government agencies to utilise location tracking has enabled spatial databases to be updated automatically in real time when equipment and facilities are installed or constructed in the field. This has also made possible the development of numerous consumer-oriented LBS applications, as explained in Section 4.2.

3.2.2 *In-situ* Location-based Data Collection and Processing

The huge growth in computer and telecommunications markets have resulted in the development and production of state-of-the-art electronic chips that are now deployed in a large variety of consumer products. At present, it is relatively easy to attach a micro radio chip to electronic devices and communicate with them using wireless technology. The *Sensor Web*, which was conceived at the National Aeronautics and Space Administration (NASA) Jet Propulsion Laboratory (JPL) in 1997, took advantage of cheap consumer-market chips to create a system of location-based data collection platforms that has the potential to revolutionise the way spatial data are collected and deployed (Delin, 2002).

The fundamental building blocks of the Sensor Web are called *pods* (Figure 12.4a). These are electronic devices that can be installed on space- and air-borne as well as terrestrial carriers or stations whose positions are either fixed or mobile. Each pod contains one or more sensors which communicate, usually via wireless, with its local neighbours (Figure 12.4b). The data collected by one pod in a Sensor Web are shared and used by other pods. This means that the information in a Sensor Web is not purposefully routed or focused towards the end user, but rather it is intended for a close web of multiple numbers and types of sensors that continuously collect, monitor, aggregate, analyse and control an environment through coordinated efforts. End users are also able to communicate with a Sensor Web through special *portal pods* in order to extract knowledge from the raw data collected and then respond, adapt and react accordingly.

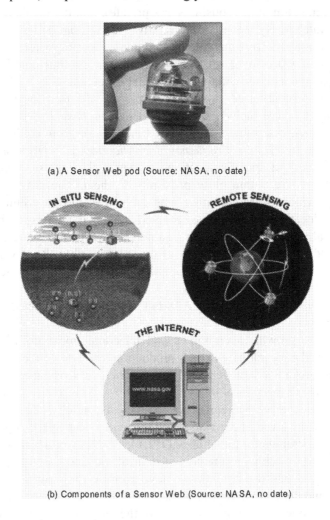

(a) A Sensor Web pod (Source: NASA, no date)

(b) Components of a Sensor Web (Source: NASA, no date)

Figure 12-4. Configuration and components of a sensor web

The growing importance of Sensor Webs and other sensor-based real-time data collection technologies has prompted the OGC, in conjunction with NASA, the National Geospatial-Intelligence Agency (NGA), and the United States Environmental Protection Agency (EPA), to develop a standard XML encoding scheme for metadata describing sensors, sensor platforms, sensor tasking interfaces, and sensor-derived data. The proposed specifications of this encoding scheme include (Botts, 2004):

- *Sensor Model Language* (SensorML). This language facilitates the development of general models and XML encodings for describing the geometric, dynamic and radiometric properties of sensors. (SensorML is also under review as part of ISO TC211 Project 19130).
- *Observation and Measurements.* This includes the general models and XML encodings for sensor observations and measurements.
- *Sensor Collection Service.* This is a service by which a client can obtain observations from one or more sensors/platforms.
- *Sensor Planning Service.* This is a service by which a client can determine the feasibility of collecting data from one or more mobile sensors/platforms and submit collections or processing requests to sensor/platform systems.
- *Web Notification Service.* This is a service that allows a client to conduct asynchronous dialogue or message interchange with one or more other services.

Sensor Webs have opened new horizons for spatial information applications in many fields, including environmental monitoring, ecological surveillance, micro-climatic measurement in agriculture, as well as automated alert generation in homeland security and national defence. However, the real-time processing of spatial data collected by Sensor Webs has an accompanying and substantial set of new challenges in spatial database design and implementation. As the data collected by sensors represent the process or evolution over time of the objects that they monitor, it is necessary for spatial database systems to have sophisticated spatio-temporal data processing capabilities not generally found in the paradigm of spatial database design and operation that is prevalent in current applications. These capabilities include the need to handle the near real-time information flow among the pods, the continuous updating of the spatial database, as well as integration with conventional spatial data analysis and modelling methods.

3.3   Spatial Data Delivery

The advent of the Internet has played a significant role in revolutionising the nature and methods of delivering spatial data (see Chapter 10). The term "portal" was originally used to denote a catalogue of hyperlinks that assisted in the navigation of or search for Web resources. It was no more than a Web site that acted as a gateway to provide a single access point to multiple Internet resources. Web portals today operate more like a "dashboard" that consolidates multiple data sources, applications and other resources that reside in both local and remote domain servers into a single Web page. Some of the new generation of portals have the ability to make use of artificial intelligence to understand the preference and privileges of individual users. They are customisable by using building blocks of mini-applications called *portlets*, each of which has its own self-contained user interface and performs a specific function such as database query, data analysis or visualisation (Lowe, 2004).

To assist the spatial technology community in implementing standards-based spatial portal solutions, the OGC (2004) has released a Geospatial Portal Reference Architecture as a resource guide for rapid development and informed acquisition of portals and portal-exploiting applications. When implemented using this reference architecture, a spatial portal provides a single secure point-of-entry into an organisation's information resources for applications hosted on a variety of computing platforms, databases and development languages. Similarly, applications developed using the reference architecture are able to plug and play with spatial data and services both within an organisation and also from other organisations throughout the world.

The debut of spatial Web portal sites (for example, GeoConnections in Canada, GeoSpatial One-Stop in the United States, INSPIRE in Europe, and the ESRI Geography Network) reflect the trend of using spatial database systems as a form of public services, and as part of the spatial data infrastructures that exist at organisational, national and global levels. In this context, the concept of Web services introduced in Chapter 6 is particularly important. The Web services model combines the notions of service and infrastructure in a tripartite relationship involving data users, data providers, and data brokers (or metadata information providers).

Using nothing more than a Web browser, a user of Web services can access published spatial data, as well as discover, access, integrate and apply a wide variety of heterogeneous spatial data resources and processing functions distributed in local and remote servers and/or workstations throughout the Web environment. The concept of Web services is explained from the perspective of spatial database applications in the next section.

4.   TRENDS OF SPATIAL APPLICATIONS

The integration of spatial and non-spatial information technologies has resulted in several significant trends in spatial database applications. These include the emergence of spatial Web services and the diversification of conventional domain-specific solutions.

4.1 Spatial Application Service Providers and Web Services

Organisations have implemented spatial database technology mainly from a systems perspective that puts data structures and database operators ahead of applications. Such a narrow approach, which dominated the earlier history of spatial information technology, focuses on how spatial database systems can be set up to serve specific business needs, rather than what the spatial databases might be used for to achieve organisational goals.

The objectives and scope of implementing spatial database systems in most organisations today have gone well beyond the realms and confines of traditional GIS. Organisations now see the implementation of spatial database systems largely as an information service (to satisfy the information requirements of internal users and external clients) and as a data infrastructure (to support organisational goals and societal needs).

The idea of developing a spatial database system as a service is realised as a business model labelled a *spatial application service provider*. This business model has its origin in the more general application and other service provider models noted in Chapter 10 that have prevailed in the computer industry for many years (Waters, 2001). In essence, a spatial application service provider is an organisation that manages and deploys access to data and applications in a centrally hosted computer, usually on a subscription or per-transaction basis, over telecommunications networks. This model is developed using Web mapping technologies provided by spatial database software vendors and a thin-client architecture that typically requires nothing operationally more demanding to operate than access to a standard Web browser. Because of the reliance on the Web as the mechanism for service delivery, the spatial application service model is now more commonly known as a Spatial Web Service (SWS) (McKee, 2003) or *Geospatial Web Portal* (Lowe, 2004)).

SWS provide a practical solution for organisations that want to use spatial information without incurring the costs of maintaining a spatial database, developing their own spatial applications and supporting database and application specialists in-house. It is constructed on a truly distributed and interoperable architecture in which the component systems of application providers, data suppliers and end users work together through standard interfaces and encoding in what has become to be known as a *service* or *value chain*. Figure 12.5 illustrates the concept of a service chain. It shows how an organisation is able to make use of the data resources and applications that are available from a variety of public and private sector Web services, and unify the resources and applications with their own data resources to generate maps and reports for a particular application such as an environmental assessment project.

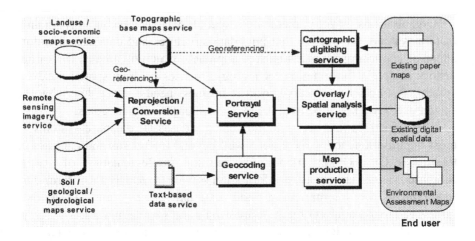

Figure 12-5. Concept of a service chain in the deployment of spatial web services in an environmental assessment project

The OGC Web Services Stage 2 (OWS-2) initiative, which was announced in early 2004, sought to develop a standards-based and vendor-neutral framework that allows distributed spatial database systems to communicate with each other. OGC Web Services specifications provide an interoperable environment to make it easier for organisations to apply geographic information through on-line business exchanges. As a result, organisations can in principle dramatically reduce the cost and time of application development, as well as offer users the ability to access plug-and-play application components. SWS in the future will become as open as the Web itself because users will be able easily to find, view, overlay, and combine different data sets residing on different spatial data servers and create spatial applications, as illustrated in Figure 12.5.

The trend of moving toward a services-oriented development strategy has prompted spatial database software vendors to change their business focus. In the past these vendors competed with one another more or less solely on the strength of their products for data acquisition and management, focusing on spatial data capture, types, operators, indexing and support for projection and coordinate transformation. In the emerging services-oriented approach to spatial information, vendors have had to adjust their competitive position in the market to emphasise the ability of their products to deliver spatial information instantly and cheaply to any point in a business process (both in terms of position in a business process chain and a location). As a result, the ability to deliver spatial information to users through SWS has become an indispensable component of the product lines of commercial database software vendors. This is shown in Table 12.3 for several prominent spatial information technology companies.

Table 12-3. Spatial web service software products from major vendors

Vendor	Spatial Web service Products	Brief descriptions
Autodesk	MapGuide Commerce, OnSite and LocationLogic.	These three products are designed for delivering spatial technology through the application service provider model, in a traditional WAN architecture as well as a wireless network
ESRI	ArcWeb Services	A collection of spatial data and capabilities that is available to users on demand when needed, accessible directly using ArcGIS or used to build unique Web-based applications. It can also be used to extend users' Web mapping and GIS capabilities by integrating them into their ArcIMS implementation
GE Smallworld	Internet Application Server (IAS)	This product is designed to enable existing GE Smallworld desktop products such as Model It to be deployed to a variety of users and Internet applications through an application service provider
MapQuest Services	MapQuest Site Advantage	An XML interface that allows users to embed street-level maps and spatial search capability of North America, Latin America and Europe in their applications
Microsoft	MapPoint Web Service	A set of APIs that developers can incorporate into enterprise and customer-oriented applications to add address lookup, map rendering, reverse geocoding, and location-aware functionality
Microsoft	MapPoint Location Server	A companion service of MapPoint Web Service that allows its users to connect real-time location information from mobile operators with mapping and routing information

(Sources of information: Technical brochures of software products and Web sites of Autodesk (www.autodesk.com), ESRI (www.ersi.com) and ESRI (2004b), GE Power (http://wwwgepower/prod_serv/products/gis_software/en/sias.htm) MapQuest (www.mapquest.com) and Microsoft (http://www.microsoft.com/mappoint/products/webservice/default.mspx)).

The recent rapid growth of SWS into pervasive consumer products signifies a new era of spatial database applications. International organisations, national mapping agencies, universities and commercial spatial data suppliers have populated the Internet with sometimes connected but still largely unconnected nodes of SWS. The connected services collectively form a network of local, national, regional and global spatial information structures that aim to make spatial data readily accessible for various purposes. For the first time in the history of spatial information technology, potentially millions of users are able to access spatial information, and obtain business or public services based on such information, without special software and training. Many of these users simply do not realise that the information and the services they obtain so conveniently are actually served by one or more huge spatial database systems working at the back-end.

SWS now affect almost every aspect of our daily life. They are used as an important part of e-government initiatives in many countries, where the public is able to use Web mapping services to find spatial information about their community, environment and other more specific aspects of spatial information such as the listings and locations of real estate properties or the real time degree of congestion on specific sections of the road network. In this context, it is now common practice for people to check weather and traffic conditions before travelling, and millions of people obtain driving directions from commercial Web services such as Microsoft's recently released Windows Live Local (powered by Virtual Earth), Google (including Google Earth), MapQuest, MapPoint (which incorporates MapBlast following its acquisition by Microsoft for use in the Microsoft Network (MSN) Web Portal), and Expedia.

SWS have powered the move toward the integration of spatial information with enterprise and consumer applications in various forms of LBS, and all indicators suggest that this application area will grow substantially over the next half decade. LBS, which grew rapidly and then faded somewhat, have been reborn, especially with the challenge mounted by Microsoft to the popularity of Google Maps and Google Earth. The future role of LBS seems to be central to the continued growth of Web mapping service tools and functions, as explained in the next section.

4.2   Location-based Services

In Section 2.3 LBS were introduced as a new paradigm of using spatial information that is distinguished from conventional domain-specific spatial database applications by their purpose, scope and method of delivery. LBS make use of information about the location of a mobile device in order to deliver personalised, localised and real-time business and information services. This is accomplished by combining the functionalities of three enabling technologies, namely (a) GIS/spatial database systems, including positioning by GPS or ground-based methods, (b) mobile wireless computing, and (c) the Internet, and especially the SWS that are based on it (Figure 12.6). In addition, LBS are characterised by a tight coupling with business and enterprise IT applications such as customer relationship management, and mobile resource and workforce management.

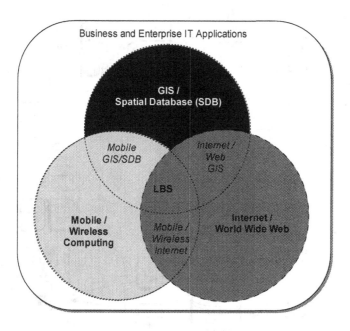

Figure 12-6. Enabling technologies of location-based services (LBS)

LBS can be hosted using the internal spatial information resources of the service provider, or alternatively using a combination of internal resources and SWS as shown in Figure 12.7. In order to reduce the cost of creating and managing very large spatial databases that serve no more than the function of spatial referencing, it is common for LBS providers to take advantage of SWS by integrating the two types of spatial information services. For example, navigation assistance, vehicle tracking, and location-aware billing can all be developed as special LBS on top of Web mapping services (WMS).

As a service-oriented application, LBS are designed to handle remote and simultaneous access by hundreds or even thousands of users. This is in sharp contrast to conventional GIS and spatial database systems where typically only a handful of users perform relatively complex spatial queries and operations on a desktop client/server environment. In order to maintain a high level of service in a concurrent real-time user environment, LBS demand much more stringent technology and performance standards than those required by conventional spatial database applications.

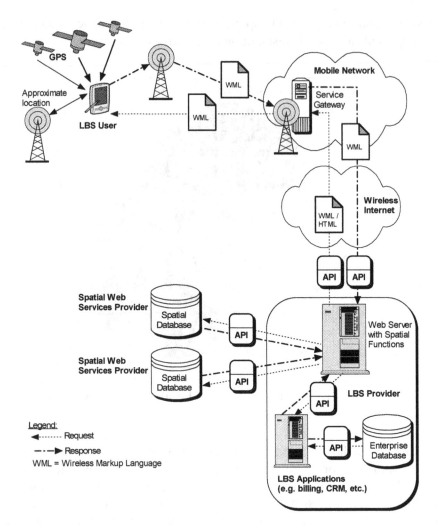

Figure 12-7. Typical systems architecture of location-based services (LBS)

This means that spatial databases designed for LBS are typically characterised by the following features:

- *High performance.* The ability to provide sub-second responses to multiple queries over the Internet using a wireless connection.
- *Scalability.* The ability to support thousands of concurrent users, potentially terabytes of spatial data, and a wide variety of business applications.
- *Reliability.* Assurance of near-100% up-time 24 hours a day, 7 days a week.

- *Open architecture.* The ability to support all data and communications standards and protocols to ensure interoperability with different hardware devices and software applications.
- *Geographic coverage.* The ability to reach users from any location using either wired or wireless devices.
- *Severity.* The ability to ensure confidentiality and integrity of the information being delivered by leveraging database transaction constraints and rules.
- *Ease of use.* The use of intuitive interface design to minimise the amount of training required of the users and, where possible, the deployment of artificial intelligence to understand user preferences and automate the required services accordingly.

LBS are an evolving technology and are being used in an increasing number of areas. Services that are currently available can be generally divided into the following categories:

- *Safety services.* These include services such as the Enhanced 911 (E-911) system that is capable of identifying the location of the user in case of an emergency.
- *Consumer information portal services.* These services include traffic information, navigation assistance, yellow pages and travel/tourism information obtained over the Internet, and the delivery of "local" news, weather reports and forecasts, driving directions, and traffic information, as determined by the location of the mobile phone and computer.
- *Enterprise mobile resources management.* These include vehicle tracking, logistic planning, utility asset inspection and management, and mobile workforce management.
- *Telematics services.* These include services that utilise GPS technology to obtain real-time location information to provide driving direction and vehicle tracking functionality in fleet management.
- *Enterprise location software and services.* These are characterised by the integration of spatial information and BI applications such as enterprise resource planning, consumer relationship management, and market analysis (see Section 4.3).
- *Location-aware transaction services.* These services include location-sensitive or location-triggered billing and product promotion and advertising.

Several standards initiatives and projects are now underway that will set the course of LBS development in the future. These include the inter-operability program for LBS spearheaded by the OGC (2004) in partnership with the W3C, ISO, the *Open Mobile Alliance* (OMA) and the *Internet Engineering Task Force* (IETF), among other organisations. While the OMA and IETF focus their interoperability efforts on locational positioning, the OGC is concerned with standards for LBS interfaces such as geocoding, web mapping, navigation assistance, yellow page search and real-time traffic data acquisition. It is obvious that LBS are not only popularising the use of spatial information, they have also set the trend of moving further away from the proprietary spatial data formats and interfaces that hindered the development of spatial information technology in the past.

4.3   Emerging Domain-specific Applications

The application arena of spatial information has witnessed tremendous diversification outside the conventional realms of land and resource management. Growth of spatial database applications in multiple areas, including but by no means limited to health care, law enforcement and public safety, and business intelligence, have become areas of innovation in their own rights. Since the user base of these application domains is cumulatively much larger than land and resource management, it is reasonable to expect that major growth of spatial decision support applications will occur in these areas in the future.

4.3.1   Health Care Applications

The relationships between place, location and health have been explored for more than two centuries in the medical sciences. Medical geography, also known as health geographics, is the branch of human geography that deals with spatial aspects of the status of public health and health care systems. Advances in spatial database systems and the growing availability of relevant data have given public health officials, medical practitioners, insurance providers, epidemiologists and biomedical researchers access to powerful tools that can enhance knowledge of the factors that affect public health, while helping to manage health care systems in a cost cost-effective manner (Boulos, 2004; Lang, 2000).

Generally speaking, spatial information can be applied to two principal areas of health care:

- The spatial study of disease incidence and distribution, also commonly called *landscape epidemiology*, which explores, describes and models the spatial-temporal occurrence and spread of infectious and environmentally related diseases.

- The study of health care systems, which analyses and models the delivery of and access to health care services by populations with respect to their identified health care needs and demographic characteristics.

Numerous applications in these two areas have focused on a varierty of issues concerning, for example:

- Identification of problems and issues associated with inequalities in the delivery of health services among people from different social classes and living in different regions.
- Location and allocation of health care resources, including the management and planning of health care facilities, resources, and practitioners with respect to patient access and types of required medical services.
- The efficient dispatch and routing of ambulance trips.
- The status of public health and health care systems and developing preventive health care measures.

Spatial database systems have proven to be extremely useful in managing and analysing the huge amounts of health care and medical information that exist, using spatial location as the common link. A spatial database approach to health information management enables researchers to investigate effectively the long term impacts of global environmental change, population growth, poverty levels, urbanisation and other human activities on the physical well-being of populations. Over the last half century, the World Health Organisation (WHO) and its affiliates as well as national health care organisations, notably the United States Centre for Disease Control and Prevention (US-CDC), and national governments have collected a wealth of public health statistics. The spatio-temporal analysis of this information and translation of findings into effective public health policies are and will continue to be an extremely challenging and critical application domain of spatial information in the years to come.

4.3.2   Law Enforcement and Public Safety Applications

In the United States, law enforcement and public security are now commonly referred to as *homeland security*. More broadly this concept embraces policing, criminal justice, border patrol, emergency management, intelligence services and aspects of investigation, inspection and enforcement within various government departments and agencies. Since much of the information used in law enforcement is related to various aspects of geography, spatial database systems have become a standard part of the information management infrastructure of law enforcement and public safety organisations.

Figure 12.8 illustrates a variety of areas in which spatial information is applied in law enforcement and public security organisations, using policing as a specific example.

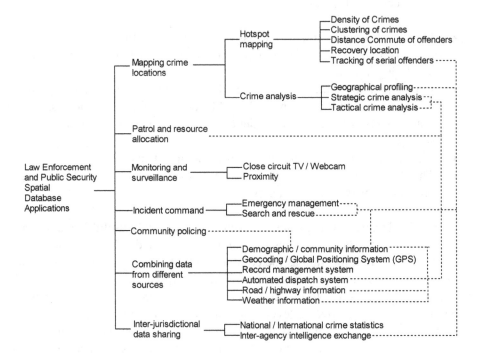

Figure 12-8. Law enforcement and public security applications of spatial information: the example of policing with links among the various application areas

The major application areas of law enforcement and public security applications can be described as follows:

- *Mapping of crime or incident locations.* This class of applications is generally divided into two subclasses:

    - *Hotspot Mapping.* The use of colour-coded pushpins on wall maps to indicate locations of crimes and incidents is a standard practice that has been used by police departments for a long time. Spatial database systems now allow police departments to manage and organise crime and incident data more effectively by means of versatile digital hotspot maps that utilise a database of reported crime and incident locations in conjunction with digitised maps of the affected areas. Since the data are in digital form, it is relatively easy to turn incident location data into useful criminal information, such as the spatial density and clustering of crimes. Moreover, digital capture of incident data allow numerous aspects of criminal activity to be investigated, for example measuring the distance between the residences of offenders and crime locations and the identification and tracking of serial crime locations.

- *Crime Analysis.* Geographic profiling (also called criminal geographic targeting) uses real-time crime data spanning several days to identify short-term crime patterns that have common characteristics (for example, description of suspects, date and time of crimes, and types of weapons used). In addition, profiling uses spatial information to develop an investigation and problem-solving approach to law enforcement (Rossmo, 1999). Strategic crime analysis is a form of research-focused analysis that aims to determine long-term crime trends and forecasting with respect to changing demography and community dynamics.

- *Patrol and resource allocation.* This form of information is used to generate patrol briefing maps, to optimise the design of beat routes, and to set the allocation of resources for patrol officers and criminal investigators.
- *Monitoring and surveillance.* Spatial information allows identification of locations for installing close circuit televisions or webcams for crime prevention, investigation and collection of crime data. Spatial database systems and Web map servers are also used to generate proximity maps to monitor and manage the residential locations and activities of registered sex offenders and offenders under house arrest.
- *Incident command.* Spatial database systems provide real-time geographic information for emergency management and search and rescue operations. These systems are also used as the platform for combining information from other sources, such as road and weather conditions, to improve decision making and assist in carrying out planned actions.
- *Community policing.* This actively involves members of the community in the development of crime prevention strategies through public education, neighbourhood watch and similar programs. Such an approach to policing depends heavily on a thorough knowledge of local communities. Spatial database systems provide the necessary tools for police officers to manage community information and use it to support the implementation of community policing strategies and practices.

- *Combining data from different sources*. Conversion and import of demographic and community data, such as population characteristics, economic status, racial and ethnic diversity, activities of social groups and volunteer organisations, are necessary for crime analysis and the development of community policing strategies. Geocoding of address-based crime reports and the use of GPS allow crime and incident locations to be acquired rapidly. GPS is also used to convey the locations of patrolling police cruisers to command centres so they can be deployed optimally in the event of emergencies. Supporting automated dispatch systems are also used in this capacity to provide the shortest or fastest routes for police cruisers to reach the location of a reported crime or incident. In addition to the above information sources, road and highway information from a city's works and engineering departments can be used to develop traffic control plans and deploy highway patrol assignments. In this context, weather information can be integrated with other data resources to assist in emergency planning and management such as diversion of traffic and search and rescue operations.
- *Inter-jurisdictional information sharing*. Spatial database systems provide the means for reporting and aggregating crime location statistics at local, regional, national and international levels. These systems also facilitate the exchange of criminal information among police departments from different jurisdictions across municipal, county, provincial/state and international boundaries.

The use of spatial database technology has revolutionised the ways law enforcement and public security organisations collect, manage and use crime information. Spatial information is the foundation of knowledge-based and intelligence-led law enforcement characterised by targeting offenders, effective management of information resources, and the ability to correlate serial crimes and incidents. It allows law enforcement and public security organisations to develop and adopt preventative measures rather than employ reactive measures in dealing with criminal activities, to work in tandem with local communities to reduce crime and disorder, and to cooperate with organisations in different jurisdictions.

4.3.3 Business Intelligence Applications

Although the concept of business intelligence (BI) is relatively new in and of itself, the idea of using computer-based analytical methods to support business planning and operations has been used, in one guise or another, for decades. The concept of BI has largely replaced conventional terms such as business decision support, executive information systems (EIS) and management information systems (MIS) in the prevailing jargon. BI applications combine business data with analytical tools to present complex and competition-oriented information to decision makers in order to improve the timeliness and quality of information this is used in decision processes.

BI enables decision makers to identify the strengths and weaknesses of an organisation, market trends, emerging technologies and best practices, and the regulatory environment in which organisations have to compete. It also helps decision makers predict possible actions of competitors, and the implications of these actions if and when they materialise. BI is essentially a specific form of data mining and decision support systems, as defined in Chapter 11, that allows organisations to adopt a proactive rather than a reactive approach to business planning and operations.

The software tools used in BI are typically client-side applications capable of analysing legacy business data residing in the back end server of a data warehouse. The results of an analysis, typically in the form of spreadsheet results, tabular reports, pie charts and bar graphs, are collectively called a *BI dashboard* (see Figure 12.9a). The term "dashboard" is used in this context because it contains everything necessary to "drive" a business. Users are able to change the granularity of data analysis with respect to classification (the number of classes into which attribute values are divided) and temporality (yearly, seasonal, monthly, weekly and daily summation and tabulation) for different decision scenarios and purposes.

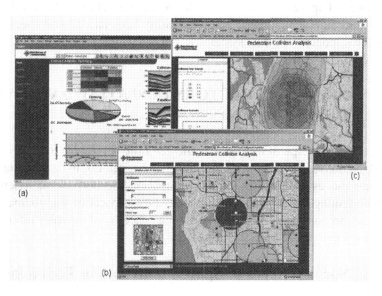

Figure 12-9. Examples of a BI dashboard: (a) conventional dashboard components; (b) and (c) spatial dashboard components (Source: Gonzales, 2004)

Advances in spatial database technology and the growing ability to integrate spatial and business transaction data from different sources through the use of Web services have led to the development of spatial BI tools. The trend toward the use of spatial BI has also been driven by the realisation that conventional BI applications are limited in their ability to explore the spatial dimension of business data. Consequently, new generation spatial BI tools make use of spatial analytical techniques to help users identify the geographical characteristics of customers, incidents and business relationships. The results are presented by means of various 2- and 3-dimensional visualisation techniques in the form of a spatial dashboard to depict distribution, clustering, spatio-temporal trends and inter-relationships that would otherwise be very difficult or impossible to detect, describe and explain (Figure 12.9b and c).

Ideally, a spatial BI tool is product-independent so that it can seamlessly interoperate with different BI client computers, back-end database or data warehouse servers, and Internet map servers from various spatial database vendors. This allows the design of BI systems that are capable of transparently integrating spatial and non-spatial business data, business and spatial analytical techniques, and visualization technology into a single decision support environment for executive, managerial and operational users. At present, spatial capabilities are more an optional component than a standard feature of BI systems. Given the growing understanding of the importance of spatial information in the business world and the trend toward spatial and non-spatial data interoperability, spatial BI will become an integral component of all BI systems in the near future.

5.   PEOPLE AS FACTORS IN SPATIAL DATABASE DEVELOPMENT

A recurrent theme in this book is the centrality of people to all aspects of spatial databases. People will continue to be central to spatial database development, as without the human dimension the purpose of assembling and maintaining spatial databases is lost. Two aspects of this theme are of particular importance, namely the human needs of spatial information technology and public participation GIS.

5.1   Societal Needs of Spatial Information Technology

The societal needs of spatial information technology can best be considered as a three tiered hierarchy. The top level of the hierarchy contains global spatial databases that are significant in terms of international strategic needs in times of crisis. Regional databases that meet strategic regional needs fall in the central tier, and at the bottom are local databases that address procedural (query-based), day-to-day needs that enable communities to be more efficient and equitable. Each successively lower tier should have upward linkages with the data and databases that operate at sequentially higher levels in the needs hierarchy.

Global spatial databases have existed for more than twenty five years. They continue to be developed and are disseminated by a number of international organisations and institutions, prominent among which are the agencies of the United Nations including the United Nations Environment Program (UNEP), the Food and Agriculture Organisation (FAO), and the World Health Organisation (WHO). Thematically, global spatial databases can be divided into four classes that each applies to a different dimension of global concerns, namely environment, food and agriculture, population and poverty, and health. Clearly, each of these classes is highly interrelated with the others and the needs that spatial information technology addresses within each class are similarly interlinked. For example, human population and poverty dynamics are inextricably linked with health status, which in turn is linked with the condition of the environment as well as availability of adequate food and potable water.

The increasing use of more accessible and interactive Internet-based map servers, and greater capabilities by users to access and utilise global datasets, has magnified the needs for reliable, accurate, accessible and standardised spatial data for strategic/critical decision making and management purposes. These needs are profound, especially during the current era of severe global weather events, increased security threats due to international terrorism, and an acceleration in the need to manage global natural resources. In the future, there are clear and elevated needs to harmonise global spatial data and associated information resources, to identify more clearly user needs for online global data services, to improve user education, to improve capacity building and support, and to improve user access through increased interoperability and the use of open standards.

As the potential for using spatial information technology to address global human needs continues to grow, so too do the problems related to inconsistent data integration and visualisation, variable data quality and documentation, uncoordinated growth in the presence of different versions of essentially the same data sets, unnecessary duplication of effort, complex restrictions on data re-dissemination and use, and incomplete or incorrect citation and attribution of data. Addressing these and other problems will be one of the most significant challenges of the spatial information technology industry in the coming years. Efforts described to this end in a report edited by de Sherbinin and Chen (2005) suggest that positive initial steps are being made. In particular, they note seven areas of collective responsibility at the global scale, that set the human needs agenda for productive use of global data by spatial information technology, namely:

- To make global-scale data and derived information as widely accessible and usable to all types of users as possible, while recognising the intellectual property rights of the underlying data sources.
- To promote the appropriate use of these data and information resources among all types of users, through provision of suitable metadata and documentation, expert guidance, outreach to key user communities, and other means.
- To improve the quality, comprehensiveness, and usability of global-scale datasets and derived information through collaboration with the relevant data sources and managers, the scientific community, diverse data users, and key sponsors.
- To improve the capacity of data sources, data managers, and data and information users in developing countries to contribute to and benefit from global-scale data and information resources.
- To promote efficient and seamless integration of global-scale data development, management, and access with corresponding local, national, and regional data programs, initiatives, and networks.
- To establish effective coordination with other related data and information efforts including the development of national and global spatial data infrastructure, ongoing intergovernmental data programs, relevant international efforts to develop and implement open standards, and present and future international scientific initiatives.
- To ensure the long-term stewardship of these data including their long-term preservation and access.

The diversity of organisations working in developing and maintaining spatial databases is more extensive at the regional than global level, as regional databases include national *spatial data infrastructures* (SDI), spatial databases assembled by multiple national government ministries, and the resources of private sector data-focused companies. Typically, regional level organisations focus on human needs that concern aspects of economic and social security and trade, while also addressing issues that have global ramifications stemming from environmental conditions in specific regions, such as Antarctica and the Amazon basin among others. Regional level human needs for spatial information technology are no less important than the global stage, and may actually be more significant given the geographic proximity of member nations and the shared basis of regional cooperation.

Essentially the same problems and needs noted above apply at the regional level, with the additional complexity of multiple independent jurisdictions and governments that use different standards and practices for geospatial data, and that also have different policies on spatial data access. In Europe for example, there are multiple datums and geographic projections in use for spatial data between member nations of the European Union. This runs contrary to the objectives of organisations such as EUROGI, COGI and INSPIRE (see Chapter 5) that are seeking to harmonise geospatial standards between countries. Moreover, it makes merging coincident data sets cumbersome and may serve to impede rather that advance the ability of spatial information technology to address important regional issues that relate, for example, to watershed and environmental management where physical systems do not coincide with the administrative boundaries of nations.

Regional co-operation to improve the capacity of spatial data resources is an essential prerequisite to merging human spatial data needs with emerging developments in spatial information technology. In this regard, several challenges exist to overcoming current regional spatial information technology level difficulties. Specifically, the following issues must be addressed in the future:

- The variable quality of most of the key terrestrial data and information.
- The generation of information that is genuinely demand-driven and need responsive.
- National policies that restrict spatial data access and resist standardisation.
- Poor data and information management, update, assembly, and assimilation.
- Weak mechanisms for data and information sharing, including collaborative inter-jurisdictional analysis.
- Uncoordinated investment in end-to-end institutional capacity.
- Unharmonised data and information collection and dissemination methods and standards.
- Continued collection of unused data and information.

Both the global and regional needs of spatial information technology noted above are applicable at the local level. At this level authorities often operate as spatial information technology silos, sometimes using data that are collected by higher levels of the human needs hierarchy while responding to localised needs of constituent populations and ecosystems within their jurisdiction. In this context, entities such as school boards, health districts, municipal planning offices, conservation authorities, land valuation and transfer offices, as well as a myriad of private sector companies typically operate completely independently of one another, duplicating substantial effort in data assembly and processing. Local-level needs mostly address issues related to procedural spatial data use as well as less frequent special projects. The focus of much of this work is on day-to-day operational use as well as improving the quality of life for local residents.

While nearly everyone within a local community is indirectly affected by or directly uses some form of spatial information technology, often without knowing it, on a regular basis, most of this use is purely of a service-oriented nature. Hence, local needs in spatial information technology focus on facilitating these types of uses and, especially, in building the capacity and identity of local communities to become self-managed. In this context, one of the greatest human needs of spatial information technology in the coming years is to make it easier for the public to obtain access to and use spatial data. The basic human needs for communication and independence are central to this and are reflected in the popularity of mobile services such as cellular telephones to access LBS.

Steps to improve the spatial orientation of local communities by promoting greater community planning are important in this regard. In fact, the challenge to move spatial databases into the public realm is one the major drivers of innovation at the present time. *Public participation GIS* (PPGIS) has become a major focus of academic research and the objective of engaging the public more fully not only in the use of spatial data but also in planning and decision processes that involve questions of land use and management has become a consideration of substantial importance. In this regard, the human need to be involved in important decisions that affect community cohesion and quality of life is one of the most important needs that the spatial information technology industry can address.

5.2   Public Participation GIS and Spatial Databases

The development of spatial databases and spatial information technology discussed in earlier chapters has deviated away from proprietary, large-scale hardware and software environments toward increasingly distributed and interoperable platforms using open standards and integration of networking and the Internet as central means of facilitating data access. This new direction has also changed the predominant view of users of spatial databases as highly trained specialists with strong computing skills, toward the inclusion of non-specialist users, including members of the general public. Hence, attention has begun to focus on developing user-friendly, accessible and highly functional or adaptive interfaces, such as the BI dashboard discussed in Section 4.3. The specialist user/developer/manager still remains centrally important to the success of spatial databases, however the non-specialist user has taken on a new importance.

While the enterprise-level management of spatial databases remains largely within the more traditional and specialised domain of spatial information technology (Keating et al., 2003), public access to and use of spatial databases has opened a new horizon of possibilities that is both broad in its application as well as innovative in its orientation. The growth in what is commonly known as PPGIS or community GIS has coincided with the popularisation of mapping and map use on the Internet (see Section 4.1).

This trend, plus the already strong interest in PPGIS within the geo-spatial research community has evoked considerable debate over what constitutes "the public" and what constitutes "participation", how participation can be facilitated through the use of spatial data, and what roles PPGIS should play in the future development of spatial information technology.

5.2.1 The Public and Participation

Schlossberg and Shuford (2005) describe the public aspect of PPGIS as ranging from every resident in a neighbourhood engaged in community asset mapping to every citizen in the United States interested in viewing census data on-line. This spectrum of public involvement with GIS, while clearly very general, is still somewhat restrictive. Broadly speaking, "the public" in the context of PPGIS can encompass any person or collection of people anywhere who can access and use a spatial database for any purpose outside of the context of tasks that are required of GIS professionals through paid employment. In other words, the spirit of PPGIS is to provide access to spatial information technology and data for any member of the public. This conceptualisation of the public evokes the underlying origins of the concept of PPGIS, which promotes a view of spatial information technology as a means of facilitating input or voice in processes for individuals and groups who have traditionally been voiceless. This input can be achieved both through distributed access to spatial data using the Internet (Carver et al., 2000) and/or co-located spatial data use where participants work together face-to-face (Jankowski and Nyerges, 2003).

Schlossberg and Shuford (2005) note that a complete understanding of the concepts of public and participation can only be achieved by considering their relationships in the form of matrix, such as that shown in Table 12.4. This allows the impacts of the public and participatory dimensions in spatial problem solving through the use of GIS to be understood both at the start and end of a decision process, as well as in specifying the various classes of each dimension. At the start, various approaches to participation can be adopted depending on who is targeted and what the participation goal is. At the conclusion of a decision process, all parties can establish whether the use of PPGIS has produced desired results by systematically evaluating project outcomes relative to spatial data use.

Table 12-4. Intersection of public and participation within PPGIS

		Simple	Public Dimension			*Complex*
		Decision Makers	Implementers	Affected Individuals	Interested Observers	Random Public
Simple	Inform					
	Educate					
Participation Dimension	Consult					
	Define Issues					
	Joint Planning					
	Consensus					
	Partnership					
Complex	Citizen Control					

Source: Schlossberg and Shuford (2005)

Integration of the public and participation dimensions allows administrators and planners considerably flexibility in crafting practical instances of PPGIS. The inescapable reality is that with spatial data increasingly appearing in the public realm via the Internet, and with the move toward participatory democracy in decision processes at various levels of government, PPGIS is no longer an optional consideration. Rather, it is rapidly becoming and important concern for spatial information technology that requires new interfaces and tools to be built around spatial databases.

5.2.2 Future Developments of PPGIS

The future use of PPGIS approaches for community organisations wishing either to self-organise around the use of spatial databases for their community or to participate in spatial planning decision processes is likely to focus on one or some combination of four general models among others identified by Leitner et al. (2002), namely:

- Community-based, neighbourhood GIS centres used by communities for their own self management purposes.
- University-community based partnerships.
- Publicly accessible GIS facilities available at tertiary institutions or in public libraries, community centres or municipal offices.
- Internet map servers that provide data from all three tiers identified in Section 5.1.

Independent of the specific approach that communities use to engage in the use of PPGIS, the challenges of educating the lay public about geography and determining how exactly to facilitate public spatial knowledge building and expression of ideas about space and place through the use of GIS and digital mapping technologies remain substantial obstacles. Several writers have noted that most PPGIS case studies reported in the literature do not use advanced GIS functions for spatial analysis (Weiner et al., 2002), but rather rely on the use of GIS more to present digital map views of areas of interest. Use of spatial databases for this purpose is likely to become more pronounced with the increased use of Internet map services, to the point where the Internet will become the dominant PPGIS platform. However, this is likely to vary from place to place depending on the availability of Internet connections and useable spatial data, as well as appropriate software tools for public users to display and manipulate map views in Internet browsers.

One of the key objectives of PPGIS developments is to encourage broad-based use of the technology by members of the public independent of their level of affluence and/or technological readiness to utilise the tools of modern information systems. However, this is far more easily said than done. In some parts of the world, field methods of PPGIS may not actually involve the IS component of the acronym at all due to technological constraints, and manual methods of spatial data and map use may achieve the objectives of PPGIS without the use of digital spatial information technology at all. Moreover, many technical and procedural challenges of implementation need to be overcome and the whole notion of public empowerment that is central to the objectives of PPGIS needs to be understood better. For example, GIS technology can disempower groups through government agency policies or private sector data producers making it difficult or impossible to obtain reasonably priced and reasonable quality spatial data. Further, empowerment may serve only reinforce existing power structures that relate to privilege and influence, rather than providing opportunity for the voiceless to participate in spatial decision making.

Known weaknesses of PPGIS need to be addressed in future developments. For example, more emphasis needs to be placed on incorporating local knowledge into spatial databases through use of qualitative narratives and representational interactions of participants with various aspects of their local environment. Methods that capture qualitative perceptions and expressions of localities need to be devised so that members of the public can interact with spatial data in ways that they are comfortable with, rather in ways that are dictated by the nature of conventional spatial data models and user interfaces. This will require investing thought and programming expertise in radical changes in the ways that people interact with maps, especially in on-line environments. New approaches to intelligent user interface design are required that adapt to the technological skills and knowledge of participants revealed through their use of a system. Finally, effort needs to be investigated in developing methods that are capable of verifying and validating data inputs provided through PPGIS.

6.   SPATIAL DATABASE RESEARCH TRENDS

This section concludes Chapter 12 by first reviewing current spatial database research trends and then extending this discussion to consider the likely areas of development in the foreseeable future.

6.1   Current Trends

A dominant theme in spatial databases in recent years is their incorporation into the mainstream IT database industry. As noted elsewhere in this book, spatial extensions both to Oracle (Oracle Spatial and Oracle Locator) and IBM (DB2 Spatial Extender and Informix Spatial DataBlade) databases allow users of these tools to define and manipulate spatial data using standard GIS-based functions as well as SQL-based *object-relational queries* and *spatial queries*. Hence, the integration of IT infrastructure has strengthened the relationship between spatial and other forms of data, as well as provided spatial databases with corporate database management tools to achieve improved scalability, better reliability, better versioning and improved maintenance functions, as well as enhanced performance, among other things. As a result, the move away from systems and applications that focus on solving single problems using highly specific spatial data models and specialised databases to a more general approach has produced a much broader array of services that are utilised across multiple job functions and departments.

In turn, this development has initiated a change from the previous single database, single application focus to logically consolidated, consistent, open and large enterprise data hubs. The consolidation of spatial and IT data management within one infrastructure has allowed workflows within organisations to be consolidated, while also avoiding duplication of effort and reducing the overall costs of deployment and data maintenance. Hence, the general trend of change toward a service focus for spatial data in modelling business processes provides information when and where it is needed.

The trend toward IT integration has helped to accelerate the speed and lower the cost of spatial data collection, aggregation and dissemination within and between organisations. Spatial data now need only be stored once and maintained centrally within an enterprise database which provides multiple views, applications and access points to distributed users. Alternatively, databases can, themselves, be distributed across various servers resident on a network that is internal (intranet) or external (extranet and Internet) in its connectivity and degree of openness to users, depending on the nature of the data and the application.

Hecht and Kucera (1999) itemised nineteen research trends in geographic information services at the end of the 1990s. Several of these trends remain current, suggesting that while the pace of technology change and diffusion is rapid, some fundamental issues persist in terms of their importance. These trends, supplemented by more recent research trends, are summarised as follows:

- *Spatial data acquisition and integration.* The substantial growth in the collection of spatial data in recent years from multiple sources (especially GPS and senor-based devices) has required the development of new approaches to integrating or blending data from multiple sources within a common database schema (also referred to as *data conflation*). The huge volumes of data that are now feasible and increased data storage capacity have required rethinking the design and function of databases, as well as developing new algorithms.
- *Distributed and mobile computing.* The move toward integrating GIS services and spatial data with mainstream IT has required investigation of bandwidth use and development of spatial data compression techniques for transmitting large volumes of data across wireless networks for LBS-type applications. Moreover, the development and adoption of spatial metadata standards is an important and growing area of interest to improve browse and search functions of spatial databases.
- *Geographic representation.* As noted in several chapters in this book, previous methods for storing and accessing spatial data were not designed or intended to address the increased volume, complexity and robustness that are required to integrate diverse data, including

multi-media, field-based and temporal data into a common database schema. Advances have been made in developing new representational schemas for these data that allow spatial geometry and other data types to be stored as attributes of database tables, coupling spatial feature objects with other types of data, especially continuously distributed data.

- *Interoperability of geographic information.* Considerable advances have been made in the last half decade, especially by the OGC and ISO to specify and standardise the formal semantics of spatial data, and to use these standards to foster interoperation across open interfaces, especially in the context of global (but in some cases also local) distributed networks, such as the Internet. This has produced frameworks, open architectures and reference architectures that have been adopted by virtually all spatial data platforms.

- *Open formats and GML.* In the past it was important to know whether data were in a specific vendor's format. With the development and adoption of open standards and interfaces, the need to know these formats is no longer an important issue. When used together, XML and GML encoding now make it possible to resolve many of the former difficulties associated with incompatible data models.

- *Geographic data infrastructures.* Changes in government information policies and practices over the past decade have served to promote the development of robust spatial information infrastructures. This in part is a function of the realisation that spatial data represent both an asset and a commodity of value. Hence, the economics of spatial information relative to their access and use figure prominently in discussions surrounding spatial databases. This recognises the importance of establishing standards within data infrastructures, while giving increasing attention to the integration of locally generated spatial data in order to encapsulate local knowledge within spatial databases.

- *Spatial data and society, including PPGIS.* A number of general research questions relating to spatial data and society have galvanised in recent years around the concept of PPGIS. Also, related to this research activity are larger questions that concern the influence of spatial information technologies on legal, economic and institutional structures within society, as well as on interpersonal relationships and spatial behaviour.

- *Spatial analysis, modelling, and uncertainty.* In addition to the many operational functions of spatial databases, their use for numerical modelling and analysis cannot be neglected. The development of new spatial analysis techniques and the need for advancing understanding of uncertainty and how it can be modelled in spatial analysis reflect this research trend. Moreover, the propagation of spatial uncertainty within database operations is ever more important as spatial data continue to penetrate beyond the traditional use domains into new areas.

- *Spatial query optimisation.* The widening net of spatial data applications, especially in BI-type uses, requires that new techniques be developed for optimising database queries. Traditional query methods, with large and complex spatial databases are inefficient in their execution. Work has been underway for some time to extend SQL syntax to treat spatial objects and continuous field model queries (Egenhofer, 1994; Laurini et al., 2004), and this research as well as the development of efficient spatial indexing is likely to increase in importance in the future.
- *Use of geometry in database queries.* Geometric operations are now becoming standard functions in conventional relational, object-relational and object-oriented databases that support spatial data. However, the need for more efficient methods of database storage of complex geometries is required and this will be an important area of research activity in the future.
- *Incorporation of time varying data into spatial databases.* Although no consideration is made of time in the implementation of most GIS, and scant attention was paid to temporal process modelling within the spatial data world until relatively recently, spatial data clearly vary, often quickly and substantially, through time. Hence, research on spatio-temporal data and associated database schemas have become more prominent, especially in the context of automated data creation and collection methods.
- *Use of spatial data on the Web.* This aspect of spatial databases was discussed in detail in Chapters 10 and 11 and it continues to attract increasing attention as Internet-ready spatial databases become more prominent. Associated with this trend are issues concerning database security, client-server architectures, and distributing both data and their processing across multiple servers and algorithms.
- *Knowledge discovery and spatial data mining.* Spatial data mining is a relatively new field that extends conventional data mining approaches. The need to utilise spatial relations in knowledge discovery through use of mining operations (such as spatial drill down, roll-up and drill across or slice and dice) is especially important for enterprise-level strategic decision making. In this context, standardisation of data models, data archiving, and the approaches used for knowledge discovery will serve to consolidate research advances in related areas, such as those mentioned above.
- *Improve geospatial interoperability between systems, data types and databases.* There is an important need to create system components that

communicate with each other and in which the underlying format and types of data are transparent to the user. This need is especially important in the context of applications that relate to customer response BI-related management, logistics, and LBS using wireless devices.

In addition to the trends noted above, a number of newly emergent issues, many of which stem from the influence of general IT trends on spatial databases, are appearing on the spatial information landscape.

6.2   Research Frontiers on Spatial Database Systems

An article in the influential journal Nature in 2004 noted that the United States Department of Labor identified geotechnology as one of the three branches of technology with the greatest growth potential in that country (Gewin, 2004). Much of this growth does not originate from within the conventional GIS industry, but rather from external trends that are impacting on GIS and will continue to do so in the future. In particular, the past two years have seen the emergence of a number of *disruptive technologies* within the spatial information technology industry (for example, the paradigmatic shift of Oracle version 10g to open and interoperable standards, the use of *grid computing* for spatial databases, the appearance of Google Maps and Google Earth, the counter release of Microsoft's Virtual Earth, and so on). These new technologies have challenged prevailing paradigms and introduced innovations that are sufficiently profound in their impact to change fundamental aspects of future development paths. The disruptions instigated by these developments were long overdue in the GIS industry, which had continued to evolve along a very narrow path for many years.

The resulting effect of the recent innovations is that the current spatial database landscape is considerably different relative to its configuration even one or two years ago. Certainly, the fusion of mainstream IT with spatial IT has established a fundamental launch pad for a host of future developments in the coming half decade. However, it is unlikely that there will be further radical departures from current growth areas and trends and the industry, as well as the research community, will continue to invest time and effort in the areas of interest and activity discussed above. Six general trends in spatial databases and spatial IT form the core areas of development in the foreseeable future.

6.2.1   Large Spatial Data Stores Including Warehousing

Integrated content databases, including approaches from data warehousing, database federation and data marts, satisfy the need for robust and standards-based platforms. In these environments spatial data management resides directly within the IT infrastructure, rather than isolated in a GIS that may have proprietary data access and formats. Hence, integrated content storage offers advantages in robustness, security, and scalability of the infrastructure within which the data reside.

The rapid collection of very large volumes of spatial or locationally-referenced data through the adoption of sensor Web technology as well as the use of *point cloud data* collection methods, GPS, RFID, and airborne or terrestrial *LIDAR* has created the need to develop new approaches to handling massive data storage as well as efficient data processing and visualisation. These needs, in turn, have lead to the recent adoption of enterprise-level grid computing and other methods for dealing with the previously unheard of volumes of spatial data that are now feasible. As with any other IT project the business problem driving the adoption of grid computing must be thoroughly understood and a business case made before embarking on the creation of this type of solution.

6.2.2 Improved Visualisation

As grid computing may provide a solution to the management and use of massive spatial data stores, the heterogeneity of data types and the growth, especially, in multi-media and 3-D data will require the development of new approaches to data visualisation. In particular, there is likely to be more seamless integration of 3-D and continuous data models for creating and displaying complex terrain as well as models of entire cities. Current constraints in network bandwidth and data streaming across networks remain substantial obstacles to implementing breakthroughs in these areas. However, this will change in the future with the adoption of gigabit and higher network speeds. Moreover, BI dashboard-type interfaces on the Web will likely facilitate linked views of information presented in maps, graphs, reports as well as 3-D visualisations that can be generated by easy-to-use query and report generation interfaces.

6.2.3 Improved Service-oriented Architectures

The fast growing trend toward adoption of *service-oriented architectures* (SOA) within mainstream IT and their integration with real time or near real time sensors for operational and decision support applications will also serve to drive future innovations in spatial databases and their management. In the same way that grid computing seeks to optimise the use of underutilised computing resources, a SOA-based approach seeks to develop collections of services that communicate with one another. Clearly, this concept is not new as it is at the heart of previous innovations such as DCOM and the ORB-type architectures discussed in Chapter 3. However, placing a SOA within the context of Web services, is a considerably newer concept.

Web map services within the context of spatial SOA are likely to become of elevated importance in the next half decade as Internet mapping, the use of XML and GML, and service-oriented functions associated with the growth of wireless technologies and LBS become more widely used. The presence of these newer systems architectures and standards will have a fundamental influence on the way that organisations go about building their internal systems and failure to keep up with the rapid change in technology may mean that lags will result in reducing the interoperability of internal and external systems. This, in turn, may impact on the overall viability of organisations that require rapid and frequent access to externally derived spatial data.

6.2.4   New and Improved Spatial Data Standards

The adoption of open spatial data standards within the IT industry is especially important given the incorporation of spatial data into mainstream database management. Such standards are needed for data formats, metadata, semantics, data content, and service interoperability. The degree of connectivity between systems that is sought can only be achieved by industry-wide participation in an open and consultative process that seeks to produce universally agreed upon and freely distributed standards for defining spatial features such spatial object classes and object names, setting standard rules for object behaviours and map topologies, specifying open programming interfaces, and other aspects of spatial data and metadata.

Without standards in place and widely used, the potential for spatial data interoperability and many of the above areas of future development in spatial databases and their management will be severely impeded. Numerous organisations, especially the OGC and ISO, have made significant advances in this regard, however a great deal remains to be done, especially in establishing broad-based industry acceptance of the standards that are currently in place and contributing to the development of new standards such as the OGC simple features for geography markup language (SFGML) and extensions to other standards, such as WMS and WFS, that are likely to grow in importance in the future.

6.2.5   Use of the Web

The impact of the Internet on spatial databases, as with all areas of IT, is already substantial and will continue to grow in the future. Spatial data are increasingly available on the Internet either for interactive on-line exploration through Internet map servers using WMS and WFS for spatial data browsing and analysis, or for download and use with a desktop GIS. The adoption of the Internet for spatial data access and use has possibilities that are currently difficult to imagine not only in terms of real time and near real time data access, but also in terms of collaborative interaction with live spatial databases which reduces, but not removes, the need for data resources and workers to be near one another.

One of the key benefits of Web services is their ability to reuse existing data and applications and deliver them as components or composite services. In this context, there is substantial growth in the number of companies using open online mapping APIs to create applications using existing protocols and standards such as SOAP, XML, Java and the relatively newer asynchronous Java and XML (AJAX) approach to Web page management. It is likely that this trend will continue to grow and spatial data content will become embedded within a host of, as yet, unimaginable applications. The recent formation of the Open Source Geospatial Foundation (OSGeo), which is similar in concept and function to existing open source foundations such as the Apache Foundation, underlines the emergent importance of open source geospatial Web technologies.

Recent developments in the above areas have revolutionised the current spatial database landscape. In the future much less time will be spent on building spatial data, and a data-centric environment that places emphasis on the need to understand spatial databases and their management rather than focus on technical aspects of the spatial data themselves will characterise the future spatial information industry. Virtually all enterprise-level database products now contain sophisticated geospatial capabilities, and all indicators suggest that these functions will be consolidated and grow in the future. Hence, it is likely that within the next five years the majority of applications that use spatial technologies will not be regarded as traditional GIS. While the need for specialised GIS applications will remain intact for the foreseeable future, their pre-eminence in the world of spatial data processing may decline into niche areas unless a convincing case can be made to challenge the spatial data management functions of mainstream database products.

7.   SUMMARY

This chapter has summarised the current state of spatial databases with a view toward likely future trends in the evolution of spatial databases and their management. Discussion first focused on trends in spatial information technology. Technology integration was discussed in Section 2, followed by a discussion of spatial data trends in Section 3. Spatial data usability and interoperability were initially discussed, followed by a review of advances in real-time spatial data processing, innovations in location tracking and sensor webs and the delivery of spatial data to end users. Section 4 discussed trends in spatial data applications including spatial application service providers and Web services, followed by a review of the continued emphasis being placed on location-based services. Applications in health care, law enforcement and public safety and business intelligence were briefly reviewed. Section 5 discussed the human factor in spatial database development in terms of a hierarchy from global to local concerns including the emergent role of public participation GIS relative to future developments. The chapter and the book concluded in Section 6 with a synoptic review of spatial database research trends and frontiers.

8.   REFERENCES

Anselin, L. (1998) "GIS Research Infrastructure for Spatial Analysis of Real Estate Markets", *Journal of Housing Research*, Vol. 9, Issue 1, pp. 113-133.

Anselin, L., Florax, R.J.G.M. and Rey, S.J. (Eds.) (2004) *Advances in Spatial Econometrics: Methodology, Tools and Applications*, Dordrecht, Netherlands: Springer.

Batty, P. (2004a) "Future Trends and the Spatial Industry, Part 1", *Geospatial Solutions*, Vol. 14, No. 7, pp. 36-40.

Batty, P. (2004b) "Future Trends and the Spatial Industry, Part 2", *Geospatial Solutions*, Vol. 14, No. 9, pp. 32-35.

Bédard, Y., Merrett, T. and Han, J. (2001) "Fundamentals of Spatial Data Warehousing for Geographic Knowledge Discovery", in *Geographic Data Mining and Knowledge Discovery* by Miller, H. and Han, J. (Eds.), London, UK: Taylor and Francis.

Botts, M. (ed.) (2004) *Sensor Model Language (SensorML) for In-situ and Remote Sensor*, Wayland, MA: Open Geospatial Consortium.

Boulos, M.N.K. (2004) "Towards Evidence-based, GIS-driven National Spatial Health Information Infrastructure and Surveillance Services in the United Kingdom", *International Journal of Health Geographics*, Vol. 3.

Bruce, C.W., Hick, S.R. and Cooper, J. P. (Eds.) *Exploring Crime Analysis: Readings on Essential Skills*, Overland, KS: IACA Press.

Buehler, K. (2004) "Interoperability: It's Mission-Critical", *GeoIntelligence*, Mar/Apr 2004, pp. 36-40.

Carver, S., Evans, A., Kingston, R. & Turton, I. (2000) "Accessing Geographical Information Systems over the World Wide Web: Improving public participation in environmental decision-making," *Information, Infrastructure and Policy*, 6, pp. 157-170.

Cooklev, T. (2004) *IEEE Wireless Communication Standards: A Study of 802.11, 802.15, and 802.16*, New York, NY: IEEE Standards Press.

CGDI (2001) *Canadian Geospatial Data Infrastructure: Architecture Description*, Ottawa, ON: CGDI Architecture Working Group.

Craglia, M. and Onsrud, H.J. (2003) "Workshop on Access to Geographic Information and Participatory Approaches in Using GIS: Report of Meeting and Research Agenda", *URISA Journal*, Vol. 15 (APA II), pp. 9-15.

Crompvoets, J. and Bregt, A. (2003) "World Status of National Spatial Data Clearinghouses", *URISA Journal*, Vol. 15 (APA I), pp. 43-50.

Datz, T. (2004) "What You Need to Know Service Oriented Architecture", *CIO Magazine*, January-15, 2004 Issue.

de Sherbinin, A. and R.S. Chen (2005) (eds.) *Global spatial data and information user workshop: Report of a workshop*, 21-23 September 2004.Palisades, New York: Socio-economic Data and Applications Center, Center for International Earth Science Information Network, Columbia University.

Delin, K.A. (2002) "The Sensor Web: A Macro-instrument for Coordinated Sensing", *Sensor*, Vol. 2, pp. 275-280.

Egenhofer, M. J. (1994) "Spatial SQL: a query and presentation language", *IEEE Transactions on Knowledge and Data Engineering*, 6, 1, 86-95.

Engelhardt, J. (2004) "Mobil GIS: Handhelds Get Rugged", *Geospatial Solutions*, Vol. 14, NO. 9, pp. 36-39.

ESRI (2003a) *Spatial Data Standards and GIS Interoperability* (An ESRI White Paper), Redlands, CA: Environmental Systems Research Institute, Inc.

ESRI (2003b) "Geospatial One-Stop Portal Is Key to President's E-Government Strategy", *Federal GIS Connections*, Redlands, CA: Environmental Systems Research Institute, Inc.

ESRI (2004a) *Versioning* (An ESRI Technical Paper), Redlands, CA: Environmental Systems Research Institute, Inc.

ESRI (2004b) *An Overview of ArcWeb Services*, (An ESRI White Paper), Redlands, CA: Environmental Systems Research Institute, Inc.

GeoConnections Secretariat (2004) *A Developer's Guide to the CGDI: Developing and Publishing Geographic Information, Data and Associated Services*, Ottawa, ON: GeoConnections Secretariat.

Gewin, Virginia (2004) "Mapping opportunities", *Nature*, 427, 376-377.

Gonzales, M.L. (2004) "Spatial Business Intelligence: The Spatial & Visualization Components for Effective BI", Eveleigh, NSW: Integeo.

Gonzales, M.L. (2005) "Components of a BI Dashboard: Spatial Data & Visualization", *DM Review*, March 25 Issue.

Goodchild, M.F. and Janelle, D.G. (2004) *Spatially Integrated Social Science*, New York, NY: Oxford University Press.

Grejner-Brzezinska, D. (2004) "Positioning and Tracking Approaches and Technologies", Chapter 3 of *Telegeoinformatics: Location-based Computing and Services*, by Karimi, H.A. and Hammad, A. (Eds.), Boca Raton, FL: CRC Press.

Hecht Louis and Kucera Barbara (1999) *"Toward improved geographic information services within a digital government"*, Report of the NSF Digital Government Initiative, Board of Trustees of the University of Illinois, National Computational Science Alliance, Champaign Il.

Huff, D.L. (2003) "Parameter Estimate in the Huff Model", *ArcUser*, October-December, 2003, pp. 34-36.

Hunter, G.J., Wachowicz, M. and Bregt, A.K. (2003) "Understanding Spatial Data Usability", *Data Science Journal*, Vol. 2, pp. 79-89.

Jankowski, P. and Nyerges, T. (2003) "Toward a Framework for Research on Geographic Information-supported Participatory Decision-making", *URISA Journal*, Vol. 15 (APA I), pp. 9-17.

Karimi, H.A. and Hammad, A. (Eds.) (2004) *Telegeoinformatics: Location-based Computing and Services*, Boca Raton, FL: CRC Press.

Keating, Gordon N., Rich, Paul M., and Marc S. Witkowski, (2003) "Challenges for enterprise GIS", URISA Journal, 15, 2, 23-36.

Kraak, M.-J. and Van Driel, R. (1997) "Principles of Hypermaps", *Computers and Geosciences*, 23, 4, pp. 457-464.

Lang, L. (2000) *GIS for Health Organisations*, Redlands, CA: ESRI Press.

Laurini, R., Paolino, L., Sebillo, M., Tortora, G. and Vatiello, G. (2004) "A Spatial SQL Extension for Continuous Field Querying", *Proceedings*, IEEE 28th Annual International Computer Software and Application Conference (COMPSAC 2004), Hong Kong, China.

Leitner, H., R.B. McMaster, S. Elwood, S. McMaster, and E. Sheppard, (2002) "Models for Making GIS Available to Community Organizations: dimensions of difference and appropriateness, in Craig, W., T. Harris and D. Weiner (Eds.), *Community Participation and Geographic Information Systems*, (London: Taylor and Francis), 37-52.

Lopez, X. (2004) "Location-based Services", Chapter 6 of Karimi, H.A. and Hammad, A. (Eds.) *Telegeoinformatics: Location-based Computing and Services*, Boca Raton, FL: CRC Press.

Lowe, J.W. (2004) "Geospatial Web Portals", *Geospatial Solutions*, Vol. 14, No. 7, pp. 42-45.

McKee, L. (2003) *The Spatial Web* (An OpenGIS White Paper), Wayland, MA: Open Geospatial Consortium.

Malerba, D., Esposito, F., Lanza, A., Lisi, F.A. and Appice, A. (2003) "Empowering a GIS with Inductive Learning Capabilities: The Case of

INGENS", *Computers, Environment and Urban Systems*, Issue 27, pp. 265-281.

Malinowski, E. and Zimányl, E. (2004) "Representing Spatiality in a Conceptual Multidimensional Model", *Proceedings*, 12th ACM International Symposium on Advances in Geographical Information, Washington, DC.

McLafferty, S.L. (2003) "GIS and Health Care", *Annual Review of Public Health*, No. 24, pp. 25-34.

NASA (no date) "NASA Sensor Webs Project Briefings" available at http://sensorwebs.jpl.nasa.gov.

OGC (2003) *OpenGIS Reference Model*, Version 0.1.2, by Buehler, K. (Ed.), Wayland, MA: Open Geospatial Consortium.

OGC (2004) *OpenGIS Location Service (OpenLS)* by Mabrouk, M. (Ed.), Wayland, MA: Open Geospatial Consortium.

Onsrud, H.J. and Craglia, M. (2003) "Introduction to the Second Issue on Access and Participatory approaches in Using Geographic Information", *URISA Journal*, Vol. 15 (APA II), pp. 5-7.

Rivest, S., Bédard, Y. and Marchand, P. (2001) "Toward Better Support for Spatial Decision Making: Defining the Characteristics of Spatial On-line Analytical Processing (SOLAP)", *Geomatica*, Vol. 55, No. 4, pp. 539-555.

Rossmo, K. (1999) *Geographic Profiling*, Boca Raton, FL: CRC Press.

Schlossberg, Marc and Shuford, Elliot (2005) "Delineating "Public" and "Participation" in PPGIS, *URISA Journal*, 16, 2, pp. 15-26.

Smith, J., Mackaness, W., Kealy, A. and Williamson, I. (2004) "Spatial Data Infrastructure Requirements for Mobile Location Based Journey Planning", *Transactions in GIS*, Vol. 8 Issue 1, pp. 23-32.

Smith, T., Peuquet, D., Menon, S. and Agarwal, P. (1987) "KBGIS II: A Knowledge-based Geographic Information System", *International Journal of Geographical Information Systems*, Vol. 1, No. 2, pp. 149-172.

Tang, W. and Selwood, J. (2003) *Connecting Our World: GIS Web Services*, Redlands, CA: ESRI Press.

Turban, E., Aronson, J.E. and Liang, T.-P. (2005) *Decision Support Systems and Intelligent Systems*, 7th Ed., Upper Saddle River, NJ: Prentice Hall.

Waters, N. (2001) "Internet GIS: Watch Your ASP", *GeoWorld*, Vol. 14, No. 6, pp. 26-28.

Wehr, A. and Lohr, U. (1999) "Airborne Laser Scanning – An Introduction and Overview", *ISPRS Journal of Photogrammetry and Remote Sensing*, Vol. 59, pp. 68-82.

Weiner, D., Harris, T.M., Craig, W.J. (2002) "Community participation and geographic information systems" In Craig, W.J., Harris, T.M. and

Weiner, D., (eds.), *Community Participation and Geographic Information Systems*, London, Taylor and Francis, pp. 3-16.

Wilson, J.D. (2000) "Mobile Technology Takes GIS to the Field", *GeoWorld*, Vol. 13, No. 6, pp. 32-36.

Wilson, J.D. (2001) "The Next Frontier – GIS Empowers a New Generation of Mobile Solutions", *GeoWorld*, Vol. 14, No. 6, pp. 36-40.

Glossary of Terms

Chapter 1	**The Current State of Spatial Information Technology**
Computational intelligence	A branch of the study of artificial intelligence that seeks to use learning, adaptive, or evolutionary algorithms to create programs that are, in some sense, intelligent. Computational intelligence research either explicitly rejects statistical methods (e.g., fuzzy logic), or tacitly ignores statistics (e.g., most neural research).
Spatial data type	A form of data based on the mathematical concepts of metric, *topology* and order. Constituent *data types* provide a fundamental abstraction for modelling the geometric structure of *objects* in space, their *relationships*, properties and *operations*. Their definition is to a large degree responsible for a successful design of spatial data *models* and the performance of spatial *database systems* and exerts a great influence on the expressive power of *spatial query* languages.
Spatial datamart	A small, single-subject area spatial *data warehouse* subset that provides decision support to information users from a specific department or business *function* of an organisation.
Spatial indexing	A mechanism to facilitate access to a spatial database by means of stored coordinates in two-dimensional space. There are many different indexing options, such as R-tree, quadtree and B-tree, each of which has its strengths and weaknesses depending on the specific data format and applications.

Spatial operator

A suite of data processing *functions* and processes that can be achieved through the use of Structured Query Language (SQL) to query and retrieve selected database contents, join database *tables* according to specific spatial and non-spatial criteria, and generate the results of processing in specific formats.

Chapter 2

Concepts and Architecture of Database Systems

Application programming interface (API)

A means of executing database procedures and queries through a set of routines, protocols and tools that can also be used for building general software applications.

B-tree index

"Balanced" index format for rapid search and information retrieval within a *database system* that is composed of one or more levels of branch blocks and a single level of leaf blocks. The branch blocks contain information about the range of values contained in the next level of branch blocks.

Client/server computing

Client/server computing describes the *relationship* between two computer programs in which one program, the *client*, makes a *service* request from another program, the server, which fulfils the request. Although the *client*/server idea can be used by programs running on a single computer, it is commonly applied across a network. In a network, the *client*/server *model* provides a convenient way to interconnect programs and share resources that are distributed efficiently across different locations.

Client-side extensions

Components of a Web-browser running on a *client* computer that can include *plug-ins* as well as Java, JavaScript and other routines that allow sites to extend the behaviours and interfaces of their *services* with portable user-interface elements that integrate transparently into the browser's existing interface.

Custodian system

A large-scale, subject-oriented data store set up to serve the long-term information *needs* of an organisation or a community of users. National, provincial, and regional natural resource inventories and topographic mapping systems are typical examples of custodian *database systems*.

Data sharing

A primary feature of a *database management system* (DBMS) that allows the same data resource to be shared by multiple applications or users. It implies that the data are stored in one or more servers in a network and that there is some software locking mechanism that provides concurrent access while preventing the same set of

	data from being changed by two people at the same time.
Data warehouse	A construction that differs from a conventional database in that it is not only a data repository and an information management tool, but it also represents a new approach to managing and thinking about data. Geared towards the business intelligence requirements of an organisation by integrating data from the various operational systems. It also contains historical information that enables *analysis* of business performance over time.
Database transaction	A more complicated process than a *database query* because of the need to handle possible conflicts caused by concurrent transactions (that is, when two or more users access the *database system* and attempt to change the same values at the same time).
Domain constraints	A database integrity constraint that sets the *permissible values* that can be stored in the columns of a *table* in a relational database, such as numeric, character or string, Boolean, date and time, and user-defined. Differs from other types of *constraints* that RDBMS have provided for a long time. These include *referential constraints* (to define primary and foreign *keys*) and unique *constraints* (to prohibit duplicate entries).
Entity	A real world referent that can also represent an abstract concept contained in the rows of a *table* in the relational database *model*. It has a unique name, which is always singular and is used for the *entity type* and entity class. Also called a data *object* or simply an *object*.
Entity instance	Each row in a relational *table* for a specific *entity* represents an instance of this *entity* - one land parcel in a *table* containing all land parcels in a jurisdiction.
Fat client model	A configuration of *client/server computing* that does most of the work on the *client*-side than on the server.
Fat server, thin client model	A configuration of *client/server computing* that places more processing emphasis on the server than on the *client*.
Information system	An interconnected set of information resources under the same direct management control that shares common functionality. An information system normally includes hardware, software, information, data, applications, communications, and people.

Spatial database systems	A form of *database system* including Geographic Information Systems (GIS) that are designed to process location-based data pertaining to land resources, the physical and biological environment, as well as human activities associated with resources and the environment.
Thin client configuration	A three-tier architecture where the data processing processes are moved from the *client* to the application server tier. This shift of workload away from the *client* means that there is no longer any need for using costly high-end computers as *client* machines.
Three-tier client/server architecture	An extension of the two-tier architecture where the *functions* of each tier are dependent on individual implementations with the tiers usually configured as a *client*, an application server and a *database server*, which is also often known as a *thin client configuration*.
Universal access architecture	A *database system* that allows users within an organisation (in an *intranet*) as well as users outside of the organisation (in an *extranet*) to access its databases dynamically.
Chapter 3	**Database Models and Data Modelling**
Cardinality	The number of rows in a *table* or the number of indexed entries in a defined index that participate in a *relationship*.
Common object request broker architecture (CORBA)	A language-independent *middleware* technology standard that manages communication and data exchange between *objects* in *object*-oriented programming and databases. Endorsed by the *OMG* for the *Object Management Architecture* of *client*/server-based distributed *database systems*. More commonly known by its acronym, CORBA.
Computer-aided software engineering (CASE)	A structured approach developed to automate the systems development activities in a *SDLC* and, by extension, in a *DBDLC*. Often used by its acronym CASE.
Database development life cycle (DBDLC)	A generic description of the processes of developing a database in six phases, namely database initial study, database design, implementation and data loading, testing and evaluation, operationalisation, and maintenance and monitoring. Often referred to by its acronym DBDLC.
Domain	The smallest unit of data representation and set of valid values that a column or *data type* can take on in a relational database *table*. Smallest in this case means that the value of a *domain* that is not divisible into smaller components.

Entity-relationship model	A method of conceptual data modelling that depicts the real world *objects* to be stored in the database and their respective characteristics placed into a framework that describes how one *object* is related to another in the database and any *constraints* that may exist in the *relationships*.
Key	A set of one or more columns in a database *table* whose values, in combination, are required to be unique within the *table*. A primary *key* ensures that no two records in a database contain the same value for that field. It is the field that uniquely identifies the record. The primary *key* must be unique, stable, minimal and non-*null* under all conditions. A foreign *key* is the primary *key* of one *data structure* that is placed into a related *data structure* to represent a *relationship* among those structures. Foreign *keys* resolve *relationships*, and support navigation among *data structures*.
Metamodel	In the context of UML, this is a set of definitions that describe in fairly precise syntax the underlying meaning of each *element* used in visual modelling and the *relationships* among all elements in a four-layer architecture. Also a *model* that explains a set of related *models*.
Normal form	A relational *table* is said to be a particular *normal form* if it satisfied a certain set of *constraints*. Currently five *normal forms* are defined, but only 3 are practically used (1NF, 2NF, 3NF). A relational *table*, by definition, is in first *normal form* if all values of the columns are atomic in that they contain no repeating values. A relational *table* is in second *normal form* if it is in first *normal form* and every non-*key* column is fully dependent upon the primary *key*. Third *normal form* requires that all columns in a relational *table* are dependent only upon the primary *key*.
Normalisation	The process of reducing a complex *data structure* into its simplest, most stable structure. In general, the process entails the removal of redundant *attributes*, *keys*, and *relationships* from a conceptual data *model*.
Object management group (OMG)	An industry-sponsored standardisation organisation formed in 1989 to establish industry guidelines and detailed *object* management *specifications* that serve as a common framework for *object*-oriented application development across different hardware platforms and operating systems.

Schema   The structure of a *database system* described in a
                                      formal language supported by the *database
                                      management system* (DBMS). In a relational
                                      database, this defines the *tables*, the fields in each
                                      table, and the *relationships* between fields and
                                      tables that are stored in a *data dictionary*.

Systems development life   A generic description of the process of developing
cycle (SDLC)   a *database system* in six phases, namely planning,
                                      analysis, design, building, implementation and
                                      maintenance. Often referred to by its acronym
                                      SDLC.

Unified modelling   A non-proprietary, diagrammatic language for
language (UML)   modelling, designing and visualising object-
                                      oriented systems. It can be used for modelling
                                      hardware (engineering systems) and is commonly
                                      used for *business process modelling* and
                                      organisational structure modelling. Commonly
                                      abbreviated to UML.

Chapter 4   **Spatial Databases and Spatial Database Systems**
Geodatabase   The spatial database *model* of ESRI's ArcGIS
                                      software that stores various types of spatial data,
                                      topology, *attibute* data and *metadata* all using a
                                      single DBMS. In such a database, spatial data that
                                      share the same *attibutes* (that is, data of the same
                                      feature class) are stored in a single *table*.

Geographic object   An identifiable discrete real world feature or
                                      phenomenon represented by a point, a line or a
                                      polygon, or a square grid cell, or a picture element
                                      (pixel), nestled within a tessellation or rectangular
                                      grid of equally sized pixels.

Geometry object model   A hierarchy of *spatial data types* proposed by the
                                      Open Geospatial Consortium (OGC) that allow
                                      spatial *features* to be represented in a database. In
                                      the *geometry object model*, the word "*geometry*" is
                                      used to represent a spatial feature as an "*object*"
                                      having at least one *attibute* of a *geometric type* in a
                                      database.

Geo-relational data model   A spatial data *model* where data are abstracted into
                                      a series of independently defined *layers*. Each of
                                      these *layers* represents a selected set of associated
                                      spatial *features* such as roads, soil types, land
                                      cover, land parcel and drainage.

Open geospatial   A non-profit organisation that groups more than
consortium (OGC)   250 commercial, governmental, other non-profit
                                      and research organisations worldwide, encouraging
                                      and prescribing *standards* for GIS data processing
                                      and exchange.

Primal-dual multi-valued vector map	This concept is a mathematical graph structure that is derived from partitioning of a 2-dimensional plane into discrete and discernable features. A dual graph is created from the planar graph by adding a vertex for each region and a universe vertex. Edges are added between region vertices that share a common region boundary in the underlying planar graph, and all parallel edges are deleted. In the resulting constructions vertices can take on multiple values representing geographic coordinates as well as qualitative and quantitative attributes that have database representation. Such a spatial data model allows any real world feature as well as its geometric, topological and descriptive attributes to be modelled using the same construction.
Workspace management	A technique used by Oracle Corporation to handle long transactions as well as using workspaces as a virtual environment to isolate collections of changes to production data, histories of changes to data, and maintaining multiple data scenarios for "what if" modelling and *analysis*.
Chapter 5	**Spatial Data Standards and Metadata**
De jure standards	These *standards* come from the literal meaning of *de jure* (according to the law) and are generally known as public or industry *standards* that are established by public bodies. These *standards* are endorsed and disseminated by official *standards* organisations.
Extensibility	In the context of *metadata* this allows for *profiles* to be developed so that particular *needs* of a given application can be accommodated without unduly compromising the functionality of the base *metadata schema*.
Implementation specifications	Technical details of how to use *standards*. An implementation *specification* specifies how the conceptual framework described by an abstract *specification* can be put into practice by using specific interface requirements, data encoding methods and structures, programming language constructs.
Industry de facto standards	From the literal meaning of *de facto* or "as a matter of fact", these *standards* are generally created by a specialized product or company with market dominance. They may be widely used and implemented, but controlled by a single vendor or group.
Namespace	A formal collection of terms managed according to a policy or algorithm that provide the mechanism

to ensure global uniformity in the vocabulary used by a *metadata* standard.

Open standards — *Standards* developed according to the ability of stakeholders to participate voluntarily, without royalty and without discrimination in *standards development*, the use of consensus in the review and *standards* approval process, as well as public access to all development documents and ultimately to the completed *standards* themselves.

Proprietary standards — A standard that is exclusively owned by and unique to an individual or organisation, the use of which generally would require a license and/or a fee.

Public domain standards — Open or *public domain standards* that are not owned by a particular organisation. Such *standards* are developed by members of a user community collectively rather than by an identified organisation.

Spatial metadata — A special type of *metadata* that is associated with a spatial database, a spatial data set, or a particular class or instance of spatial *features*.

Standards — A document or collection of documents, usually but not always published, that establishes a common reference framework, language, terminology, accepted practices and levels of performance, as well as technical requirements and *specifications*, that are used consistently for the development and use of products, *services* and systems.

Standards development — The process by which national and international standardisation agencies, such as the International Organisation for Standardisation (ISO), prepare and eventually declare the existence of a standard. In the case of the ISO, this follows six stages: proposal, preparation, committee discussion, enquiry, approval and publication.

Standards organisation accreditation — The predetermined minimal criteria or conditions set by an accrediting board, agency or body that must be satisfied during the process of an accreditation evaluation. The result of this process is usually the awarding or not of a status of being "accredited" according to the *rules* and expectations that are set by the accrediting body.

Chapter 6 — **Spatial Data Sharing, Warehousing and Database Federation**

Broker — A *service metadata portal* that registers the *services* of server *nodes* and facilitates the discovery of these *services* by *client* computers across a network.

Component object model (COM)	The *object* request *broker specification* and implementation developed by Microsoft to provide a framework for integrating software components in the Windows environment running on the same computer. *Interoperability* of the *component object model* is achieved by defining *application programming interface* (API) components using a binary structure that allows interoperation no matter which programming language a component is written in.
Data mart	A subject-specific *data warehouse* that is usually set up to meet the information *needs* of users of a particular department or functional unit within an organisation.
Database federation	A database architecture for data *integration* in which *middleware*, consisting of a relational *database management system* that provides uniform and simultaneous access to several heterogeneous *data sources*. This may remove or reduce the need for creating separate *data marts*.
Database heterogeneity	Describes the inconsistencies between spatial data sets that result from different geographic projections and datums, errors in measurement, different definitions of precision, fuzziness of spatial *objects*, and variations in the use of terminology and nomenclature.
Distributed COM (DCOM)	The distributed *component object model* is a set of Microsoft concepts and program interfaces in which *client* program *objects* can request *services* from server program *objects* on other computers in a network. DCOM is based on the *component object model*.
Enterprise data warehouse	A repository of data derived from operational *data sources* within an organisation that that can be defined and shared across the whole enterprise along the lines of common dimensions to be used for *analysis*. Typically contains hundreds of gigabytes or terabytes of data.
Federated database system	This type of *database system* is an ordinary *database management system* that is configured to serve as the control centre of a *database federation*. It shields database users from the need to know what the sources are, where they are stored, how they are modelled and managed, what hardware and software they run on, how they are accessed.
Interoperability	In terms of databases, this refers to the transparent exchange and *integration* of data drawn from different databases independent of format or

	origin. From a more general computing perspective, the capability of two or more hardware devices or two or more software routines to work harmoniously together.
Java database connectivity	An *application programming interface specification* for *Java applets*, servelets and applications to access data stored in databases, spreadsheet and text files, used to connect a user application to a *data source*, regardless of what *database management system* is used to control the database.
Object linking and embedding (OLE)	The *de facto* standard for data access in the Microsoft Windows environment that allows users to access and exchange data between Microsoft software products, for example creating a PowerPoint presentation with a linked or embedded Access database *table*.
On-line analytical processing (OLAP)	A category of software technology that enables analysts, managers and executives to gain insight into data through fast, consistent, interactive access to a wide variety of possible views of information that is transformed from raw data to reflect the real dimensionality of the enterprise as understood by the user. This is typically implemented in a multi-user *client*/server mode and offers consistently rapid response to queries, regardless of database size and complexity.
On-line transaction processing (OLTP)	A class of program that facilitates and manages transaction-oriented applications, typically for data entry and retrieval transactions, that uses *client*/server processing and *broker*ing software that allows transactions to run on different computer platforms in a network.
Ontology	A controlled vocabulary that describes real world *objects* and phenomena and the *relations* between them in a formal way. A grammar exists for using the vocabulary terms to express *relationships* within a specified *domain* such as spatial databases. The vocabulary is used to make queries and assertions. Ontological commitments are agreements to use the vocabulary in a consistent way for knowledge sharing.
Open database connectivity (ODBC)	An open standard *application programming interface (API)* for accessing one or more database from an application. Initially created by the SQL-Access Group in 1992 and extended and deployed by Microsoft as a standard *call-level interface* that allows *client* and server applications to exchange and share data without the need to know anything about each other.

Service	The foundation of networked computing. Network *services* are installed on one or more servers that provide user authentication at log-on, directory *services*, dynamic *host* configuration (DHCP), *domain name services*, e-mail *services*, printing, and Web hosting, among many other types of network computing *services*
Simple object access protocol (SOAP)	Based on *XML*, this is the standard for transmitting *messages* between computing *nodes* and *Web-services messages*. Also known as the service-oriented access protocol.
Web services protocols	Software components that can be accessed over the *Internet* through *standards*-based protocols that can include desktop application programs, Web browsers, *Java applets*, and software running on mobile devices such as cellular phones.
Web services-based architecture	A generic *standards*-based computing *model* that provides an open and interoperable environment for *data sharing* over the *Internet*.
Chapter 7	**User Education and Liability of Spatial Data Services**
Civil law	A unified legal system that allows, with maximum precision, conclusions to be drawn from its basic principles.
Common law	Past decisions of courts of law that are used as precedents or case law by judges in their rulings. The *rules* of *common law* have the authority of statutory laws unless they are specifically overridden by legislation.
Copyright	Protects against the right to copy. *Copyright* laws are the legal instruments used to protect the *intellectual property* of the creators of original literary, dramatic, musical and artistic works. The term "work" in the context of *copyright* laws suggests the use of a certain degree of skill and effort, originality and creativity to create the IP that is to be protected.
First sale doctrine	This prevents the owner of the *copyright* in a work from controlling subsequent transfers of that work. There is an exception to this doctrine with respect to two types of work, namely computer programs and sound recordings, which cannot be disposed of at the will of licensees.
Intellectual property	The legal definitions that govern individual and/or organisational ownership of data, databases and software in spatial database implementation. The 'property' refers to an abstract construct that is created through human intelligence or inventions in art, industry, science and technology.

Intellectual property right	A concept that embraces the various intangible rights that are granted by law to enable *intellectual property* owners to restrict others from using their ideas or inventions without prior approval.
Liability	The condition of being liable or answerable by law or equity for any discretion from the prevailing legal code.
Notification of copyright	The form of the *copyright* notice that is used for visually perceptible copies (those that can be seen or read, either directly (such as books) or with the aid of a machine (such as films) and those that refer to sound recordings (such as compact disks or cassettes)) of original works. Usually contains three elements, the *copyright* symbol ©, the year of first publication, and the name of the owner of *copyright* of the work. In the United States this was mandatory prior to March 1, 1989. On this date and thereafter use of *copyright* notification is optional.
Patent	A set of exclusive rights granted by a government to a person for a fixed period of time in exchange for the regulated, public disclosure of certain details of an invention. The person applying for a patent does not need to be the inventor who created or authored the invention.
Sui generis	Latin term used in *copyright* law that protects the content of a database. Literally translated means 'of its own right' and this is interpreted as being independent of any *copyright* already existing on the database.
Tort	A legal wrong. *Tort* law refers to that body of law where the plaintiff is the victim of an alleged wrong and the unsuccessful defendant is directed by the court to pay damages to the plaintiff (the usual remedy) or else to desist from the wrongful activity.
Transferability of right	A covenant that allows any or all of the *copyright* owner's exclusive rights or any subdivision of those rights may be transferred by contract to third-parties. The transfer of exclusive rights is not valid unless that transfer is in writing and signed by the owner of the rights conveyed by such owner's duly authorized agent. Transfer of a right on a nonexclusive basis does not require a written agreement.
TRIPS agreement	An international agreement introduced by the World Trade Organisation that sets minimum *standards* for most forms of *intellectual property* regulation (although length of time for compliance

is longer for developing countries). Acronym for Trade-Related *Intellectual Property Rights* agreement that introduced *intellectual property* law into the international trading system for the first time, and remains the most comprehensive international agreement on *intellectual property* to date.

Chapter 8 — **User Needs Assessment and Multi-user Spatial Solutions**

Continuous quality assurance
Best practice-oriented component of *database system* design and implementation that uses the problem solving/problem prevention cycle in modern manufacturing industry that emphasises the detection of errors before they occur and the prevention of problems before and after delivery of the system to users.

Data flow diagram
An approach used in *business process modelling* that depicts the movement of data between processes, data stores (that is, data in a database or in temporary memory) and interfaces (that is, sources and destinations of data).

Data structure-oriented approach
An approach used in *business process modelling* that focuses on the identification of information entities and the actions that are applied to them. The end product of *data structure*-oriented process modelling is a *structure chart* that depicts the procedural logic of application programs in a basic format consistent with the layout of structured computer programs.

Joint application development (JAD)
An approach introduced by IBM in the late 1970s as a means of bringing together systems developers and users with different backgrounds and opinions to explore the design requirements of new computer systems in a productive and creative environment. Now this technique is formalised as a focus group-like sequence of workshops designed to elicit, verify, decompose and prioritise these *needs* in an analytical and structured manner.

Needs management
The front end of the product development process. Involves many aspects of activity including collecting and understanding *needs*; understanding regulatory, safety and legal issues; reconciling internal and external requirements; decomposing all requirements into logical groups for *analysis* purposes; mapping of requirements onto solutions; validating proposed solutions with use cases, test cases, and designs; provide traces between solutions and results to assess the impacts of change relative to the satisfaction of *needs*.

Object-oriented approach	An approach used in *business process modelling* to create a representation of a real-world problem *domain* and map it into a solution *domain*. This approach is used in conjunction with *object*-oriented design and *object*-oriented programming to form a set of *software engineering* activities for the construction of a *object*-oriented systems.
Rapid prototyping	Strictly defined as a special class of machine technology that quickly produces *models* and prototype parts using an additive approach so that both its designers and users can learn more about a *user need* or proposed solution to a particular need. More generally defined as a process that creates parts in an additive, layer-by-layer manner.
Chapter 9	**Project Management for Spatial Database Implementation**
Gantt chart	A horizontal bar chart developed as a production control tool in 1917 and frequently used in *project management*. It provides a graphical illustration of a schedule that helps to plan, coordinate, and track specific tasks in a project. *Gantt chart*s may be simple versions created on graph paper or more complex automated versions created using *project management* tools such as Microsoft Project.
ISO 9000 certification	A series of *standards* established in the 1980s by countries of Western Europe and now used world-wide as a basis for judging the adequacy of the *quality control* systems of companies. Revised to ISO 9001 in 2000 to promote a process based approach to increase the effectiveness of the quality management system in translating customer requirements to customer satisfaction.
Organisation chart	A flow chart or diagram that shows the departments of an organisation and the lines of responsibility that exist between them. Usually arranged hierarchically to illustrate the lines of reporting upward to senior management. Often referred to as an 'Org Chart'.
Project binder	The *project binder* documents a project as it proceeds. It contains all of the important documentation for the project. All of the material in the binder can be revised as the project progresses, but nothing should be replaced. If a revision occurs, this should be documented with a new date and explanation, and added to the binder. The binder is a living and evolving document describing the life of a project.
Project management life cycle (PMLC)	The tasks that are required to manage a project. This is always the same, regardless of the *project life cycle* that is being used. Project tasks and

	project management tasks are concurrent and ongoing and are associated by *project management* deliverables.
Quality assurance	Work done to ensure that quality is built into work products, rather than defects. This is by identifying what quality means in context, specifying methods by which its presence can be ensured, and specifying ways in which it can be measured to ensure conformance.
Quality control	The measurement of both products and processes for conformance to quality requirements (including both the specific requirements prescribed by the product *specification*, and the more general requirements prescribed by *quality assurance*). It identifies acceptable limits for significant quality *attibutes*, identifies whether products and processes fall within those limits (conform to requirements) or fall outside them (exhibit defects), and allows reports to be filed.
Request for a proposal (RFP)	An announcement distributed by a contractor to potential contractees defining the terms of reference for required work. This is used to assess competing bids on the work that is called for in the RFP.
Responsibility chart	A chart or *table* that lists tasks that need to be performed along the left side and the names project team members on the top. This forms a matrix of tasks against names in which a cross or check mark is placed under each member's name across from the task that he or she is responsible for. Also used in the context of RASCI responsibility *models* (responsible, accountable, supportive, consulted, informed) for workflow management.
Revision control	The management of multiple revisions of the same unit of information. It is most commonly used in engineering and software development to manage ongoing evolution of digital documents like application source code, art resources such as blueprints or electronic *models* and other critical information that may be worked on by a team of people. Changes to these documents are identified by incrementing an associated number or letter code, termed the revision number, revision level, or simply revision and associated historically with the person making the change. Also known as version control.
Total data quality management (TDQM)	A comprehensive and structured approach to organisational management from the customer's point of view that seeks to improve the quality of

products and *services* through ongoing refinements in response to continuous feedback. This may be defined separately for a particular organisation or may be in adherence to established *standards*, such as the International Organisation for Standardisation's ISO 9001 series. It originated in the manufacturing sector and has since been adapted for use in almost every type of organisation including schools, highway maintenance, hotel management, and churches.

Chapter 10

Web-enabled Spatial Database Systems

Application service provider (ASP)

An organisation that deploys, *hosts* and manages access to packaged applications for multiple parties from a centrally managed facility. The applications are delivered over networks on a subscription basis. This delivery *model* speeds implementation, minimizes the expenses and risks incurred across the application life cycle, and overcomes the chronic shortage of qualified technical personnel available in-house. Equivalent of a *spatial application service provider.*

Domain name

The word sequences a user enters in the address bar of an *Internet* browser to visit a specific *web site*. Each *domain name* is assigned to an *IP address*. A *Domain Name System* is used to translate IP addresses into words.

Domain name server (DNS)

This server runs the system that matches the URL of a website with its numeric IP address. Whenever a user requests a web page, the web browser consults the *domain name server* to find out what the numeric translation of the URL is. This is necessary because computers only understand the numeric IP address, whereas humans prefer to use meaningful and more memorable text.

Domain name system (DNS)

A general purpose distributed, replicated, data query *service* that runs on the *domain name* server. This system links *host IP addresses* with *host names*. The style of *host* names used in the *Internet* is called the *domain name*. Some important *domains* are .com (commercial), .net (network *operations*), .edu (education in the United States). Most countries also have *domains* such as .ca (Canada) and .nz (New Zealand).

GML

A non-propriety dialect of *XML*, based on the *OGC Abstract Specification*, that makes possible the full *integration* of geographic information into daily business applications in the enterprise computing environment that is increasingly *XML*-based. Acronym for Geography *Markup Language*

Host	Any computer that has full two-way access to other computers on the *Internet*. A *host* has a specific "local or *host* number" that, together with the network number, forms its unique *Internet Protocol (IP) address*. Point-to-Point Protocol access to an access provide assigns a unique IP address for the duration of any connection to the *Internet* and the connecting computer is a *host* for that period. In this context, a *host* is also a *node* in a network. Also, a *host* is a computer with a *Web server* that serves the pages for one or more *Web sites*. A *host* can also be the company that provides this *service*, which is known as hosting.
Hypertext transfer protocol (HTTP)	Commonly abbreviated to HTTP, this is the protocol that sits above *TCP/IP* to enable access to and exchange of resources residing in computers anywhere on the *Internet* by using a *Web* browser. Formally, HTTP is a stateless (that is, transient and temporary) remote procedure call.
Internet mapping server	A Web-enabled spatial database component running from a server connected to the *Internet* that is responsible for spatial processing in response to *client* requests, also known as a *Web mapping engine* or Web map server.
Internet Protocol address (IP)	A 32-bit number that identifies each sender or receiver of information that is sent in *packets* across the *Internet*. The *Internet* Protocol part of a *TCP/IP* transmission includes the *IP address* in the *message* (actually, in each of the *packets* if more than one is required) and sends it to the *IP address* that is obtained by looking up the *domain name* in the *Uniform Resource Locator* (URL) or in the e-mail address. At the other end, the recipient can see the *IP address* of the Web page requestor or the e-mail sender and can respond by sending another *message* using the *IP address* it received.
Octet	An *octet* is 8 bits. It is equivalent to a byte, as long as the byte is also 8 bits. Bytes range from 4 - 10 bits, but *octets* are always 8 bits.
OGC abstract specification	A common *model* of geography developed by the *Open Geospatial Consortium* and agreed to by the vast majority of spatial database software vendors in the world. In effect a *de facto standard* for spatial data.
SVG	An *XML*-based *specification* was created by the *World Wide Web Consortium* specifically for describing 2-dimensional graphics in three forms, namely graphic *objects*, namely vector graphic

shapes, images and text. Acronym for scalable vector graphics.

Transmission control protocol/internet protocol (TCP/IP)

A two-layer program that is the basic communication language or protocol of the *Internet*. It can also be used as a communications protocol in a private network (either an *intranet* or an *extranet*). With direct access to the *Internet* a computer is provided with a copy of the TCP/IP program just as every other computer that you may send *messages* to or get information from also has a copy of TCP/IP.

Web mapping engine

Back-end spatial data processing *functions*, also known as a Web map server, that are collected as a constantly running program that "listens" for requests requiring spatial *operations* from *client* computers, such as selecting specific *features* within a user-defined window, identifying *attibutes* of specific *objects*, overlaying *layers*, creating buffers, and performing *spatial joins* on data *tables*.

Web mapping server implementation specification

A *specification* from the *OGC* to set up a *Web-enabled mapping system* that contains a set of common interfaces for *client* computers (called a *Web* map viewer) to query, request and display spatial information from remote spatial databases (called map servers). Often abbreviated to the acronym WMS.

Web server

A computer that delivers *Web* pages. Every *Web server* has an *IP address* and possibly a *domain name*. For example, the URL http://www.fes.uwaterloo.ca/crs/gp555/index.html sends a request to the server whose *domain name* is fes.uwaterloo.ca. The server then fetches the page named index.html from the path \crs\gp555\ and sends this to the user's browser. Any computer can be turned into a *Web server* by installing server software and connecting it to the *Internet*.

XML

A *World Wide Web Consortium* (W3C) initiative that allows information and *services* to be encoded with meaningful structure and semantics that computers and humans can understand. This language can easily be extended to include user-specified and industry-specified *markup* tags. Acronym for eXtensible Markup Language.

Chapter 11

Spatial Data Mining and Decision Support

Active server pages (ASP)

A *specification* that enables database-driven *Web sites*. Web pages that have an .asp extension (instead of an .html or .htm extension) are rendered 'actively' using updated information from the

	database. This enables instant updating and easier content management. It can also present security problems because it allows information to be accessed and viewed in real time. Abbreviated to ASP.
ActiveX controls	Similar to a Java applet. However, unlike *Java applets*, ActiveX controls have full access to and are limited to the Windows operating system for sharing information among different applications. This gives them potentially much more power than *Java applets* within this environment.
Concept hierarchy	A hierarchically organised collection of domain concepts used in data mining. The organising relationship is "part-of".
Decision trees	A process used in *data mining* from a training data set in a top-down, general-to-specific direction by recursively partitioning the data set until a completely discriminating tree is obtained. The discriminating tree is then "pruned" to generalise branches that are too specific with respect to a predefined threshold.
Inductive databases	Systems that allow collaborative decision support in a distributed network environment by deploying the *data mining* capabilities of *database systems* to work on databases belonging to different organisations at different locations (or different databases at the same location for the same organisation).
Intelligent decision support system	A person-computer system with specialised problem-solving expertise. The expertise consists of knowledge about a particular *domain*, that allows the system to understand problems within that *domain*, as well as "skill" at solving aspects of these problems. The system is able to operate somewhat independently of the user based on decision *models* programmed into the *model* base. Also, known as a knowledge-based system.
Internet Inter-Object Request Broker Protocol (IIOP)	A protocol that makes it possible for distributed programs written in different programming languages to communicate over the *Internet*. It is a critical part of the *Common Object Request Broker Architecture* (CORBA). Using CORBA's *Internet* Inter-*Object* Request *Broker* and related protocols, an organisation can write programs that will be able to communicate with their own or other organisation's existing or future programs wherever they are located and without having to understand anything about the program other than its *service* and a name.

Proxy server	A server that acts as an intermediary between a user and the *Internet* so that an organisation can ensure security, administrative control, and caching *service*. A *proxy server* is typically associated with or part of a *gateway* server that separates the organisation's network from the outside network and a *firewall server* that protects the organisation's network from outside intrusion.
Spatial data mining	A special application *domain* of *data mining* that has its theoretical foundation in *data mining* and relies heavily on general *data mining* techniques to handle the *attibute* component of spatial data.
Spatial on-line analytic processing (SOLAP)	A software platform designed to run from a *client* workstation linked to a *data warehouse* to support rapid and easy spatio-temporal *analysis* and exploration of data using a multidimensional approach and multiple levels of aggregation. Data must be visualised and accessible as cartographic displays as well as in *tables*, reports and visual summaries such as graphs. Abbreviated to SOLAP.
Unsupervised machine learning	Exploration-oriented *data mining* that aims to detect aspects of the properties of a data set in a concise and summary manner. Algorithms in this category of *data mining* make no assumptions or hypotheses about the target data set and attempt to find associations, clusters and trend in the data independent of any pre-defined objective. Also referred to as descriptive *data mining*.
Chapter 12	**Trends of Spatial Database Systems**
BI dashboard	A dashboard is a visual display of the most important information (key performance indicators) needed to achieve one or more objectives, consolidated and arranged through simple visual graphics such as gauges, charts and *tables* within a web browser on a single screen so the information can be monitored at a glance.
Data conflation	The process of transferring with minimal manual intervention *attibutes* from a source spatial data layer based on one level of accuracy and precision to a target *layer* of a different precision and accuracy.
Disruptive technologies	A new technological innovation, product, or *service* that eventually overturns the existing dominant technology in the market. A disruptive technology comes to dominate an existing market by either filling a role in a new market that the older technology could not fill or by successively moving up-market through performance improvements

	until finally displacing the market incumbents.
Enterprise GIS	An integrated, multi-departmental system composed of interoperable components that provides broad access to geospatial data, a common infrastructure upon which to build and deploy GIS applications, and significant economies of scale. It is characterised by significantly reduced data redundancy, improved accuracy and integrity of geographic information management, efficient and timely *data sharing*, improved enterprise-wide knowledge management and decision support capabilities, a high level of *interoperability* between GIS and non-GIS applications, more effective use of departmental GIS skills and resources, and reduced overall GIS maintenance and support costs.
Geospatial Web portal	A single sign-on, user-customisable *Web site* that integrates and displays spatial data from multiple independent sources. Use of *XML* content exchanged by HTML allows *integration* and presentation of spatial data from scanned paper records, isolated files, *geodatabases*, GIS software, non-spatial databases, and non-spatial applications (such as content management systems). Also known as a geospatial Web *service*.
Global positioning systems (GPS)	A worldwide satellite navigational system formed by 24 satellites orbiting the earth and their corresponding receivers on the ground that was developed by the United States Department of Defence. The satellites orbit the earth at approximately 12,000 miles above the surface and make two complete orbits every 24 hours. In addition to military purposes it is widely used in marine, terrestrial navigation and *location-based services*. Commonly referred to as GPS.
Location-based services	A *service* based on current geographic location that delivers locational information between mobile and/or static users via the *Internet* and/or a wireless network. Location is determined by user entry or a GPS receiver, but most often the term implies the use of a radiolocation *function* built into a cellular network or handset that uses triangulation between the known geographic coordinates of the base stations through which the communication takes place.
Multiple Criteria Analysis (MCA)	Generic term given to a family of analytic techniques and procedures by which concerns about multiple conflicting criteria can be formally incorporated into decision analysis. Also referred to as multi-criteria decision analysis.

Object-relational query	A form of *database query* that requires mapping or linking *tables* in a relational database with SQL queries from *object*-oriented programming. This approach is at best an intermediate step between purely relational databases and *object*-oriented *database management systems* which removes the need to convert an SQL form to execute a query.
Open mobile alliance (OMA)	An organisation that is the focal point for the development of mobile *service* enabler *specifications*, which support the creation of interoperable end-to-end mobile *services*. The OMA drives *service* enabler architectures and open enabler interfaces that are independent of the underlying wireless networks and platforms. In addition, the OMA creates interoperable mobile data *service* enablers that work across devices, *service* providers, operators, networks, and geographies.
Point data cloud	3D digitized data that define a part or object that is obtained from a measurement device such as a laser range scanner
Portlets	Reusable Web components run from modular programs that do simple, specific jobs. Each has its own self-contained *user interface* within a Web *portal's* overall page layout. They also support application-to-application communication, allowing developers quickly to create business-specific composite applications from a library of *portlets*.
Public participation GIS (PPGIS)	There are almost as many definitions for this concept as there are writers. Generally refers to increased involvement of the public in the definition and *analysis* of questions tied to location and geography. The concept seeks to overcome the limitations of present GIS technologies and to address barriers in the institutional settings within which GIS is practiced. A public participation approach seeks to achieve an expanded framework of communication and discourse that opens opportunities for the public to use spatial data and provide their perspective on planning and land management issues.
Sensor Web	A new type of GIS of intra-communicating spatially distributed and synchronous sensor *pods* that can be deployed to react and adapt to environments as well as monitor them. The components of the *sensor Web* may be ground-based, aerial, or space-based.

Spatial application service provider (Spatial ASP)

An organisation also known as a commercial *service* provide that *hosts* spatial data processing software applications on one or more servers that customers rent on a subscription basis and access over the *Internet* or via a private connection. A *Web* browser, acting as a *universal client* interface, provides the means of access.

Index